T0271039

Metal Ions in Biochemistry

Metal Ions in Biochemistry

Second Edition

Pabitra Krishna Bhattacharya
Prakash B. Samnani

CRC Press
Taylor & Francis Group
Boca Raton London New York

CRC Press is an imprint of the
Taylor & Francis Group, an **informa** business

Second edition published 2021
by CRC Press
6000 Broken Sound Parkway NW, Suite 300, Boca Raton, FL 33487-2742

and by CRC Press
2 Park Square, Milton Park, Abingdon, Oxon, OX14 4RN

© 2021 Taylor & Francis Group, LLC
First edition published by Narosa Publishing House 2005
Second edition published by CRC Press 2021

CRC Press is an imprint of Taylor & Francis Group, LLC

Library of Congress Cataloging-in-Publication Data
Names: Bhattacharya, P. K., author. | Samnani, Prakash, author.
Title: Metal ions in biochemistry / edited by
P.K. Bhattacharya and Prakash Samnani.
Description: 2nd edition. | Boca Raton : CRC Press, 2021. |
Originally published: Metal ions in biochemistry / P.K. Bhattacharya. 2005. |
Includes bibliographical references and index.
Identifiers: LCCN 2020034827 | ISBN 9780367622251 (hardback) |
ISBN 9781003108429 (ebook)
Subjects: LCSH: Metals in the body. | Bioinorganic chemistry.
Classification: LCC QP532 .B53 2021 | DDC 572/.51—dc23
LC record available at https://lccn.loc.gov/2020034827

ISBN: 9780367622251 (hbk)
ISBN: 9781003108429 (ebk)

Typeset in Times
by codeMantra

Dedicated to the Department of Chemistry, Faculty of Science, the Maharaja Sayajirao University of Baroda, Baroda, which gave us the opportunity to learn and practice the new area of bioinorganic chemistry.

Contents

Preface to First Edition

It was earlier believed that the biomolecules are only carbon compounds, and hence, carbon compounds were called organic compounds. The elements, other than carbon, were termed inorganic, implying that they have no role in life processes. But it has now been realised that the inorganic elements, specially the metal ions, have a very significant role in biochemistry. This book aims at understanding the nature of binding of the metal ions with the biomolecules and the consequent effect on the activity of the biomolecules.

Most of the books dealing with bioinorganic chemistry discuss primarily the biochemical process, with suggestion of probable role of the metal ions. In this book, the metal ions present in animal systems have been first identified, and then their role in biochemistry has been investigated. This is mainly to understand how the principles of inorganic chemistry and specially coordination chemistry work in biochemical systems.

The first chapter presents an introductory discussion on cell biology and the biomolecules. The coordinating sites of the biomolecules have been located, and their nature of binding with the metal ions has been discussed. The various roles of the biomolecules bound to metal ions have been classified.

The second chapter of this book deals with the thermodynamic and kinetic properties of metal complexes, and the concepts have been extended to understand the role of metal bound biomolecules.

Chapters 3–10 embody a detailed discussion on the different roles of bulk and trace metal ions in biochemical processes, based on the principles of inorganic chemistry.

The last two chapters of this book are devoted to the toxic effects of the metal ions, diseases caused due to deficiency or excess of metal ions in animal systems, and also the role of metal salts and metal complexes as drugs.

This book is mainly written for the postgraduate students and research workers in chemistry and biochemistry. However, it will also be useful to the students and research scholars in related disciplines, like biotechnology and medicine, for getting introduced to this new area.

I hope the readers of this book will enjoy understanding how the biochemical processes are governed by the principles of coordination chemistry, and I will be happy to know their reactions.

A list of books and research papers has been suggested for background studies and to advance further the knowledge in the subject.

I have to express my gratitude first to Prof. Daryle. H. Busch, in whose laboratory in Ohio State University, Columbus, OH, in 1976–1977, I had the opportunity to fortify and advance my comprehension of the then fast developing subject of bioinorganic chemistry.

I am also thankful to my research students Dr. Debjani Chakraborty, Dr. Prakash B. Samnani, Dr. Sujit Dutta and Dr. Daksha Patel for the help in the preparation of the manuscript and to Sri. Ninad Mehta and Mrs. Revathi Ganesh for bringing the manuscript in the typed form.

Finally, I have to express my appreciation to my wife, daughter and son and their lovely families, for the encouragement and the moral support extended by them during the completion of the work.

Pabitra Krishna Bhattacharya

Preface to Second Edition

We are thankful to the readers for warmly receiving the first edition of the book *Metal Ions in Biochemistry*. We regret that the book was not available in the market for some time, due to the delay in the publication of the second edition. But we are sure that the readers will find this second edition of the book worth their waiting.

In this book, which is revised and enlarged, attempt has been made to remove all typographical and symbolic discrepancies in the first edition. New information has been added to the chapters on alkali and alkaline earth metals, copper, zinc, iron, cobalt and metal ion toxicity. Chapter 10, on trace metal ions, has been reorganised, and new information on the role of aluminium and lanthanide elements has been added. The role of vanadium and chromium in biochemistry has been discussed in more detail, especially emphasising on their antidiabetic activity.

A new feature of the book is Chapter 13, giving the details of the role of nonmetal ions in biochemical reactions.

We will look forward to receiving views of the readers on the second edition.

Pabitra Krishna Bhattacharya
Prakash B. Samnani

Acknowledgements

We are thankful to Prof. B. V. Kamath, former Professor and Head, Department of Chemistry, the Maharaja Sayajirao University of Baroda, Vadodara, India, for going through the first and second edition of the book thoroughly and making useful suggestions for revision. We are thankful to our families for bearing the inconveniences with patience and encouraging us.

Authors

Pabitra Krishna Bhattacharya retired as Professor and Head, Department of Chemistry, the Maharaja Sayajirao University of Baroda, Vadodara, India, in 2001 working in the university for 34 years in the capacity of Reader and Professor. Prior to that, he served in Kurukshetra University, Thanesar, India, University of Rajasthan, Jaipur, India, and Vikram University, Ujjain, India.

Prof. Bhattacharya obtained MSc and PhD degrees from the University of Saugor, India. He was awarded senior fellowship under Fulbright Hayes Programme to carry out postdoctoral work in Ohio State University, Columbus, OH, during 1976–1977. He also worked at the University of Sheffield, UK, in 1987 under the INSA Royal Society Exchange Programme.

Prof. Bhattacharya's teaching and research interests involved quantum chemistry, spectroscopy, coordination chemistry, bioinorganic chemistry and homogeneous catalysis. Thirty-seven students obtained their PhD degree under his supervision. He presented papers and delivered lectures at various Indian and foreign conferences. He was invited as a speaker at International Conference in Coordination Chemistry, held at Gera, Germany, in 1990.

Prof. Bhattacharya was awarded the Prof. P. Ray Memorial Medal of Indian Chemical Society in 2000, Platinum Jubilee Lecture Award of Science Congress in 2001 and Silver Medal of Chemical Research Society of India, Bangalore, India, in 2002. He worked as a member of various committees, more notable being IUPAC and DST, programme advisory committee.

Prof. Bhattacharya is the author of three books: *Group Theory and Its Chemical Applications* (Himalaya Publishing House, Mumbai, India, 1st ed. 1987, revised 1996, reprints 1997, 1999, 2003); *Metal Ions in Biochemistry* (Narosa Publisher, Delhi, 1st ed. 2005); and *Threshold of Inorganic Chemistry* (Himalaya Publishing House, Mumbai, India, 1st ed. 2016).

Prakash B. Samnani obtained his MSc and PhD degrees from the Maharaja Sayajirao University of Baroda, Vadodara, India. After a 2-year stint as Research Officer in Analytical Division of the Advanced Research Center of Sun Pharmaceuticals, he joined the Department of Chemistry of the Maharaja Sayajirao University of Baroda, Vadodara, India, where he is presently working as Professor.

With an industrial experience and more than 25 years of teaching and research experience in chemistry in the university, he has a number of research publications to his credit. He has published research papers in the area of analytical chemistry and homogeneous catalysis in journals of national and international repute. He has also authored five books: *Experiments in Chemistry* (Anmol Publications Pvt Ltd., New Delhi, India, 2007); *Chem Companion - Graduate Chemistry Practicals*, Zenith Publications Pvt Ltd., New Delhi, India, 2007); *Teaching of Chemical Analysis* (APH Publishers Ltd., New Delhi, 2008); Quantitative Chemical Analysis (Ria Publishing House, Anand, India, 2014); and *Threshold of Inorganic Chemistry* (Himalaya Publishing House, Mumbai, India, 2016). He has completed research projects funded by UGC, CSIR and DST.

His teaching and research interests include inorganic chemistry, analytical chemistry, especially chromatographic methods, environmental chemistry and homogeneous catalysis.

1

Structure of Cells and Introduction to Bioinorganic Chemistry

The living matter is, almost entirely, made up of cells. Human body consists of nearly 10^{11} cells, whereas amoeba has single cell. The average diameter of the cell is 10^{-3} cm, and the average weight is 10^{-6} g (Figure 1.1).

Except the red blood corpuscles of some mammals, the cells contain a nucleus. Its area is one-third of the cell and is darker in colour. Each nucleus contains a definite number of chromosomes, which are highly linearly organised DNA–RNA–protein complexes, called chromatin. The daughter cell formed by the mitotic division of a cell contains the same number of chromosomes as the parent cell. The interior of the nucleus has a dense and more spherical region, mostly made of RNA, called nucleolus or nucleoplasm.

The nucleus is separated from the surrounding cytoplasm by a 75 Å thick membrane. The macromolecules, like RNA, are synthesised inside the nucleus and then passed on to the cytoplasm through the holes in the nuclear membrane.

The cytoplasm, surrounding the nucleus, is a jelly-like substance. The cytoplasmic fluid contains water, carbohydrates, phosphates, enzyme proteins, and alkali metal and chloride ions. The cytoplasm also consists of rigid ellipsoidal structures, called mitochondria and granular, globular or rod-like bodies, called Golgi bodies. The mitochondria are next to nucleus in size and are the powerhouse of the cell, being the sites for production of intracellular energy. The biological processes, such as oxidation, respiration, formation and breakdown of the proteins and synthesis of fats and glycogens, take place in different parts of the mitochondria. The Golgi bodies are specialised for the production and secretion of hormones, proteins, etc.

The cytoplasm also consists of lipoprotein membranes, in the form of tubules and vesicles, which act as interconnecting channels between nuclear membrane and cell membrane. Such complex network is called endoplasmic reticulum. There are millions of granules of ribonucleoproteins, attached to endoplasmic reticulum. These granules are called ribosomes. The Golgi bodies collect the proteins and other secretions, produced by the ribosomes, and pass these outside the cell.

Besides these, there are vesicles in the cytoplasm, filled with watery fluids and also vacuole-like cytoplasmic bodies called lysosomes. The lysosomes are third in order in size after nucleus and mitochondria

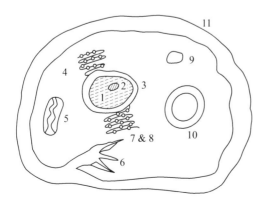

	1. Nucleus
	2. Nucleolus
	3. Nuclear membrane
	4. Cytoplasmic fluid
	5. Mitochondria
	6. Golgi body
7 & 8.	Endoplasmic reticulum with granules of ribosomes
	9. Vesicles
	10. Lysosome
	11. Cell membrane

FIGURE 1.1   Animal cell.

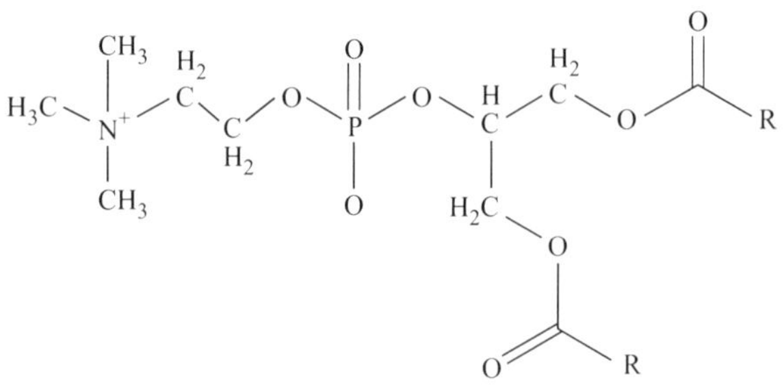

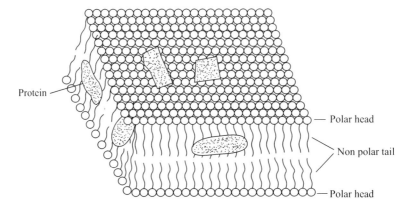

FIGURE 1.2   Lipid structure.

Protein

Polar head

Non polar tail

Polar head

FIGURE 1.3   Cell membrane.

and are surrounded by a membrane. They mainly store hydrolytic enzymes useful for the digestion of ingested micronutrients. In living cells, these hydrolytic enzymes remain confined inside the membrane. On the death of the cell, the lysosome membrane breaks, and the hydrolytic enzymes are released, leading to rapid disintegration of the dead cell. Thus, the lysosomes act as scavengers in quick elimination of the dead cell.

The cells are surrounded by 70 Å thick biomolecular membrane. It is strong and insoluble, and thus protects the cell and maintains its structure. The cell membrane can discriminate between useful and harmful substances. It allows access to nutrients and prevents the entry of toxic substances.

The cell membrane is made of phospholipids, coated on both the sides with proteins. The general structure of the lipid molecule is shown in Figure 1.2.

The phospholipids have a hydrophobic and a hydrophilic end, and exhibit surfactant properties. In the cell membrane, the lipids have a bilayer structure, with the polar heads of the lipids pointing outwards. Depending on the hydrophobic or hydrophilic nature of the protein, it occupies different parts of the lipids. The protein is considered to float like icebergs in the "fluid mosaic model" of the cell membrane (Figure 1.3).

Biomolecules and Their Metal Coordination Behaviour

Seventy to ninety per cent of most form of life is made up of water. The H bonding between water molecules leads to molecular association, resulting in its unusual physical properties, such as high melting point (m.p.) and boiling point (b.p.), high specific gravity and specific heat, and high latent heat of fusion and evaporation. The dipolar water molecule can form only weak bonds with the metal ions.

Besides water, other molecules present in biosystems are as follows:

i. **Sugars:** They serve as source of energy in biochemical systems. Oxidation of glucose in the respiration liberates energy and sustains life. Hexose sugar ($C_6H_{12}O_6$) occurs in the form of aldohexoses (glucose, galactose, mannose being more important, having six-membered ring structure) and ketohexose (fructose, having five-membered ring structure) (Figure 1.4).

   Various hexoses condense, forming glycosidic linkages, and result in polysaccharides ($(C_6H_{10}O_5)_n$), e.g. starch and cellulose.

   Pentose sugar ribose ($C_5H_{10}O_5$) and deoxyribose ($C_5H_{10}O_4$) are the constituents of RNA and DNA, respectively (Figure 1.5).

   Sugars have weak binding tendency with metal ions. Weak complexes are formed by the coordination of sugars from −OH sites. Mainly, sugar complexes of hard acid metal ions, like Ca^{2+}, are known. The structure of Ca(II) complex of inositol is shown in Figure 1.6.

   Allose, lactose and ribose form relatively more stable complexes, but are not abundant in nature. The gel formation of the polysaccharide alginic acid, in the presence of Ca, is attributed to the binding of the Ca(II) ions to the −OH and −O groups of the two constituent molecules, β-D-mannuronic acid and α-L-gluconic acid, of the polysaccharide.

ii. **Amino acids:** Amino acids are the basic constituents of the polypeptides and proteins. Essentially, they have an acidic −COOH group and a basic −NH$_2$ group and hence in neutral solution exist as zwitterion (Figure 1.7a).

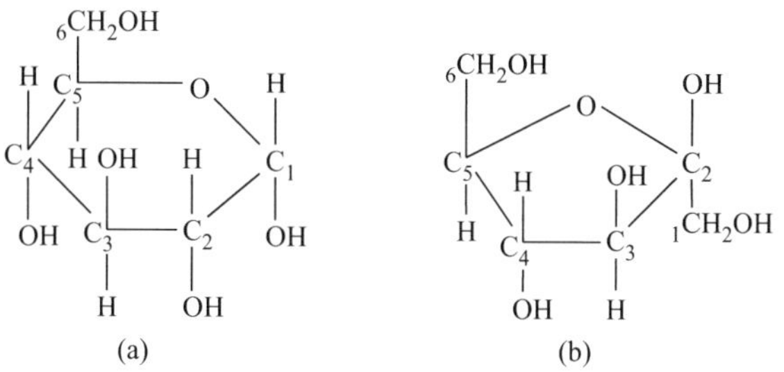

FIGURE 1.4   (a) Aldohexose. (b) Ketohexose.

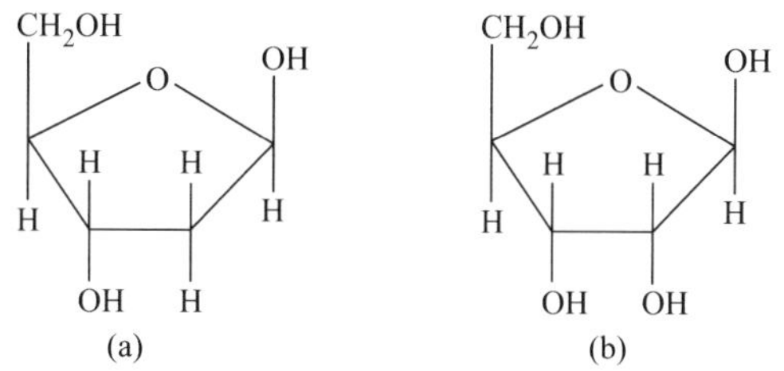

FIGURE 1.5   (a) Deoxyribose. (b) Ribose.

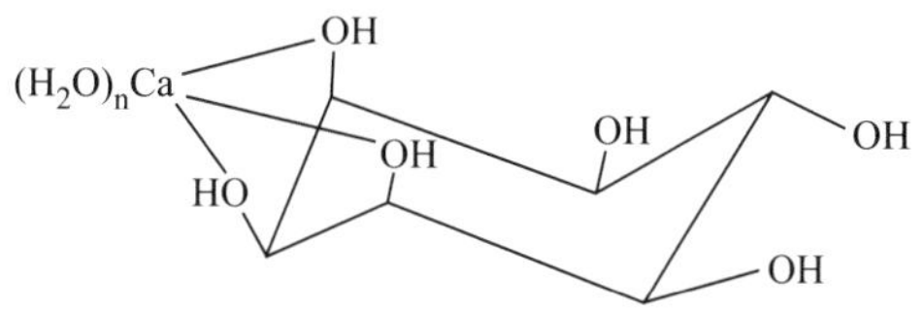

FIGURE 1.6   Calcium complex of inositol.

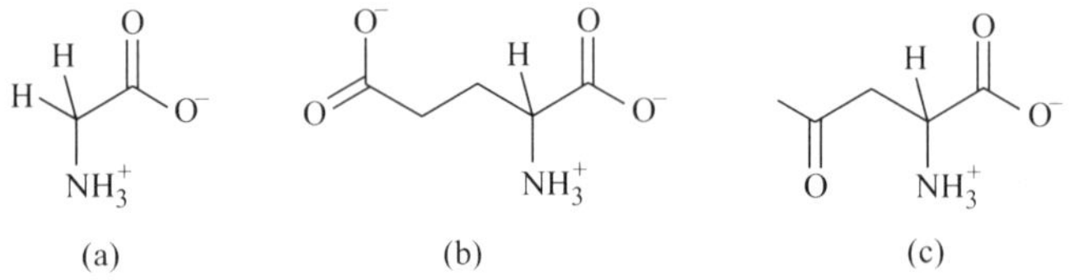

FIGURE 1.7   (a) Glycine. (b) Glutamic acid. (c) Aspartic acid.

In the case of polycarboxylic amino acids (glutamic acid, aspartic acid), there are additional acidic sites (Figure 1.7a, b and c), whereas the amino acids histidine and arginine have additional basic imidazole or secondary amine sites, respectively (Figure 1.8a and b).

The ionic properties of amino acids are responsible for their buffering activity in biological systems.

Amino acids are strongly coordinating molecules, binding to the metal ion from NH_2 and COO^- sites, resulting in the formation of chelate rings.

The formation constants of the metal complexes K_{ML}^M should increase with the basicity of the amino acid, that is proton binding capacity of NH_2 site.

$$M + L \rightleftharpoons ML \quad K_{ML}^M = \frac{[ML]}{[M][L]} \tag{1.1}$$

If the amino acid has additional coordinating sites, it can form more stable complex. For example, aspartic acid has an additional acidic carboxylic group. It forms more stable complex, as there is coordination from two carboxylates and NH_2 site, resulting in greater stability (Figure 1.9a).

Cysteine, with an additional SH coordinating site, is ambidentate in character (Figure 1.9b). It can coordinate from NH_2 and SH sites or from NH_2 and COO^- sites. It has, however, been confirmed that the coordination is of the former type, as the formation constant of Zn(II) cysteine complex is comparable with that of mercaptoethylamine ($SH-CH_2-CH_2NH_2$) complex. As mercaptoethylamine has only NH_2 and SH sites, cysteine must also be coordinating from NH_2 and SH sites, only.

Histidine is also an ambidentate ligand (Figure 1.10). At low pH, it coordinates from NH_2 and COO^- sites (alanine type), whereas at higher pH, it coordinates from NH_2 and imidazole

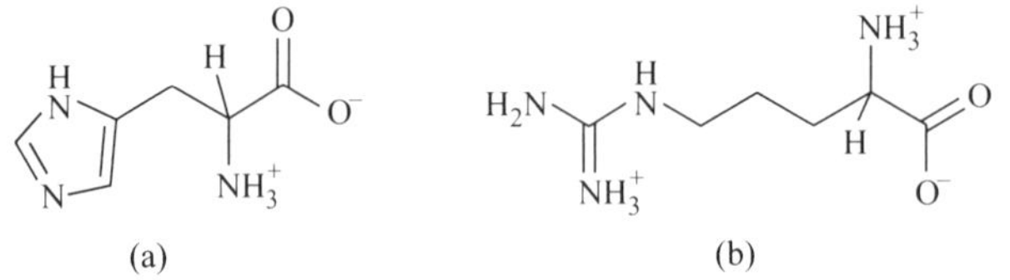

FIGURE 1.8   (a) Histidine. (b) Arginine.

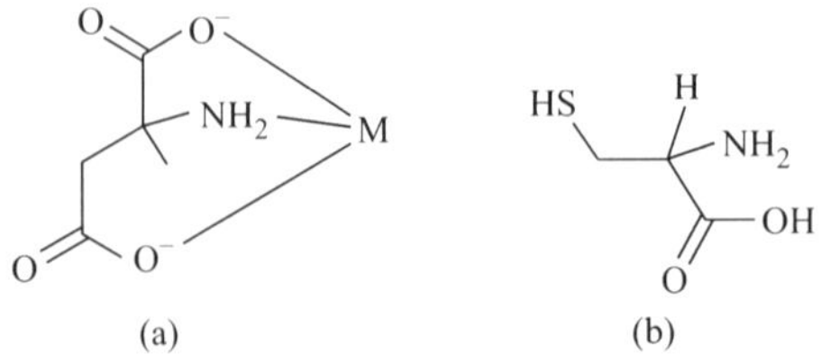

FIGURE 1.9   (a) Aspartate. (b) Cysteine.

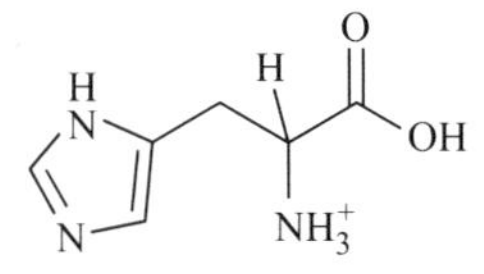

FIGURE 1.10 Histidine.

N (histamine type). The electronic spectrum also supports the coordination sites. At low pH, spectrum of Cu-histidine is comparable with that of Cu–alanine complex, whereas at high pH, Cu-histidine spectrum is comparable with that of Cu–histamine complex.

iii. **Dipeptides:** They are formed by the condensation of amino acids. They are named starting with the amino acid on the left of the peptide linkage, that is the N terminal end (Figure 1.11).

X-Ray crystallographic study of the peptide bond shows that it is planar. This is because of the following (Figure 1.12) resonance stabilisation. Both C=O and C–N bonds involve sp^2 hybridisation and have intermediate double- and single-bond character.

At low pH, the dipeptide coordinates through NH_2 of the amino end and CO of the peptide bond, retaining the planar structure of the chelate ring. The COO^- group remains uncoordinated. At high pH (>4.5), the dipeptide coordinates from the NH_2 group, and the peptide nitrogen (N^-), generated by the deprotonation of the peptide NH. Deprotonation is essential, because the coordination of lone pair of NH should make the nitrogen atom tetrahedral, with consequent loss of planarity in the chelate ring. Hence, resonance stabilisation is lost. Planarity can be retained by the deprotonation of NH, resulting in planar N^-. This also facilitates a weak coordination of carboxylate group. The equilibrium between the two forms is shown in Figure 1.13.

The fact that the coordination site changes at higher pH has been confirmed by studying the electronic spectrum of Cu-glycylglycine (1:1) solution at varying pH (Figure 1.14). At low pH, the band appears at high $\lambda > 800$ nm, due to CO and NH_2 coordination. With increasing pH, there is an increasing N^- coordination, creating a stronger field, and a higher energy absorption band is observed at 625 nm.

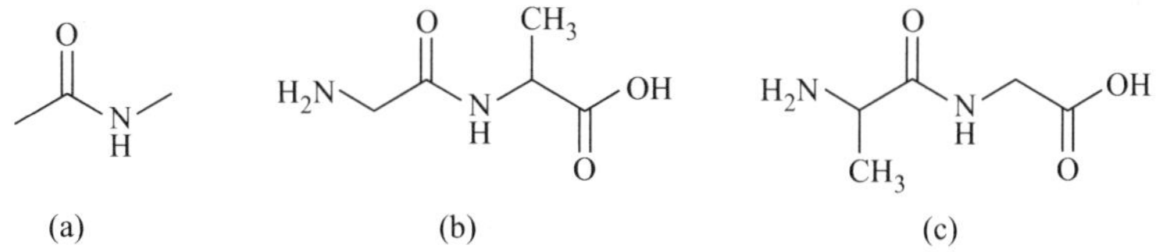

FIGURE 1.11 (a) Peptide linkage. (b) Glycylalanine. (c) Alanylglycine.

FIGURE 1.12 Resonance stabilisation of peptide bond.

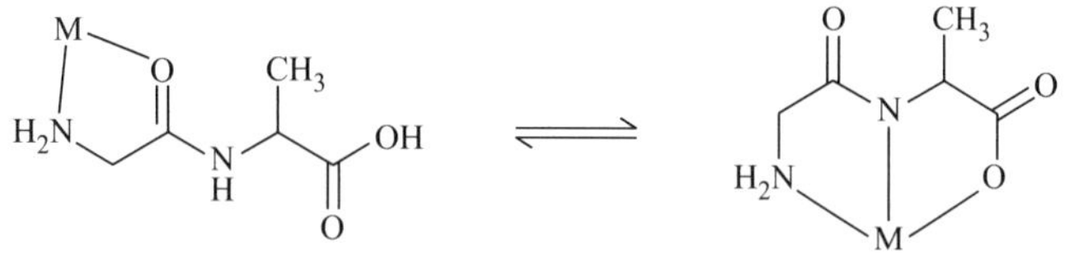

FIGURE 1.13 pH dependent binding of dipeptide.

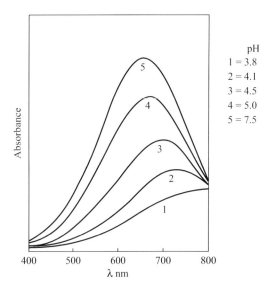

FIGURE 1.14   Electronic spectrum of Cu-glycylglycine solution: pH effect.

Cyclic voltammetric studies also confirm the formation of two different species at different pH. At low pH, with relatively weaker field, due to NH_2 and amide CO coordination, Cu^{2+}/Cu^+ appears at more positive potential. However, at higher pH with N^-, NH_2 and COO^- coordination, the ligand field is stronger, and hence, there is an increase in electron density around the metal ion. Therefore, the reduction takes place at more negative potential, and the reduction peak shifts to more negative side.

Polypeptides and Proteins

Condensation of several amino acids results in the formation of a polypeptide and proteins are nothing but long-chain polypeptides. The constituent amino acids are called the amino acid residues. The polypeptides and proteins have the same structure, with amino acid residues joined by peptide group –CONH-.

The end containing the amino acid residue with –NH_2 is called the amino end, and the end containing the amino acid residue with –COOH is called the carboxylic end. Every amino acid residue in the peptide chain has an attached side group, like –CH_3 in alanine, –$CH(CH_3)_2$ in valine, –$CH_2C_6H_5$ in phenyl alanine and imidazole in histidine. The side chains affect the properties of the proteins.

This long-chain peptide structure is called the primary structure of the protein (Figure 1.15). There is a sequential array of the amino acid residues in the protein chain, which determines the properties of the proteins.

The proteins differ from polypeptides in having higher molecular weights (over 10,000 daltons). They have a more complex structure, as the peptide chains in proteins are arranged in space. As the peptide bond is rigid due to planarity, two peptide chains cannot be connected through peptide bond and consequently branching of polypeptide chains is not possible. However, the peptide chains can be interconnected by hydrogen bonds (N–H•••O) (Figure 1.16a) or disulphide linkages (S–S) (Figure 1.16b), formed due to biological oxidation of cysteine residue to cystine.

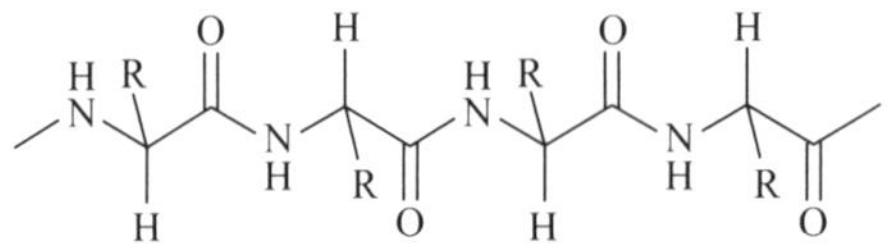

FIGURE 1.15   Protein primary structure.

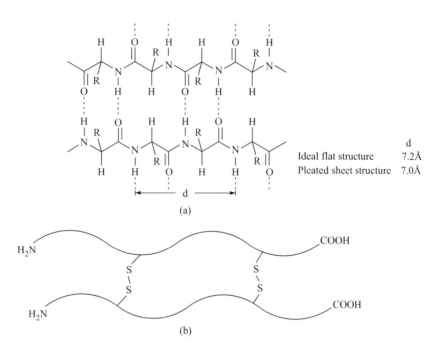

Ideal flat structure   7.2Å
Pleated sheet structure   7.0Å

(a)

(b)

FIGURE 1.16   (a) H – bonding and (b) disulphide linkages between peptide chains.

The simplest two-dimensional structure that can be thought of is the flat sheet in which the peptide chains lie side by side, held by H bonds between two neighbouring chains (Figure 1.16a). In this ideal flat structure, the distance between alternate amino acid residues should be ~ 7.2 Å. However, because of the crowding of the side chain R on the amino acid residues, ideal flat structure is not possible. If the side groups R are small or medium in size, they can be accommodated with slight contraction of the peptide chain. So chains can lie side by side, held to each other by H bond. The contraction of the peptide chain results in a shorter distance between the alternate amino acid residues than in the flat sheet, and a pleated sheet structure is formed. This is called beta arrangement, as found in silk fibre, with a distance of 7.0 Å between the alternate amino acid residues.

However, if the side chains are very large, for accommodating them, each peptide chain becomes coiled to form a helix like a spiral staircase. There is a hydrogen bonding between different coils of the same chain, and this holds the helix together (Figure 1.17a). This structure is found in the α-keratin of unstructured wool, hair, horn and nails. It was suggested by Pauling and Corey in 1951 that the five coils in the helix of the polypeptide chain of keratin are made of 18 amino acids, that is, there are 3.6 amino acid residues per turn of the helix and a repeat distance of 1.5 Å between the amino acid residues, thereby providing room for the side groups. This is called α-helix structure and is of significance in protein chemistry (Figure 1.17a).

The α-helix and β-sheet structures represent the secondary structure of the protein. When wool is stretched, there is uncoiling of the helical structure of the single chain of the protein. The hydrogen bonds within the helix are broken, and new hydrogen bonds are formed between the adjacent chains, leading to sheet structure. Thus, α-keratin changes to β-keratin.

The contraction of muscles is considered to involve the reversible change of α-helix structure of the fibrous protein of muscle (called myosin) to β-structure.

There is a third type of secondary structure of protein. In collagen, present in tendon and skin, helical chains get bound by interhelix hydrogen bonding of β-type. Thus, a three-strand helix is formed. This combines the characteristics of α-helix and β-sheets. On dissolving in water, collagen changes to water-soluble gelatin. The interhelix H bonds are broken, and the single chains form H bonds with water, making the collagen soluble.

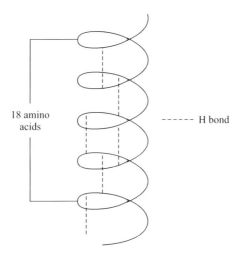

FIGURE 1.17a   α-helix structure of keratine.

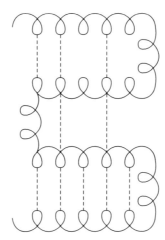

FIGURE 1.17b   Coiled structure of protein coils.

Some protein molecules contain a nonpeptide part, called prosthetic group. The protein gets isolated only along with the prosthetic group. For example, there is an Fe(II) protoporphyrin (haem) prosthetic group in haemoglobin (to be discussed in Chapter 5). The proteins, with metal complex as prosthetic group, are called metalloproteins. Protein structure may be further modified due to binding with a sugar (glycoprotein), lipid (lipoprotein) or fatty acids or due to phosphorylation of the amino acid residue.

Pauling has suggested that there can be further coiling of the coils of the protein, like the coiled coils in the lamp filaments (Figure 1.17b). Between the layers of coiled chains, there are weak interactions due to H-bonding, electrostatic bonds or disulphide bridges. This constitutes the tertiary structure of the protein.

Quaternary structural feature is observed in multimeric proteins. A definite number of identical or distinct component protein units of the multimeric protein can be held together by electrostatic interaction. Such interactions are observed between the component haem proteins of tetrameric haemoglobin (to be discussed in Chapter 5).

The proteins are very specific in their physiological reactions. It has now been realised that the activity of the protein does not depend only on the sequence of amino acid residues or the prosthetic groups present, if any. It also depends on the secondary and subsequent structural features, determining the molecular shape of the protein. Sudden change in pH or temperature can cause denaturation of the protein (irreversible loss of biological activity). This is because, on denaturation, the helical structure of protein gets uncoiled, and thus, the characteristic shape is lost, with consequent loss of biological activity.

As can be seen from the structure of proteins, they are strongly coordinating molecules. The coordination can be of three different types, from the terminal $-NH_2$ and the adjacent peptide, from two neighbouring peptide sites or terminal carboxy and the last peptide linkage. Besides this, the functional side chain of the constituent amino acids of the type histidine (imidazole N), glutamic acid (carboxylate), aspartic acid (carboxylate), tyrosine (phenolate) and cysteine (SH) also provide additional coordinating sites. Thus, the proteins are multinucleating and can take up a large number of metal ions.

Furthermore, the tertiary and quaternary structures of proteins orient the coordinating atoms at specific positions leading to a site with a specific structure suitable for particular metal ions. For example, a tetrahedral site is preferred by Zn(II) and Cu(I), and hence, they prefer proteins with tetrahedral coordinating sites. Hence, the metalloproteins, which act as enzymes, prefer specific metal ions as coenzymes. The cavities inside the proteins are also of specific size and shape, and hence, the proteins are also selective of the substrates with suitable size, which can be accommodated in the cavities of the protein. In addition to the above, in metallo-haem proteins which act as enzymes and also perform O_2 transfer process, prosthetic group has a bound metal ion. For example, haemoglobin has iron in the prosthetic group haem.

The metal ions bound to the protein are known to activate the catalytic activity of the enzyme protein. Besides this, the binding of the metal ion, at specific sites in the proteins, also helps in stabilising the specific secondary and tertiary structures of the proteins and thus affects their mechanical performance. In addition to the stabilisation of the secondary and tertiary structures of the proteins by metal binding, the interaction with metals also promotes the assembly of individual protein metal units with other such units, resulting in larger complexes. This property can have adverse effects on biological systems, leading to disorders in health such as formation of amyloid in brain, resulting in neurodegeneration. However, the metal-assisted self-assembly process of proteins has a favourable effect. It also helps in the formation of gel-like adhesive necessary for biochemical reaction or formation of structural scaffolds. The formation of metalloprotein polymers due to self-assembly with specific structures results in hard materials, required for the hardening of the tips and edges of specific parts of the body, such as nails and claws. These hard body parts of animals and worms help in several functions such as stinging, injecting venom and also drilling of soil to lay eggs.

Nucleosides, Nucleotides and Nucleic Acids

Nucleosides are comprised of purine bases (adenine and guanine) or pyrimidine bases (cytosine, thymine and uracil) (Figures 1.18a to e), bound to pentose sugar, ribose or deoxyribose from N1 pyrimidine or N9 purine base and C-1 of the sugar (Figure 1.18f).

In **nucleotides**, the sugar of the nucleoside part is linked to a mono-, di- or triphosphate from C-5′, $-CH_2$ group. The common example is adenosine triphosphate (ATP) where the base is purine (adenine) and the N-9 of the base is linked to C-1 of ribose, which is further linked to triphosphate (Figure 1.18g).

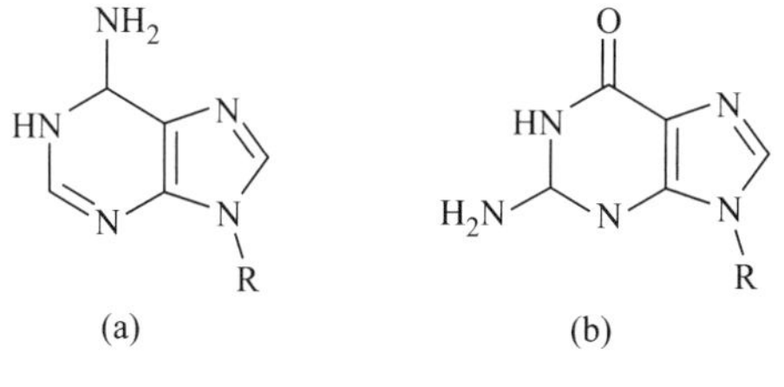

(a) (b)

FIGURE 1.18 Purine bases – one six-membered pyrimidine ring (with two nitrogen atoms) and one five-membered ring with two nitrogen atoms. (a) Adenine. (b) Guanine.

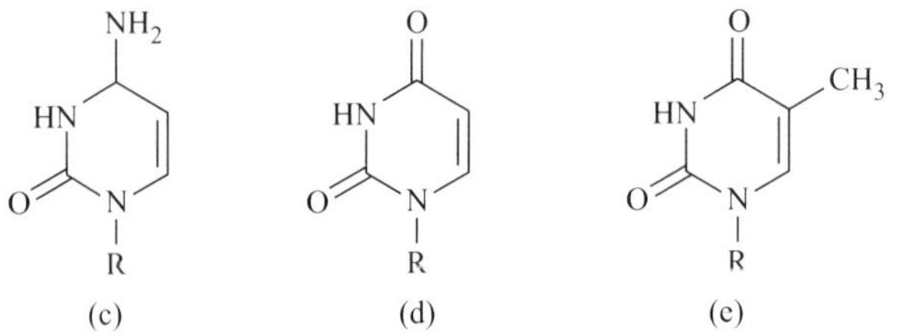

FIGURE 1.18 Pyrimidine bases – one six-membered ring with two nitrogen atoms. (c) Cytosine. (d) Uracil. (e) Thymine.

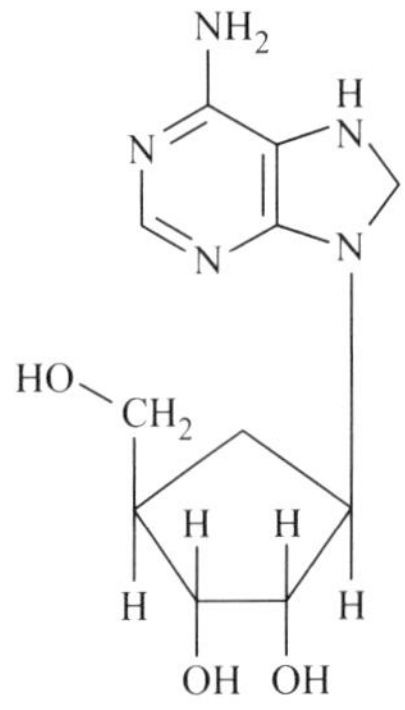

FIGURE 1.18f Adenosine nucleotide.

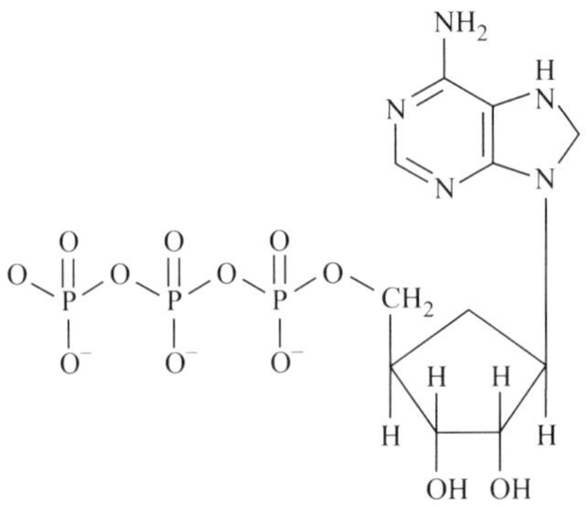

FIGURE 1.18g Adenosine triphosphate.

The nucleotides with linked monophosphate and diphosphate groups, instead of triphosphate, are called adenosine monophosphate (AMP) and adenosine diphosphate (ADP), respectively.

The **nucleic acids** are polymers of monophosphate nucleotides. They are built up by phosphodiester bond formation between the 3′ hydroxyl group of one nucleotide and the 5′ hydroxyl group of ribose of the adjacent nucleotide. In ribonucleic acid (RNA), the sugar is ribose, whereas in deoxyribonucleic acid (DNA), the sugar is deoxyribose (Figure 1.19). The bases in DNA are adenine, guanine, cytosine and thymine, whereas RNA has uracil, instead of thymine. Thus, the primary structures of both RNA and DNA consist of repeating units of nucleotides, whose sequence is the basis of the genetic code. These nucleic acids are associated with proteins called histones. This association of nucleic acid and protein is called nucleoprotein.

It was first proposed by James Watson and Francis Crick in 1953 that nucleic acids have a helical structure, with turns at every tenth nucleotide. RNA molecule has a small single helix structure.

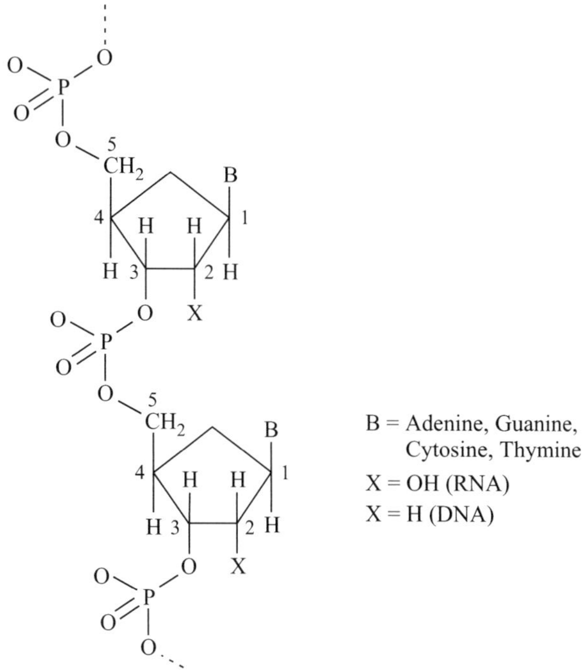

B = Adenine, Guanine,
Cytosine, Thymine

X = OH (RNA)

X = H (DNA)

FIGURE 1.19 Structure of RNA and DNA.

However, DNA has double-helix structure, as two chains of DNA are linked together by hydrogen bonding interactions between two structurally complimentary bases, that is adenine of one chain against thymine in the other, and guanine in one chain in opposition to cytosine of the other. This constitutes the secondary structure of DNA (Figure 1.20). The interaction between the bases of two helices orients the bases in specific ways resulting in the twisting of the double helix. This is called the tertiary structure of DNA.

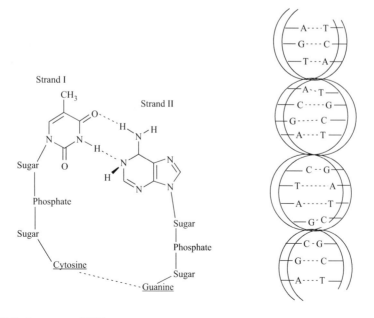

FIGURE 1.20 Helical structure of DNA.

Synthesis of DNA involves a process called replication. The double helix of DNA divides into two chains due to the breaking of the H bonds between the bases. The nucleotides of the two resulting chains combine with the single chain of another complimentary DNA, in the surrounding medium, giving rise to another DNA molecule. Thus, each DNA breaks into two new DNAs. The synthesis of the new DNA is activated by the enzyme polymerase.

The DNA molecules control the synthesis of different proteins which act as enzymes for different functions of living cells. The actual process of protein synthesis is carried out by three different forms of RNA: messenger RNA, transfer RNA and ribosomal RNA.

Metal Ion Binding with Nucleosides, Nucleotides and Nucleic Acids

The purines and pyrimidines are known to be monodentate ligands, binding with the metal ions from different N-sites. Hence, for nucleosides, base is the coordinating site, the sugar part being very weakly binding with metal ions. In nucleotides, the coordination is mainly from the phosphate. For the class A metal (alkali, alkaline earth, early transition metal ion in higher oxidation state) complexes of nucleotides, it has been observed that the formation constants are comparable with those of the inorganic triphosphates. Coordination takes place from α- and β-oxygen of the triphosphate. Simultaneous coordination of the base nitrogen is not possible, because it forms a higher membered chelate ring. However, in the case of transition metal complexes, it is observed that the nucleotide forms more stable complexes than the inorganic triphosphates. This alludes to the possibility of additional axial coordination of the base. However, the base coordination results in a multimembered chelate ring, which does not contribute much to the stabilisation. Due to the Jahn–Teller effect, which favours formation of longer chain chelate ring in the axial direction, Cu(II) complexes of ATP are more stabilised, due to axial coordination of the base part, and hence, Cu(II) forms more stable complexes with ATP than other transition metal ions.

In the case of nucleic acids, the soft metal ions prefer coordination with the base part. In the case of DNA, the coordination may be at the N7 position of guanine, N7 position of adenine, N3 position of cytosine and deprotonated N3 position of thymine and uracil. It has been observed that in most complexes, the N-site available for coordination in the double helix of DNA is of the purine base. Such coordination has mainly been studied for Pt(II), Pd(II) and Ru(II), but coordination from this site is also possible with Cu(II) and Zn(II). The metal ion can get bound to the base nitrogen atoms of the two helixes of the nucleic acid and can thus forms intrastrand bond. Such bonding of platinum of cisplatin to DNA is attributed to its anticancer activity (to be discussed in Chapter 12).

The harder metal ions can coordinate with the phosphate group of the nucleic acid. The nature of binding of the metal is important, because base coordination of the metal causes destabilisation of the helical form, whereas if the metal ion is bound to phosphate, the helical structure is stabilised.

As stated earlier, in the case of ATP, the pentose sugar site is weakly coordinating. However, RNA is known to coordinate with osmate esters from the sugar oxygen.

The study of the nature of binding of the metal ions with the nucleic acids is of significance, because the toxicity of heavy metal ions is partly due to their binding with the nucleic acids. Furthermore, the knowledge of the binding of the metal ions with nucleic acids helps in developing the metal complexes which will act as more efficient anticancer drugs.

Lipids

Oils, fats and waxes are generally known as lipids. They are the glycerol esters of long-chain fatty acids. The fats of plant origin, oils, have more double bonds, have low boiling point and are liquid in nature. The animal fats being more saturated are solids. Mixed esters of fatty acids and phosphoric acid are called phospholipids. They constitute the walls of the cell membrane. Lipids bound to proteins are called lipoproteins.

Steroids are also classified as lipids. The best-known steroid is cholesterol and is a source of hormones, vitamins and bile acids.

Lipids do not have potential coordinating groups to combine with metal ions, and there is no significant study of their metal binding tendency.

Evidence of the Presence of Metal Ions in Biochemical Systems

It was considered earlier that all the chemicals involved in biochemical processes are organic compounds. However, there is a growing realisation that biochemical processes involve metal ions. Besides the direct evidence of the presence of calcium in bone and teeth structures, alkali and alkaline earth metal ions in cells and iron in haemoglobin of blood, there are indirect evidences indicating the role of metal ions in biochemical system.

i. If we look at the environment around living organisms, it can be divided into the following four parts:

a. **Atmosphere:** a thin film of gas around the earth.

b. **Hydrosphere:** oceans, lakes and rivers.

c. **Lithosphere:** the solid part of the earth.

d. **Biosphere:** the narrow region, close to the earth's surface, which is the habitat of the living organisms.

It is known that the first three spheres have inorganic constituents. Hence, the organisms are in constant interaction with the inorganic environment, through the food chain, this inevitably leads to the incorporation of inorganic constituents in biological systems.

ii. The use of precise analytical techniques has helped in the detection of trace quantities of metal ions in biochemical systems. The enzymes, which are biochemical catalysts, were supposed to be complex proteins. It has now been shown that 70% of the known enzymes involve metal ions.

iii. Micronutrients containing metal salts are essential for plants. Deficiency of trace metals in the soil affects plant growth. Deficiency of minerals in food affects animal health. This shows that the metal ions are essential for life processes.

iv. As discussed earlier, most of the biochemically important molecules have coordinating sites and bind with the metal ion to form metal complexes. Hence, metal complex formation appears to be vital for many biological processes. Many drugs like aspirin and antifungals are coordinating molecules and may involve metal complex formation, in their therapeutic actions.

v. Various metal salts and metal complexes are used as drugs. Vitamin tablets are associated with mineral salts. In the case of acute anaemia, the patient is advised to take iron tablets and vitamin B_{12}, containing cobalt. In the indigenous Indian systems of medicines, like Ayurveda and Unani, oxides of various metals (Bhasmas) are administered for the cure of several diseases. It is advised to give a spoon of lime water or calcium tablets to a growing child to build his teeth and bones. Drinking cold water, kept in copper vessels, is also considered good for health.

vi. Some metal ions have toxic effects, due to their interference in biochemical processes.

Definition of Bioinorganic Chemistry and Inorganic Biochemistry

The understanding of the role of metal ions in biochemical processes has led to the emergence of new fields of study, called inorganic biochemistry, wherein the biochemists use the inorganic concepts and techniques to understand the role of inorganic elements in biochemical processes and bioinorganic chemistry, wherein the inorganic chemists work on metal complex systems, mimicking the biological molecules. Study on such model systems provides information of relevance to fundamental inorganic chemistry and is also of help in understanding the role of metal ions in the biological systems.

Occurrence of Metals and Nonmetals in Biological Systems

Besides the nonmetals carbon, oxygen, hydrogen, nitrogen and phosphorous, which are involved in the composition of the organic compounds of biochemical systems, such as proteins, carbohydrates, fats and nucleic acids, there are sulphur, chlorine, selenium, iodine, fluorine, boron and silicon present in animals and plants. Selenium is essential for cattle and chicken. Iodine is present in animal thyroid hormone, thyroxine and its derivatives. Iodine and bromine are accumulated in some brown algae. Silicon and boron are also essential components of small organisms.

The metal ions present in biological systems can be classified into two types:

i. **The bulk metal ions:** Na(I), K(I), Mg(II) and Ca(II), which constitute 1% of the human body weight.

ii. **Trace metal ions:** Fe(II) & (III), Cu(I) & (II), Mn (II) & (III), Zn (II), Co(II), Mo (VI), Sn, V and Ni, which constitute only 0.01% of the human body weight.

Besides these, there is some tentative evidence of the presence of chromium, cadmium and lead, which may be required at a level lower than even the trace metal ions. It is interesting to observe that all the trace metal ions present in biological systems, except molybdenum, belong to the first transition series.

Sodium, potassium and magnesium are the major components of the body fluids and the cytoplasm of the cell.

Calcium is the main component of the bones of the vertebrates and the shells of the eggs. Ca(II) and Mg(II) also bind with ATP and participate in enzymatic reactions, as components of enzyme proteins.

The transition metal ions are characterised by their ability to assume different oxidation states. Therefore, they serve as active centres for metalloenzymes catalysing electron transfer, e.g. copper containing plastocyanins, azurins, stellacyanin and Fe(II) containing cytochromes and ferredoxin, oxidation reactions and oxygenation reactions, e.g. copper (II) containing tyrosinase and ceruloplasmin and Fe(II) containing catalase, peroxidase and cytochrome P-450. Transition metal ions also constitute the compounds which help oxygen-carrying process, such as Fe(II) containing haemoglobin, myoglobin, hemerythrin and Cu(I) containing hemocyanin. Some of the transition metals such as Mn(II), Co(II) and Zn(II) act as Lewis acid and constitute active sites of enzymes which catalyse reactions like hydrolysis (peptidase and phosphatase, enzymes containing Zn(II) and Ca(II)), hydration (enzyme carbonic anhydrase, containing Zn(II)), phosphorylation (kinase, containing Mn) and carboxylation reaction (carboxylase, containing Mn).

Cobalt (III) is a constituent of vitamin B_{12}, and iron and molybdenum constitute the active sites of nitrogenase present in plants, which is responsible for nitrogen fixation.

Mg(II) and Mn(II) are essential components of chlorophyll and assist photosynthesis. Vanadium is considered to be essential for the growth of some algae and is an important component of a protein contained in a group of marine invertebrates (tunicates). It is also present in vanadin which helps in the respiration of ascidians.

Essentiality of chromium in higher animals has been suggested but has not been fully established. Although other elements such as aluminium, strontium, barium, lead, cadmium, arsenic and tin are present in biological systems, they are neither known to be essential to man nor to the other organisms. Mercury, lead, arsenic and cadmium are notorious for their adverse effects health.

QUESTIONS

1. Which cell organelle acts as control centre of the cell?
2. What is the composition of plasma membrane?
3. What does chromatin comprise of?
4. Which cells of the human body are enucleated?
5. Describe the fluid mosaic model of plasma membrane.
6. What role does glucose play in human system?
7. Name the linkages present in polysaccharides.
8. Name the types of sugar present in RNA and DNA.
9. Not many metal ions form complexes with sugars. Why?
10. Which metal ion is responsible for the gel formation of carbohydrate alginic acid?

11. Why amino acids form very stable complex compounds with metal ions?

12. What is an ambidentate ligand?

13. Give two examples of amino acids that act as ambidentate ligand.

14. Dipeptides are formed by condensation of amino acids, resulting into peptide bond, ![peptide bond structure]. X-Ray crystallography studies show that this bond is planar. How will you explain this?

15. Give experimental evidence showing that coordination of peptides with metal ions changes with pH.

16. Describe the factors responsible for α-helix structure of peptide chains.

17. Which compounds make up cell walls?

2

Thermodynamic and Kinetic
Properties of Metal Complexes

As is evident in the previous chapter, there are a number of metal ions involved in biological systems and a number of potential coordinating biomolecules. Hence, the biochemical reactions should involve the formation of metal complexes. This fact has been suitably emphasised in the following statement of J. M. Wood (Naturwiss., 62: 357 (1975)):

> If you think that biochemistry is the organic chemistry of living systems, then you are misled, biochemistry is the coordination chemistry of living systems.

It is, therefore, essential to understand the following factors, which govern the properties and reactivities of the metal complexes of the biochemicals.

    i. The strength of the metal–biochemical bonds.
    ii. The ease with which the organic component or the metal component undergoes substitution in the metal–biochemical complex.
    iii. The change in the reactivity of the biomolecule, on binding with the metal ion.

For this, it is essential first to understand the factors affecting the thermodynamic and kinetic stabilities of the metal complexes, in general.

The formation of a complex ML_n can be represented by the following equilibrium:

$$M + nL \rightleftharpoons ML_n \tag{2.1}$$

$$\beta_n = \frac{[ML_n]}{[M][L]^n} \tag{2.2}$$

β_n is known as the overall formation constant or stability constant.

The greater the value of the overall formation constant, the more stable is the complex; that is, lesser is the dissociation of the complex into M^{n+} and L, and hence, lesser is the concentration of the free metal ion in the metal complex solution.

According to Bjerrum, the formation of a complex, ML_n, takes place in steps, each step being governed by an equilibrium constant, termed stepwise formation constant:

$$M + L \rightleftharpoons ML \qquad K_1 = \frac{[ML]}{[M][L]} \tag{2.3}$$

$$ML + L \rightleftharpoons ML_2 \qquad K_2 = \frac{[ML_2]}{[ML][L]} \tag{2.4}$$

$$ML_2 + L \rightleftharpoons ML_3 \qquad K_3 = \frac{[ML_3]}{[ML_2][L]} \tag{2.5}$$

$$\vdots$$

$$ML_{n-1} + L \rightleftharpoons ML_n \qquad K_n = \frac{[ML_n]}{[ML_{n-1}][L]} \tag{2.6}$$

The overall formation constant of the reaction is shown as follows:

$$\beta_n = K_1 \cdot K_2 \cdot K_3 \cdots K_n \tag{2.7}$$

For example, in the complex ML_2, where $n = 2$, that is, for ML_2 complex,

$$\beta_n = K_1 \cdot K_2 = \frac{[ML_2]}{[M][L]^2} \tag{2.8}$$

From statistical consideration, it can be expected that $K_2 < K_1$, as the tendency of the metal ion to accept the ligand goes on decreasing, with the increase in the number of the ligands bound.

This is because when the first ligand (L) combines with a metal ion, it has more coordination positions available for bonding than available to the second L binding to ML. Hence, K_2 is less than K_1.

If the final complex, to be formed, is ML_N, the stepwise formation constant, K_n, should be inversely proportional to the number of ligands bound, that is, n, and directly proportional to the number of ligands, yet to be added, that is, $N - (n - 1) = N - n + 1$ (it may be noted that the complex ML_n is formed from the stage ML_{n-1}). This is shown as in the following equation:

$$K_n \, \alpha \, \frac{1}{n} \quad K_n \, \alpha \, N - n + 1 \tag{2.9}$$

In the case of the complex ML_2, the statistical values of K_1 and K_2 are shown in the following equation:

$$K_1 = \frac{2 - 1 + 1}{1} \quad K_2 = \frac{2 - 2 + 1}{2}$$

$$\frac{K_1}{K_2} = 2 \times 2 = 4$$

However, it is observed that the experimentally determined value of K_1/K_2 may be higher or lower than 4. K_1/K_2 value is, therefore, expressed as $4 X^2$, where X^2 is the spreading factor and accounts for the nonstatistical factors, such as electronic effect, electrostatic repulsion and steric effects, affecting the stepwise formation of the complexes.

A σ bonding L transfers a pair of electrons to the metal ion. Thus, the electronegativity of the metal ion in ML is reduced, and hence, its tendency to bind with the second ligand L is lowered. The greater the σ bonding strength of the ligand L, the lesser the K_2, and K_1/K_2 becomes greater than 4. If the ligand has negative charge, (L^{m-}), its binding with the metal ion (M^{n+}) is stronger. However, the positive charge on $(ML)^{n-m}$ is less than M^{n+}, and hence, it has less tendency to bind with the second L^{m-}. In other words, the L^{m-} in $(ML)^{n-m}$ repels the incoming second L^{m-}. Thus, K_2 is reduced. For example, K_1/K_2 value in the case of $M(NH_3)_2$ complex is less than that in the case of $M(glycinate)_2$ complex.

If the ligand has large substitutions over it, there is a steric hindrance in the binding of the second ligand with the metal ion. For example, K_1/K_2 value is more in the case of *bis*-N-alkyl ethylenediamine complex than that in the case of *bis*-ethylenediamine complex.

Conversely, K_2 is not lowered significantly, if there is an intramolecular H bonding possible between the two ligands, for example, in *bis*-dimethylglyoxime metal complex. Hence, K_1/K_2 value is lesser than 4. Besides the above examples of deviation of K_1/K_2, value from statistical expectation of 4, in some case of ML_n complexes, there are unusual changes in stepwise formation constants at specific step of complex formation.

i. If there is a hybridisational change at some stage of complex formation, the formation constant for that step is lowered. For example, in the formation of HgI_4, value of K_2/K_3 value is very large, because, when the tetrahedral $HgI_3 \cdot H_2O$ is formed from linear HgI_2, the hybridisation changes from sp to sp^3.

ii. Similarly, K_2/K_3 value is large in the formation of $Zn(en)_3$, because tetrahedral $Zn(en)_2$ has to change to octahedral $Zn(en)_3$.

iii. In the Fe(II) complexes of 2-2′-bipyridyl or 1,10-phenanthroline, the value of K_2/K_3 is less than expected, because the *bis* complex of Fe(II) is an outer orbital high spin complex, whereas *tris* complex is of low spin, inner orbital type. Thus, there is a stabilisation of the *tris* complex due to additional crystal field stabilisation energy (CFSE).

iv. In $Cu(en)_3$ complex (en = ethylenediamine), (en = ethylenediamine) K_2/K_3 value is large, because Cu(II) forms distorted octahedral complex, due to the Jahn–Teller effect (any nonlinear system in degenerate state undergoes distortion, so that the degeneracy is resolved and the system is stabilised.).

Hence, on going from $Cu(en)_2$ to $Cu(en)_3$, two ligands have to occupy one axial and one equatorial position. They are strained in spanning a longer distance along axial direction, and thus, K_3 is reduced.

Factors Affecting the Stability of the Complexes

The external factors, such as temperature, pressure, concentration and ionic strength of the solution, affect the stability constants of the complexes. If the stability constants are compared under identical external conditions, they are found to be dependent on the nature of the metal ion and the nature of the ligand.

1. **Nature of the metal ion**: According to the electrostatic model of complex formation, as proposed by van Arkel and de Boer, complex formation is due to electrostatic attraction between the metal ion and the ligand ions or ligand dipoles, arranged at positions around the metal ion, such that electrostatic attraction between the metal ion and ligands is maximum and ligand–ligand repulsions are minimum. For example, six ligands are oriented octahedrally in ML_6 complexes, whereas four ligands in ML_4 are oriented in tetrahedral or square planar way.

   The formation constant increases with increasing charge and decreasing size of the metal, that is increasing charge density over the metal ion.

   For example, in fluoride and phosphate complexes, where ML interaction is mainly electrostatic, the order of K_1 value of metal complexes is as follows:

   $$Fe^{3+} > Mg^{2+} > Li^+ \qquad \text{I set}$$
   $$Mg^{2+} > Ca^{2+} > Sr^{2+} > Ba^{2+} \qquad \text{II set}$$

In the first set, the metal ions are of comparable size, but with increase in the charge of the metal ion, charge density is more and hence Fe^{3+} forms more stable complexes than Mg^{2+} or Li^+. In the second case, for the metal ions with the same charge, complex-forming tendency decreases with an increase in size from Mg^{2+} to Ba^{2+}.

Hydration of the metal ions, resulting in the formation of weak aquo complexes, can be considered to be due to electrostatic attraction between the metal ions and negative end of the water dipole. The hydrational enthalpy changes, for the bivalent first transition series metal ions, should increase with an increase in atomic number, as there is a decrease in size with a consequent increase in the charge density. Plot of hydrational enthalpy change, ΔH, against atomic number, should give a straight line. But the plot shows a double hump shape as shown in Figure 2.1.

This can be explained in terms of crystal field theory. It is expected that there will be additional contribution to ΔH, due to the CFSE, liberated during the hydration process. In the case of Ca^{2+} (d^0), Mn^{2+} (d^5) and Zn^{2+} (d^{10}), where CFSE is zero, the hydration enthalpies are directly proportional to the atomic number, and the ΔH values follow the straight line relationship. In the other cases, there is a deviation from the straight line due to the CFSE.

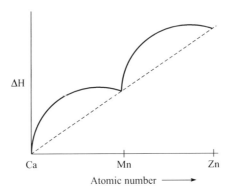

FIGURE 2.1   Plot of hydrational enthalpy change vs atomic number.

It is further known that M–L interaction is not purely electrostatic in nature, but there is a covalent interaction between the metal ion and ligand orbitals which results in donation of a pair of electrons from the ligand to the metal ion. Thus, the metal ion acts as a Lewis acid. The greater the Lewis acidity of the metal ion, the more stable the complex formed with the Lewis base ligands.

Alkali and alkaline earth metal ions, which are more basic, form less stable complexes with Lewis base ligands. Transition metal ions with higher acidity and with vacant d orbitals (which act as capture levels for the ligand electrons) form more stable complexes.

Considering the contribution of both the electrostatic interaction and the covalent interaction in the formation of complexes, Irving and Williams proposed that the stability of the metal complexes depends on the overall second ionisation potential of the metal ion and reciprocal of the ionic radius.

Second overall ionisation potential determines the electronegativity of the metal ion and hence is directly proportional to its Lewis acidity. The ionic size of the metal ion is inversely proportional to the charge density, and hence, the reciprocal of ionic radius is proportional to the electrostatic attraction of the metal ion for the ligand, that is, its Lewis acidity.

On this basis, the bivalent metal ions of the first transition series are arranged in the following order, as per their tendency to form the complexes with one specific ligand:

$$Ti(II) < V(II) < Cr(II) > Mn(II) < Fe(II) < Co(II) < Ni(II) < Cu(II) > Zn(II)$$

The above order is called the Irving–Williams order. The anamolous positions of Cr(II) d^4 and Cu(II) d^9 are due to the Jahn–Teller effect, which brings additional CFSE. Cr(II) and Cu(II), octahedral complexes, being doubly degenerate in ground state, undergo distortion and get stabilised, and hence, Cu(II) and Cr(II) form more stable complexes than the metal ions on both sides.

The order of Ni(II) and Zn(II) complexes may also vary in the case of the complexes of different ligands. This is because Ni(II) complexes have contribution from CFSE, whereas, for Zn(II) complexes, CFSE is zero. Hence, in the case of ligands, where CFSE is low and electrostatic factors predominate, Zn(II) forms more stable complex than Ni(II), whereas, in the case of ligands, wherein crystal field contribution is more, Ni(II) forms more stable complex than Zn(II).

2. **Nature of the ligand**: The ligand acts as a Lewis base in the formation of the complex. The greater the basicity of the ligand, the more stable will be the complex formed. A linear relationship between the formation constant of the complex, $\log K_{ML.m}^{M}$, and the protonation constant of the ligand, LH_n, that is, $\log K_{L.Hn}^{L} = \dfrac{[LH_n]}{[L][H]^n}$, is expected.

The greater the basicity of the ligand, the greater will be the value of the protonation constant $\log K^{1}_{LHn}$, and higher will be the stability constant of its complex with the metal ion. However, there are deviations observed from the linear relationship. This is because of the following factors.

i. **Formation of chelate ring:** If the ligand has more than one coordination site and occupies more than one coordination position around the metal, thereby forming a ring, the stability of the complex is more. The greater the number of coordinating sites on the ligand (denticity), the more the number of rings formed and hence the greater will be the stability of the complex. This is called "chelate effect". The order of the stability of the amine complexes of a metal ion is

Triethylene tetramine > Diethylene triamine > Ethylene diamine > Methyl amine
(Tetradentate)   (Tridentate)   (bidentate)   (monodentate)

Nitrilotriacetic acid (tetradentate) and ethylenediamine tetraacetic acid (EDTA, hexadentate) are also known to form very stable complexes. Though there is more than one M–L bond in a chelate, the enthalpy change involved in the formation of each metal–ligand bond, is same as in a monodentate ligand. Hence, there is no favourable enthalpy change, in the formation of the individual bond, in chelate formation. Hence, stabilisation in chelate formation is considered to be due to favourable entropy effect (to be discussed later).

ii. **Ring size:** A five-membered chelate ring, without double bond, is more stable than a six- or higher-membered ring. With increasing size of the ring, there is a strain, leading to instability.

Ethylenediamine (en) forms less stable complex than propylenediamine, but forms more stable complex than 1,3-diaminopropane (Figure 2.2).

Both en and propylenediamine form five-membered rings. However, propylenediamine, with the electron releasing CH_3 group over the C atom, is more basic than en. Hence, propylenediamine forms more stable complex than en. 1,3-Diaminopropane is similarly more basic than en, but it forms a six-membered ring and hence forms a less stable complex than en.

By similar reasoning, it can be explained that α-alanine forms more stable complex than glycine, but β-alanine forms less stable complex (Figure 2.3).

However, if there is a double bond in the ring, six-membered ring is more stable than a five-membered ring. For example, 1,3-diketone (acetylacetone) forms more stable complex than 1,2-diketone (diacetyl) (Figure 2.4).

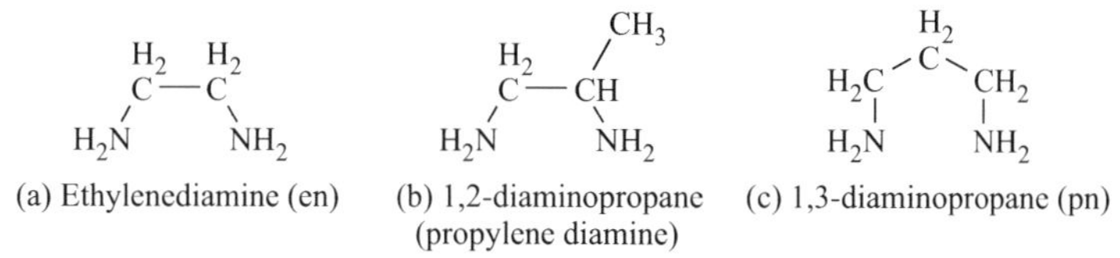

(a) Ethylenediamine (en) (b) 1,2-diaminopropane (c) 1,3-diaminopropane (pn)
         (propylene diamine)

FIGURE 2.2 Bidentate amine ligands (a) ethylenediamine, (b) 2,3-diamino propane and (c) 1,3-diamino propane.

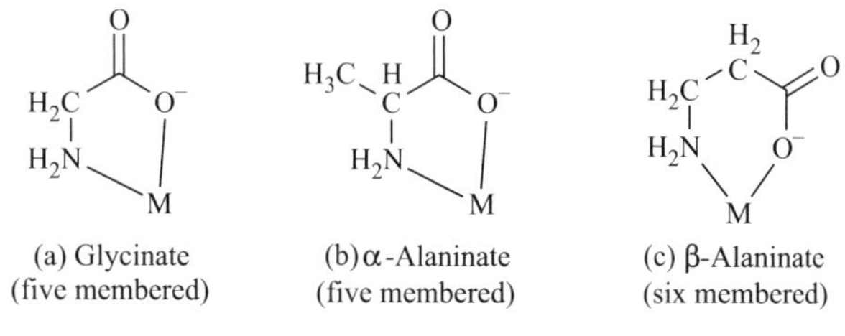

(a) Glycinate   (b) α-Alaninate   (c) β-Alaninate
(five membered) (five membered) (six membered)

FIGURE 2.3 Chelate formation between metal, M and (a) glycine, (b) α-alanine and (c) β-alanine.

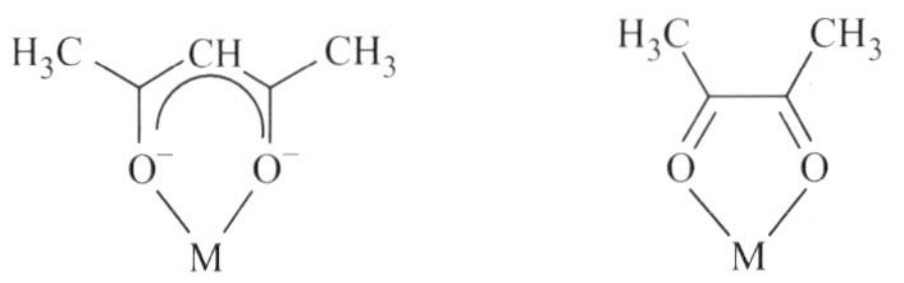

(a) Six membered chelate ring   (b) Five membered chelate ring

FIGURE 2.4 Metal chelates formed between M and diketones (a) six membered ring with acetylacetone and (b) five membered ring with diacetyl.

  iii. **Substitution on the ligand:** Substitution on the ligand may increase or decrease the basicity of the ligand, due to inductive or mesomeric effect. This in turn affects the coordinating tendency of the ligand with the metal ion.

  For example, α-alanine forms more stable complex than glycine, because the substitution of hydrogen by methyl group increases the basicity of amino nitrogen in α-alanine. Similarly, β- and γ-picoline form more stable complexes than pyridine, because the inductive effect of the methyl group makes the ring nitrogen more basic in β- and γ-picolines.

  Substitution of hydrogen by a functional group, in the compound, may introduce an additional coordinating site and thus result in the formation of stable chelate. For example, bidentate glycine forms more stable complexes than monodentate acetic acid, due to the additional $-NH_2$ coordinating site, and bidentate salicylaldehyde forms more stable complex than monodentate benzaldehyde, due to the introduction of the additional coordinating $-OH$ group, at the ortho position to the aldehydic group.

  Substitution may also lead to the destabilisation of the complex due to steric hindrance. For example, though α-picoline is more basic than pyridine, β- or γ-picolines, it forms less stable complex, due to the steric hindrance by the methyl group at the ortho position, during coordination of pyridine nitrogen with the $M(H_2O)_m^{n+}$.

  N-Alkyl-substituted ethylenediamine forms less stable complex than unsubstituted ethylenediamine, due to the steric hindrance by the alkyl group. The larger the number of substituting alkyl groups and the greater their bulk, the lesser the stability of the complex, due to increasing steric hinderance.

  In both the above cases, the hinderance is due to the groups, introduced on the front side of the ligand. This is termed forward or "F" strain. The bulky group on the back side of the ligand can also cause hindrance to the coordination of the ligand with the metal ion. For example, acetyl acetone forms more stable complex than β-diketones with the substitution of bulky alkyl group at the Y carbon atom. This is called backward or "B" strain.

  iv. **Nature of the ligand atom:** The simplest effect, that the coordinating atom has on the ligand, is that it determines the basicity of the ligand and consequently the stability of its complexes with metal ions. However, there are some more factors associated with the nature of the ligand atom, which affect the stability of the complex.

It is interesting to study the relative tendencies of nitrogen and oxygen as coordinating atoms. In the ligands (e.g. NH_3 or RNH_2), nitrogen has one lone pair of electrons, but oxygen has two pairs of electrons, as in H_2O or ROH. From the observation of the formation constants of various metal complexes, it can be inferred that the majority of non-transition metals have greater affinity for oxygen than for nitrogen. There are, however, some exceptions to this generalisation. Sidgwick has classified the metals into three classes, depending on their tendency to combine with oxygen or nitrogen.

$$O > N - Mg^{2+}, Ca^{2+}, Sr^{2+}, Ba^{2+}, Ga^{3+}, Si^{4+}, Ge^{4+}, Sn^{4+}, V^{5+}$$

$$O = N - Be^{2+}, Cr^{3+}, Fe^{2+}, Pt \text{ metal ions.}$$

$$N > O - Cu^{2+}, Ag^+, Au^+, Cd^{2+}, Hg^{2+}, V^{3+}, Co^{3+}, Ni^{2+}$$

It can be noticed that all the members of the first type are inert gas like ions. Those of the third type include transition metal ions with nearly filled d levels. The second type consists of metals with intermediate character.

In the case of the ligands, where the coordinating atoms are in the similar environment (e.g. NR_3, OR_2), interpretation in terms of electronegativity is valid. In the case of type III metals, tendency of ligands to donate electron pair makes a significant contribution to the strength of coordination bond. These metals have available vacant d orbitals (capture levels) and can accept pair of electrons from the ligand atom. With the increase in the electronegativity of the ligand atom, the tendency of the ligand to donate the electrons goes on decreasing, resulting in the lowering of the strength of M–L bond. Since N is less electronegative than oxygen, the order M–N > M–O holds good in complexes of the type III metal ions. Furthermore, these metals have filled d orbitals, and hence, there will be increasing repulsions between the non-bonding d orbital electrons of the metal ions and the non-bonding lone pairs in the orbitals of these small ligand atoms. Nitrogen has no extra lone pair of electrons. As oxygen has a non-bonding lone pair, there is a repulsion in the case of M–O bond formation. This contributes to greater stability of M–N bond, compared to M–O bond.

The type I metal ions are, however, electropositive non-transition metal ions or transitional metal ions without any d electrons or fewer d electrons. The M–L bond in such cases is more or less electrostatic in nature. The more electronegative coordinating atom, with higher negative charge density, can be expected to form more stable complexes with such metal ions. As these metal ions have no d electrons, there is no repulsion with the non-bonding lone pair of electrons of the ligand atom. Hence, such metal ions form more stable complexes with oxygen-coordinating ligands and therefore the order M–O > M–N.

Another point which needs attention is that donor atoms, which provide large ligand field splitting, form their strongest complexes with the metal ions which are particularly sensitive to ligand field stabilisation, whereas donor atoms which produce small ligand fields tend to form relatively more stable complexes with cations which are less sensitive to ligand field stabilisation. Nitrogen-coordinating ligands form stronger ligand fields, directed towards metal ions, than the oxygen-coordinating ligands. The type III metal ions, with d electrons, are sensitive to ligand field stabilisation and hence form more stable M–N bond. The type I metals, however, which have little or no contribution from the ligand field stabilisation and hence form less stable M–N bond.

The study of the relative order of M–L bond strength in complexes of the ligands, containing coordinating atoms of the same group, is another interesting field of study. It has been observed earlier that the complex-forming tendency of a ligand increases with the increase in its basicity. The non-transition metal complexes of trialkyl ammonia are found to be more stable than the corresponding trialkyl phosphine and arsine complexes, and this is in agreement with the basicity order, $NR_3 > PR_3 > AsR_3$. Thus, the non-transition metals are found to form more stable complexes with ligands containing the earlier members of a group as the coordinating atom.

However, in the case of transition metal complexes, there is a reversal in the order, that is, $AsR_3 > PR_3 > NR_3$ or Se > S > O.

Though the above pattern, of the relative coordinating tendency of the ligand atoms with metal ions, is not always valid, the metals have broadly been classified into two classes by Ahrland, Chatt and Davies. The class A consists of non-transition metals and earlier members of transition series in their common oxidation state. They form more stable complexes with the ligands having first atom in each group (N, O and F) as ligand atom. The class B metal ions consist of the later members of transition series in normal oxidation state or the earlier members of the transition series in lower oxidation states. They form more stable complexes with the ligands, having later atoms in a periodic group (N < P < As, O < S < Se < Te and F < Cl < Br < I) as the coordinating atom.

In the cases of many metals, class A and class B characters are very well defined. However, there are borderline cases, which exhibit different characters, depending on the oxidation state or the specific group of ligand atom, with which coordination takes place. Thus, the border between class A and class B types of metals is not very sharp.

Pearson, on the basis of his independent observations, classified the metals into hard and soft acids, depending on their greater or lesser tendency, respectively, to receive the electrons. He termed the ligands with higher tendency to donate their electrons as hard bases and the ligands with lesser tendency to donate the electrons as soft bases. He observed that hard acids have greater tendency to combine with hard bases and vice versa. Pearson's hard and soft acids are alternative nomenclature for class A and class B metal ions, and the terms hard and soft bases correspond to ligands containing earlier and later elements of the group as coordinating atoms, respectively.

Attempts have also been made to explain class A and class B characters of metal ions in terms of wave mechanical concept of M–L bond. According to Pauling, besides L $\rightarrow$ M σ bonding, M $\rightarrow$ L π interaction can further stabilise M–L bond. Such π bonding is possible only between transition metal d_π orbital (e.g. d_{xy}, d_{xz}, d_{yz} in octahedral complexes) and vacant p or d, π acceptor levels on the ligand atom. In the cases of complexes of pyridine, 2,2′-dipyridyl, cyanide and carbonyl, having vacant π orbitals, M $\rightarrow$ N or M $\rightarrow$ C, d_π–p_π interaction can take place, and this stabilises the complexes. However, the coordinating atoms, which are first members of a group, like nitrogen and oxygen in NR_3 and OR_2, respectively, have no suitable vacant d_π orbitals, to accept electrons from the d_π orbitals of the metal ion. On the other hand, sulphur and phosphorus have vacant 3d orbitals, which can be used for d_π–d_π bonding. Earlier transition metal ions, in lower oxidation state, or later transition metal ions in normal oxidation state, have d_π electrons, which can favourably be donated to vacant ligand d_π orbitals, leading to d_π–d_π interaction. σ and π bonds mutually stabilise each other. Hence for the complexes of these metal ions, the stability is in the order P > N or S > O. This accounts for their class B character.

In the case of non-transition metal ions and transition metal ions without available d_π electrons, back donation M $\rightarrow$ L is not possible and only L $\rightarrow$ M σ bonding accounts for the stability of their complexes. The ligands with N, O or F, as coordinating atoms, form more stable σ bonds, and hence, the above metals form more stable M–N or M–O bonds, rather than M–P or M–S bonds, respectively. This explains the class A character of the metal ions.

The concept of π bonding can also be explained in terms of molecular orbital theory. Formation of M $\rightarrow$ L π bond finds support in IR, NMR, Mössbauer spectral and X-ray studies in the case of phosphine, carbonyl and cyanide complexes. However, there are evidences to show that M $\rightarrow$ S π bonding may not be significant. Pb(II) with ns^2 configuration cannot affect significant M $\rightarrow$ S π interaction, and hence, greater stability of Pb–S bond should be attributed to some factors other than π bonding. According to Klopman, in the case of the M $\rightarrow$ L bond, with larger difference in the energies of the outer orbitals of the metal and ligand atoms, the extent of overlap is less, resulting in smaller covalent interaction. Interaction is mostly ionic in character. However, if the difference in the energies of interacting orbitals is less, there can be better overlap. The interaction of class B metal ions and soft base ligands, like S containing one, can, therefore, be expected to be more covalent in nature. Thus, the greater stability of M–S bond in class B metal complexes can be explained by considering only the stabilisation of S $\rightarrow$ M σ bond compared to O $\rightarrow$ M bond.

Jorgenson has attributed the greater stability of the M–S bond, compared to M–O, to greater polarisation of S, compared to oxygen, by the metal ion.

It is interesting to observe that above classification, of metal and ligands, has relevance in bioinorganic chemistry. The bulk metal ions, present in the body, are class A type, alkali and alkaline earth metal ions. They are hard acids.

Class B or soft acid metal ions are present in body, in trace amounts, such as Fe, Zn and Cu in enzymes.

Large excess of these metal ions can have toxic effects.

The organic ligands, of biochemical importance, are hard bases with N or O coordination, e.g. amino acids, proteins and polyphosphates. There are significant examples of M–S bond formation (from cysteine residue of the protein) in enzymes, but normally soft base ligands such as CO, CN and AsR_3 are poisonous, for the biosystems.

Thermodynamic Factors in Complex Formation

The formation constant K of the complex depends on the free energy change, as shown in the following equation:

$$\Delta G = - RT \ln K$$

$$= - 2.303 \, RT \, \log K \tag{2.10}$$

Hence, the higher the negative value of the free energy change (ΔG), the greater the stability of the complex.

Free energy change depends on the enthalpy and entropy change in the complex formation.

$$\Delta G = \Delta H - T\Delta S$$

$$\Delta H = \text{Enthalpy change} \quad \Delta S = \text{Entropy change} \tag{2.11}$$

For ΔG to be more negative, ΔH should be negative, and ΔS must be positive.

1. **Enthalpy change:** Enthalpy change is the heat evolved, in the formation of a gram mole of the complex. It can be determined by temperature coefficient method or calorimetric method. Enthalpy change depends on the strength of M–L bond. As stated earlier, according to Irving and Williams, it increases with increasing second overall ionisation potential of the metal and decreasing ionic radius of the metal ion. ΔH also increases with increasing CFSE during the formation of the complex. Furthermore, if there is π interaction in M–L bond formation, enthalpy change is more negative.

2. **Entropy change:** Both metal and ligands are solvated in solution and have specific freedom of movement. During the formation of the complex, the ligand gets bound to the metal ion and thus loses entropy. The solvent molecules around the metal ion and ligand are liberated, and thus, there is a gain in solvent entropy.

Net entropy change = Entropy gained by solvent molecules – Entropy lost by ligands

In the case of a monodentate ligand, binding with the metal ion, for each ligand getting bound, one solvent molecule is liberated from the coordination sphere of the metal ion. Hence, net entropy change is not high. However, in the formation of a chelate, the loss in the configurational entropy of the ligand is more, as it gets bound with the metal ion from more than one coordinating site. However, on binding of one ligand molecule, there is a liberation of two or more solvent molecules from the coordination sphere of the metal ion. Hence, the solvent entropy gained is more in chelate formation. Thus, the stabilisation of a chelate is due to favourable entropy change, rather than enthalpy change. If the chelate or complex formed is charged, the solvent molecules are weakly held in the second sphere of attraction. However, if the complex or chelate is neutral, the solvent molecules do not bind with the complex. Hence, the solvation entropy is more positive in the formation of a neutral complex. Thus, chelation, specially the formation of a neutral chelate, results in stability, favourably due to entropy effect.

Macrocyclic Ligand Complexes

A polydentate ligand, with the donor atoms incorporated in a cyclic backbone, is called a macrocyclic ligand. By definition, the macrocyclic ligand should have at least three donor atoms and at least nine atoms in the cyclic ring. A cyclic tetraamine shown in Figure 2.5 is an example.

A large number of macrocyclic ligand complexes are involved in biological systems. For example, chlorophyll, cytochromes and haemoglobin involve 16-membered porphyrin ring (to be discussed in Chapter 5), whereas vitamin B_{12} incorporates 15-membered corrin ring (to be discussed in Chapter 8).

It has been observed that the stability constant of a macrocyclic amine complex is more than that of the complex of the corresponding open-chained tetraamine. For example, stability constant of Cu(II)

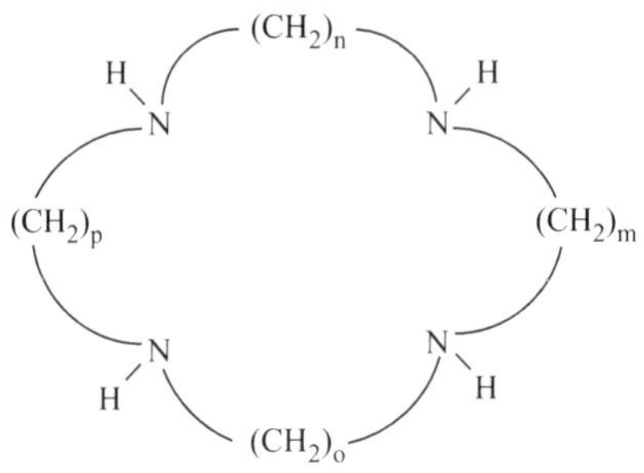

FIGURE 2.5 A cyclic tetramine – macrocyclic ligand.

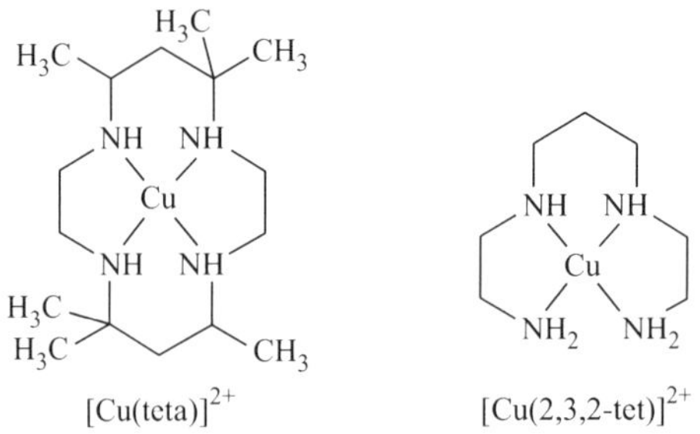

$[Cu(teta)]^{2+}$ $[Cu(2,3,2\text{-}tet)]^{2+}$

FIGURE 2.6 Macrocyclic ligand complexes of copper (a) $[Cu(teta)]^{2+}$ and (b) $[Cu(tet)]^{2+}$.

complex, of macrocyclic teta, is 10^4 times greater than that of related Cu(II) complex of 2,3,2-tet, as shown in Figure 2.6.

The greater stability of the macrocyclic ligand complex, compared to the complex of the corresponding open-chain ligand, can be explained by the consideration of the presence of one additional chelate ring. Margerum and Cabbiness attributed the greater stability of the macrocyclic ligand complex to the favourable solvation enthalpy change, in macrocyclic ligand complexation. It is expected that the desolvation of the metal ion is alike, in the binding of the tetra dentate open-chain or macrocyclic ligand. However, the macrocyclic ligand is less solvated in free state than the open-chain ligand. Hence, less energy is spent in the desolvation of the macrocyclic ligand, during its binding with the metal ion. This results in more negative, ΔH, on complexation of the macrocyclic ligand with the metal ion, leading to greater stabilisation of the complex. This effect was termed macrocyclic effect.

However, Paoletti and coworkers have indicated favourable entropy change in macrocyclic ligand complexation. During the binding of the open-chain tetra amine with the metal ion, the loss in entropy is more, because the free ligand has sufficient freedom of movement, which is lost in binding with the metal ion. However, the free macrocyclic ligand has low freedom of movement, and hence, loss of entropy is minimum when it gets bound to the metal ion. Thus, the loss in ligand configuration entropy is less in macrocyclic ligand complexation, leading to a positive entropy change. Hence, macrocyclic ligand forms more stable complex than the corresponding open-chain ligand.

Another factor determining the stability of the macrocyclic ligand complex is its cavity size. These ligands bind most strongly with the metal ion, whose ionic radius is such that it fits best in the ligand cavity. The bond strength between the ligand atom and the metal ion will be greater, when all the coordinating atoms can bind with the metal ion, without straining the ring. If the macrocyclic ring is too big, the bigger macrocycle is constrained, in binding with the metal ion, at all coordination sites. On the other hand, if the metal ion is too large and cannot be accommodated in the ligand cavity, either the ligand gets folded to coordinate with the metal ion or the metal ion stays above the plane of the ligand ring, resulting in less stable complex.

This phenomenon is illustrated in biochemical systems. In haemoglobin, bigger Fe(II) prefers a 16-membered porphyrin ring, whereas in vitamin B_{12}, smaller Co(III) prefers a 15-membered corrin ring.

Busch and coworkers have attributed the greater stability of the macrocyclic ligand complexes to the fact that in such complexes, the ligand atoms are suitably placed around the metal ion and the metal ion is simultaneously bound to number of donor atoms. Hence, the M–L bond is less susceptible to dissociation, leading to the greater stability of the complex. This factor of the stabilisation of the macrocyclic ligand complex was termed "multiple juxtapositional fixedness" (MJF).

Extensive unsaturation in the ring, as in protophyrin ring, in haem, may result in loss of flexibility in the ring, and this leads to increase in macrocyclic effect and stabilisation of the complex. There is also the possibility of interaction between metal $d\pi$ electrons and the delocalised π cloud of the rings. Thus, the M–L bond is stabilised in the macrocyclic ligand due to the M $\rightarrow$ L π delocalisation. Furthermore, due to the back donation from M $\rightarrow$ L, the lower oxidation state of the metal ion is stabilised.

Mixed–Ligand Complexes

The complexes, in which the metal ion has two or more types of ligands in its coordination sphere, are called mixed–ligand complexes. Such systems are also known as ternary complexes.

The formation of ternary complex can be considered to take place in three different ways.

One way is by simultaneous combination of two ligands with the metal ion is shown as

$$M + L + L' \rightleftharpoons MLL' \quad \beta_{MLL'}^M = \frac{[MLL']}{[M][L][L']} \tag{2.10a}$$

$$ML + L' \rightleftharpoons MLL' \quad K_{MLL'}^{ML} = \frac{[MLL']}{[ML][L']} \tag{2.11a}$$

$$ML' + L \rightleftharpoons MLL' \quad K_{MLL'}^{ML'} = \frac{[MLL']}{[ML'][L]} \tag{2.12}$$

$$\log K_{MLL'}^{ML} = \beta_{MLL'}^{ML} - K_{ML}^M$$

$$\log K_{MLL'}^{ML'} = \beta_{ML'L}^M - K_{ML'}^M$$

Another way of the formation of the ternary complex is the combination of the two binary *bis* ligand complexes.

$$ML_2 + ML_2' = 2MLL' \ldots (4)$$

On a statistical basis, in a solution containing a metal ion and two ligands L and L', the formation of the mixed-ligand complex MLL' is more favoured than the formation of binary complexes ML_2 and ML_2'. The tendency of mixed-ligand complex formation is determined by the reproportionation constant for the reaction as given in the following equation:

$$K_{reprop} = \frac{[MLL']^2}{[ML_2][ML_2']} \tag{2.13}$$

K_{reprop} should have a value of 4, from purely statistical considerations. This is because MLL' is formed by the two paths, Eqs. (2.11) and (2.12), whereas, ML_2 or ML_2' is formed by one path only, as shown in the following equation:

$$ML + L \rightleftharpoons ML_2 \text{ or, } ML' + L' \rightleftharpoons ML_2' \tag{2.14}$$

The concentration of MLL' should be double that of ML_2 or ML_2'.

$$K_{reprop} = \frac{[0.50]^2}{[0.25]^2[0.25]^2} = 4$$

$$\text{or,} \quad \log K_{reprop} = 0.6$$

Thus, log K_{reprop} = 0.6, if only statistical factors are responsible for the formation constant. However, the experimental values of the K_{reprop} constants for the mixed-ligand complexes differ from statistical expectation. The value of log K_{reprop} is either more or less than 0.6. It shows that factors other than statistical, as proposed by Bjerrum for binary complexes, are also operative in the case of ternary complex formation.

Watter and coworkers observed that for the [Cu(en)(ox)] complex, the value of log K_{reprop} was higher than the expected value of 0.6. This can be explained by the electrostatic factor, which is operative. During the formation of [Cu(ox)$_2$] from [Cu(ox)], the incoming ligand ox^{2-} is repelled by the existing ox^{2-}. In other words, [Cu(ox)]^0 has more repulsion for second ox^{2-} than [Cu(en)]$^{2+}$. Furthermore, the solvation entropy in the formation of neutral [Cu(en)(ox)] is more positive than that in the formation of [Cu(ox)$_2$]$^{2-}$. Thus, both enthalpy and entropy factors favour the formation of [Cu(en)ox], resulting in higher K_{reprop} values. (en = ethylenediamine, ox = oxalate anion).

Steric factors also affect the formation of mixed-ligand complexes. It has been observed that the steric hindrance between the two molecules in the formation of the binary complex [Cu(et$_2$ en)$_2$]$^{2+}$ (et$_2$ en$_2$ = N,N-diethyl ethylenediamine), with the bulky group on the coordinating atom, is more than that in the formation of mixed-ligand complex [Cu(en)(et$_2$ en)], and hence, the formation of the ternary complex is more favoured, and the reproportionation constant is high.

However, a high value of log K_{reprop} does not indicate absolute stability of the ternary complex. It indicates the relative stability of the ternary complex (MLL′) with respect to the binary complexes ML$_2$ or ML$_2'$. If the stability of ML$_2$ or ML$_2'$ is less, due to electrostatic or steric reasons, more of MLL′ is formed, and the value of log K_{reprop} is higher than 0.6.

Another way of quantifying the stability of the ternary complex is to determine the value of Δ log K.

$$\Delta \log K = \log K_{MLL'}^{ML} - \log K_{ML}^{M} \tag{2.15}$$

$$\Delta \log K = \log K_{MLL'}^{ML} - \log K_{ML'}^{M} \tag{2.15a}$$

It indicates the difference between the tendency of a ligand to bind with the metal ion, already bound to another ligand and the tendency of the ligand to bind with the free metal ion. It is evident from Eqs. (2.14) and (2.15) that the effects of both the ligands are mutual and of the same magnitude in the formation of the ternary complex. The binding of both the ligands is equally stabilised or destabilised, in the second step of the formation of the ternary complex.

From statistical consideration, Δ log K is expected to be negative. This is because when the first ligand (L) combines with a given multivalent ion, it has more coordination positions available for binding, than when it combines with metal already bound to another ligand L′. Hence, the order log $K_{ML}^{M} > $ log $K_{ML'L}^{ML}$ usually holds and one expects to observe negative values for Δ log K. In case, where L and L′ are bidentate ligands, for a complex having regular octahedron geometry, log $K_{(oh)}$ = −0.4. For a square planar geometry, the theoretical value of Δ log $K_{(sp)}$ should be −0.6. For a distorted octahedral ternary complex, the statistical value of Δ log $K_{(DO)}$ has been calculated to be −0.9. Hence, an experimentally determined value of Δ log K > −0.9, for a distorted octahedral ternary complex, indicates that the formation of ternary complex is favoured. A lower value indicates destabilisation of the ternary complex. This stabilisation or destabilisation, not expected from statistical consideration, is due to nonstatistical factors.

There are many nonstatistical factors that stabilise or destabilise the ternary complexes. They can be classified as follows:

1. **Effect of σ or π-acidic character of the ligand:** When a σ-bonding ligand gets bound with the metal ion, it transfers the pair of electron to the metal ion. Thus, the concentration of electrons around the metal ion increases and its electronegativity is reduced. As a result, the tendency of the metal ion, to accept the electron, is reduced. Hence, ML binds with the second ligand L′ less firmly than the binding of L′ with the free metal ion M^{n+}. Hence, Δ log K becomes more negative than expected from statistical consideration alone. However, it has been observed that in the complexes of the type [M(dipy)L], the mixed-ligand formation constant $K_{M\,dipyL}^{M\,dipy}$ is much

higher than expected from statistical consideration. 2,2'-Dipyridyl (dipy) is a bidentate ligand bound to the metal ion through N $\to$ M σ bonds. Besides this, there is also M $\to$ N π bond formation, by the back donation of electrons from the metal $d\pi$ orbitals to the vacant delocalised $p\pi$ orbitals of the ligand. This, $d\pi$–$p\pi$ interaction, does not allow the concentration of electrons on the metal ion to increase significantly, on the binding of dipyridyl. In other words, the electronegativity of $[M(dipy)]^{n+}$ is almost same as that of $[M(H_2O)x]^{n+}$. Hence, the tendency of L to combine with $[M(dipy)]^{n+}$ is almost same as to combine with $[M(H_2O)x]^{n+}$, and $K_{M\,dipy\,L}^{M\,dipy}$ becomes nearly equal to K_{ML}^M (Figure 2.7).

In the case of MAL type of complexes, where A is a π-acidic ligand, M $\to$ A π interaction results in increase in class A character of the metal ion in MA. This brings in discriminating behaviour of MA towards the secondary ligand L, coordinating through N–N, N–O$^-$, O$^-$–O$^-$. The tendency of binding is in the order O$^-$–O$^-$ > N–O$^-$ > N–N. The order can also be explained in terms of electron repulsion concept. In the formation of M–L complex, where L coordinates through O$^-$, there is a repulsion between metal $d\pi$ electrons and O$^-$ lone pair of electrons, as discussed earlier. In the ternary complex, MAL, M $\to$ A π bonding reduces π electron density over the metal ion, and hence, the lone pair of electrons over O$^-$ has to face less repulsion from d electrons, while O$^-$ coordinating L combines with MA, than when combining with M. The effect is more significant in the complexes of copper (II), so that the value of Δ log K in [Cu A Catecholate^{2-}] complex is positive.

Substitution over the π-acidic ligand, which increases its π-acidic character, makes the Δ log K of MAL complexes more positive. In the case of O$^-$–O$^-$ coordinating bidentate ligand L, substitution over L, which increases the charge density over O$^-$, stabilises the ternary complex further, making Δ log K more positive.

The study of the effect of a π-acidic ligand, in stabilising the ternary complex, is important, as in biological systems, there are instances of formation of ternary complexes, involving imidazole type of ligands with possibility of M $\to$ N π interaction. In the metalloenzymes, the electron density around the metal ion does not increase significantly on binding with the π acidic imidazole nitrogen of the histidine residue of the apoenzyme protein. Hence, the metal-apoenzyme can effectively bind with the substrates having O$^-$–O$^-$ (adenosine triphophate or ATP) or O-N (peptide) coordinating sites, resulting in the formation of stable apoenzyme-M-substrate ternary complexes.

2. **Charge on the ligands:** If one of the ligands in the ternary complex is neutral as in $[Cu(en)(ox^{2-})]$, there is no electrostatic repulsion. However, if both ligands are charged $[Cu(gly-)(ox^{2-})]$, there is a repulsion felt from negatively charged ox^{2-}, when glycine binds with Cu ox^{2-}. Hence, $K_{Cu\,ox\,gly}^{Cu\,ox}$ is much less than $K_{Cu\,gly}^{Cu}$, and therefore, Δ log K is more negative. The higher the charge on the first ligand, the greater the electrostatic repulsion and the more negative the Δ log K values. For example, Δ log K of [Cu-salicylate-glycinate$^-$] complex is less negative than [Cu-salicylate-ox^{2-}], as there is a greater repulsion between salicylate and oxalate, with two negative charges.

3. **Steric hindrance due to bulky groups:** If one of the ligands has a bulky group, there is a steric hindrance, during its coordination with the metal ion, bound to another ligand. For example, Δ log K in [M-dipyridyl-*N*-alkyl ethylenediamine] complex is more negative than in [M-dipyridyl-ethylenediamine].

4. **Tridentate character of one ligand:** Δ log K in the case of [CuAL] complex, where A = tridentate iminodiacetate and L is another bidentate ligand, is more negative than in analogous [NiAL] complex.

This is because the tridentate ligand occupies three equatorial positions. The incoming ligand L–L has to occupy one equatorial and one axial position. In Cu(II) complexes, due to the Jahn–Teller (JT) distortion, there is lengthening of bond along axial direction. The ligand L feels more strain in occupying one equatorial and one axial position, resulting in lower value of K_{CuAL}^{CuA} and a more negative Δ log K value. In Ni(II) complexes, L is not strained in occupying the axial position, as there is no JT distortion.

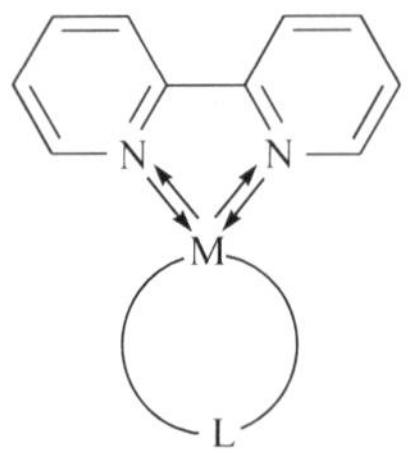

FIGURE 2.7   Mixed ligand complex of type [M(dipy)L].

5. **Intramolecular interligand interactions:** The most predominant interaction which stabilises the ternary complexes in biological systems is intramolecular interligand interaction. These are of two types:

   a. **Rigid interaction:** One of the ligands is coordinated rigidly with the metal ion, and the other coordinated ligand has a flexible non-coordinating group interacting with the rigid ligand. For example, there is a rigid interaction between a coordinated aromatic diamine and the non-coordinated base of a nucleotide as in [Cu-2,2'-dipyridyl-adenosine triphosphate] (Figure 2.11b).

   b. **Flexible interaction:** If both the ligands have free non-coordinated group, there can exist flexible interaction between them. For example, in [Cu-ATP-tryptophan] (Figure 2.11a), there is stabilisation due to interaction between the non-coordinated indole part of tryptophan and adenine base part of ATP. Similar interaction, between non-coordinated side groups of the two amino acids, stabilises the ternary complex, [Cu-phenylalanine-tryptophan].

   In both the types of interligand interactions, the nature of intramolecular forces operative may vary. They are of following types:

   i. **Hydrogen bonding:** Formation of hydrogen bonds between the parts of the two coordinated ligands, as in *bis*-dimethyl glyoxamate Ni(II), is known to exist in solid mixed-ligand complexes also. In the crystal structure of the ternary complex of [Cu-(5'-inosine monophosphate)(diethylene triamine)], interligand H bonding between oxygen of the phosphate group of 5'-inosine monophosphate (IMP) and –NH$_2$ group of diethylene triamine has been observed.

   In solution also, the interligand hydrogen bonding can stabilise the ternary complex. Stabilisation of the ternary complex [Cu-dipeptide-L] in solution is more when L = tyrosine than when L = phenylalanine, because there is an intramolecular interligand H bonding between the free COO$^-$ of dipeptide and the phenolic –OH of non-coordinated part of tyrosine (Figure 2.8).

   The greater stabilisation of the ternary complex may also be due to hydrogen bonding between two ligands mediated through a solvent water molecule; for example,

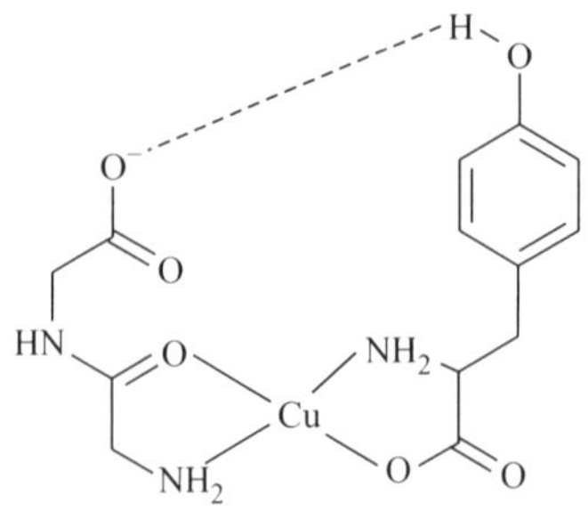

FIGURE 2.8   H-bonding in ligands of ternary complex of copper – inosine triphosphate and diethylene triamine, [Cu(IMP)(DET).

in [Cu-catecholate-dipeptide], the stabilisation has been attributed to H bonding between phenolate O⁻ and –NH$_2$ or –NH group of the peptide, through water molecule (Figure 2.9).

ii. **Ionic or electrostatic interaction:** Interligand interaction may be due to electrostatic attraction between the two ligands, for example, the positively charged protonated amine part of an amino acid and free negatively charged carboxylate part of another ligand. For example, in [Cu-Hist H⁺-dipeptide] complex, where Hist H⁺ is the histidine coordinated like alanine with free protonated imidazole site, there is an electrostatic attraction between Hist H⁺ and the non-coordinated carboxylate part of the dipeptide (Figure 2.10). Similar electrostatic interaction exists in [M-arginine-amino acid], between arginine NH$_3^+$ and the COO⁻ of the other amino acid.

iii. **Stacking interaction:** Rigid intramolecular aromatic ring stacking, between two suitable ligands within a ternary complex, was first shown to occur in the ternary complex [Cu-dipy-ATP or IMP] (Figure 2.11b). The evidence of π–π interaction between the 2,2′-dipyridyl (dipy) and the base part of ATP or IMP came from electronic spectral, NMR spectral and solution stability studies. X-ray structure determination of solids of such ternary complexes shows an intramolecular stacking between the aromatic moieties of the coordinated ligands.

Intramolecular stacking has also been observed in flexible interaction in the mixed-ligand complex of Cu(II) involving ATP and tryptophan (Figure 2.11a). UV spectral studies show a new peak at 295 nm which can be attributed to charge transfer between indole group of tryptophan and purine moiety of ATP. In ¹H NMR, positive shift of signals of H$_{(2)}$, H$_{(4)}$, H$_{(5)}$, and H$_{(6)}$ of tryptophan and H$_{(2)}$ and H$_{(8)}$ of ATP, compared to those of corresponding free ligand signals, confirms the formation of intramolecular stack between the indole residue of tryptophan and imidazole residue of adenine moiety of ATP.

In the complexes involving Cu-1,10-phenanthroline-phenylcarboxylates (C$_6$H$_5$ (CH$_2$)$_n$ COO⁻, n = 0–5), it was observed that the extent of stacking between 1,10-phenanthroline and the non-coordinated phenyl group of monodentate C$_6$H$_5$(CH$_2$)$_n$–COO⁻ depends on the flexibility

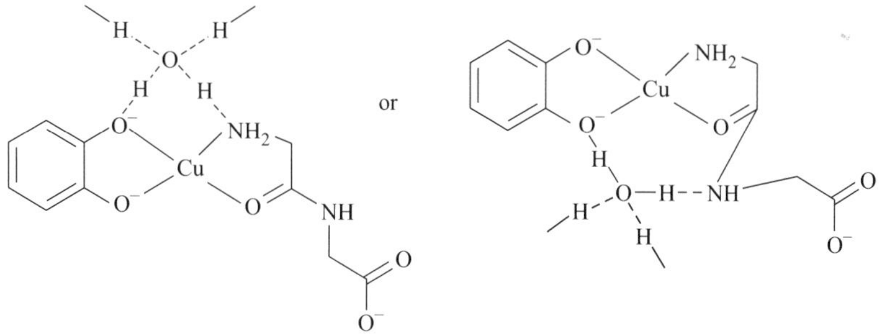

FIGURE 2.9　Water mediated H-bonding in ligands of ternary complex of copper with catechole and dipeptide.

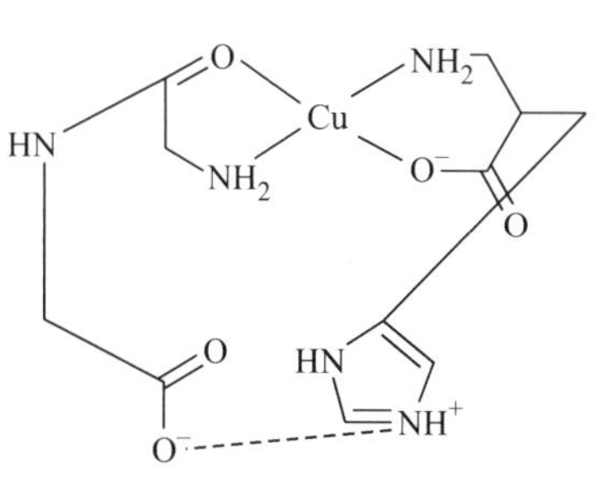

FIGURE 2.10　Interligand ionic interaction in ternary complex [Cu(Hist-H⁺)(dipeptide)] complex.

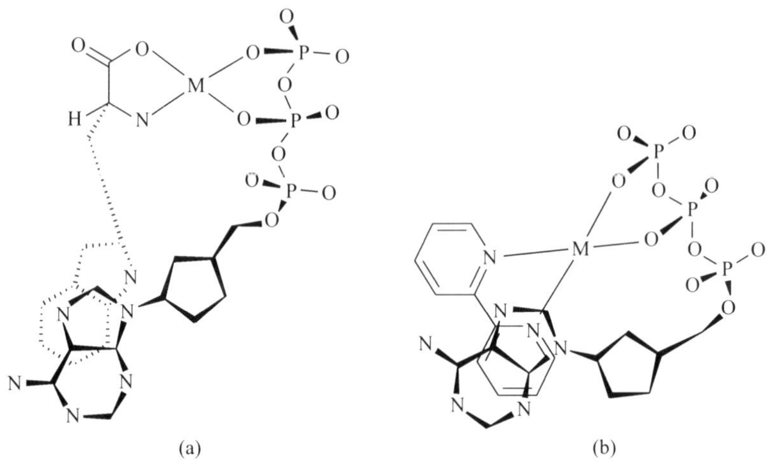

FIGURE 2.11   (a) Intramolecular aromatic ring stacking interaction between ligands in ternary copper complexes (a) flexible interaction in [Cu(ATP)(tryptophan)] and (b) rigid interaction in [Cu(dipy)(IMP).

of the coordinated monodentate ligand. The extent of stacking interaction was found to be maximum for phenyl acetate (n = 1). With increasing chain length of the phenyl carboxylate, the increasing flexibility lowers the ligand–ligand interaction. It was also observed that with the increase in the sizes of the involved aromatic ring systems, the extent of stacking increases. For example, replacement of 2-phenyl acetate by [2-(α-napthyl)] acetate promotes the interaction with 1,10-phenanthroline.

iv.   **Hydrophobic interactions:** Intramolecular hydrophobic interactions are closely related to the described aromatic ring stacking interactions. The first such interaction was shown to occur in the ternary Cu^{2+} or Zn^{2+} complexes, between metal bound, (2,2-dipyridyl, dipy) 1,10-phenanthroline and trimethylsilyl moiety of the second ligand, 3-(trimethylsilyl) propionate. The methyl signal, of the trimethylsilyl group, in the NMR spectrum is shifted to higher field in [Cu or Zn-1,10 phenanthroline-trimethylsilyl-propionate], compared to the free ligand. This supports interligand interaction.

Complexes of the type [MAL], where A = 2,2'-dipyridyl or 1,10-phenanthroline and L = amino acid, with side group, like valine and leucine, show hydrophobic interaction between the aliphatic side chain of the amino acid and the aromatic ring of dipyridyl or 1,10-phenanthroline (Figure 2.12a). Similar interaction has been suggested between non-coordinating alkyl group of the amino acid and the purine moiety of ATP in mixed-ligand complexes [M-ATP-amino acid with side group] (Figure 2.12b).

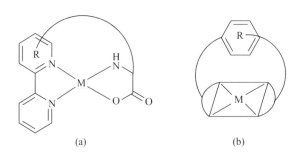

(a)   (b)

FIGURE 2.12   Intramolecular hydrophobic interaction in complexes of type (a) [M(dipy)(aminoacid)] and (b) [M(ATP)(aminoacid).

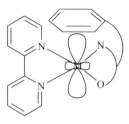

FIGURE 2.13 Ligand side group – metal interaction in ternary complex [M(dipy)(aminoacid).

The hydrophobic and aromatic ring stacking interactions are solvent dependent. Such interactions are stronger in polar solvents. Non-polar solvent molecules interact with the hydrophobic or stacking aromatic groups of the ligand and thus reduce the intramolecular ligand–ligand interaction, and there is a decrease in the stability.

Recently, it has been proposed that in complex of above type (Figure 2.12a), where the aliphatic or aromatic non-coordinated group of one ligand cannot sufficiently overlap with the second coordinated ligand, dipyridyl or analogue, there can be partial hydrophobic interaction between the two ligands, leading to the stabilisation of the ternary complex. Alternatively, it can be considered that the non-coordinated aliphatic or aromatic part of the ligand tends to occupy space near the metal ion, rather than out in solution (Figure 2.13), due to interaction between the axial metal d orbital electron and the delocalised π electron cloud over the ligand side group. This bonding tendency is referred to as ligand side group–metal interaction.

A direct favourable hydrocarbon metal interaction, by less than the van der Waals distance, has only been shown for aromatic rings, in crystal structure determination of [M-dipyridyl-amino acids with aromatic side group] complexes.

It is evident that a possible intramolecular ligand–ligand interaction, in solution, does not occur in hundred per cent molecules, but it may occur in a certain number of complex species present. This gives rise to two isomers of the ternary complex MAB, that is, an open form, where there is no ligand–ligand interaction and a closed form, where ligand–ligand interaction exists. This then leads to an intramolecular equilibrium between these two isomers which may be represented as follows:

$$K_1 = \frac{[\text{MAB (closed)}]}{[\text{MAB (open)}]} \tag{2.16}$$

The greater the closed form, the higher the value of K_1 and the greater the stability of the ternary complex. Such intramolecular interligand interactions are of great importance in biological systems. In many biological reactions, involving metalloenzymes, the metal ion gets simultaneously bound to two biomolecules (apoenzyme and substrate), leading to formation of stable ternary complexes. The biomolecules have large non-coordinating side groups, and hence, there may be non-covalent interaction between them in ternary complexes, causing stabilisation of the metal-apoenzyme-substrate complex, leading to specific and selective nature of metalloenzymes.

Mechanism of Substitution Reactions in Complexes

Substitution Reactions

In biological systems, there are reactions, in which the ligand in a metal complex may be replaced by another ligand. For example, in a metalloenzyme substrate reaction, the product formed is displaced by new molecule of the substrate, or a toxic molecule with potential coordinating group may replace the apoenzyme from the metalloenzyme, inhibiting its activities. These are ligand substitution reactions.

The progress of such reactions does not only depend on the relative thermodynamic stabilities of the binding of the ligand biomolecules with the metal ion, but also depend on the mechanism of the substitution reaction.

Since the ligands are σ bases, they are nucleophiles, and hence, the ligand substitution reactions are nucleophilic substitution reactions. The substitution reaction in an octahedral complex can be shown as follows:

$$ML_6 \rightleftharpoons ML_5X + L \tag{2.17}$$

The reaction proceeds in two ways:

$$ML_6 \xrightarrow[\text{Slow}]{-L} ML_5 \xrightarrow[\text{Fast}]{+X} ML_5X \tag{2.17a}$$

The rate determining slow step is of first order, and hence, the reaction is said to proceed through SN_1 or dissociation mechanism. There is formation of a penta-coordinated intermediate (Figure 2.14a). There is no participation of the entering group in the transition state (T.S.) and hence has no contribution to the energy of the Transition state (T.S.) (ΔG^*).

The second type proceeds through an addition mechanism:

$$ML_6 + X \xrightarrow[\text{Slow}]{} ML_6X \xrightarrow[\text{Fast}]{} ML_5X + L \tag{2.17b}$$

The rate determining step is of second order, and hence, the mechanism of the reaction is SN_2 or associative. There is a formation of a hepta-coordinated intermediate (Figure 2.14b). Again, the entering group has no role on the activation energy of the T.S. (ΔG^*).

These are, however, two limiting cases. In many substitution reactions, the incoming group can enter, as soon as the leaving group departs. There may be different extent of rupture of M–L bond and different extent of M–X bond formation in the transition state. These are concerted reactions and are called interchange processes or I processes. These are mainly classified into two types.

The process, which resembles the transition state of dissociation process, is called Id process. In this, there is a weak bonding of the metal ion to both the entering and leaving groups, but the M–X bonding is very weak, so that M–L bond rupture is complete with practically no M–X bond formation (Figure 2.15a).

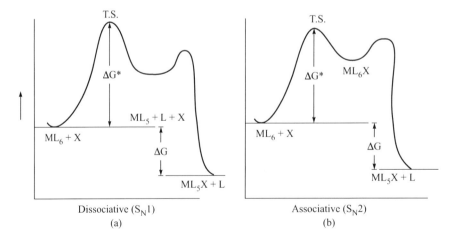

FIGURE 2.14 Energy diagram for substitution reaction in metal complexes (a) Dissociative mechanism, SN_1 (b) Associative mechanism, SN_2.

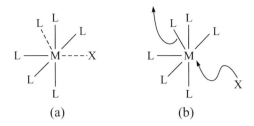

(a)   (b)

FIGURE 2.15 Transition state (TS) of (a) dissociative (Id) process (b) associative (Ia) process.

The associative interchange process resembles the transition state of associative mechanism and is termed Ia process (Figure 2.15b). In this, one of the ligands, L, tends to dissociate when the incoming ligand X enters. However, the metal ion is substantially bound to both the ligands L and X in the transition state (Figure 2.15b).

The mechanism of the reaction can be elucidated, by determining the order of the reaction. However, in most of the complexes, substitution is very fast and cannot be monitored by conventional techniques. In such cases, fast methods like stopped flow or temperature jump methods are used. In the case of Co^{3+} and Cr^{3+} complexes, the substitution reactions are relatively slower and can be studied by the slower spectrophotometric methods.

Whether the reaction proceeds through SN_1 or SN_2 path depends on the nature of the metal ion, the ligand to be substituted and the substituting ligand. The factors affecting the nature of substitution are as follows.

Conditions Favouring SN_1 Mechanism

i. The metal ion should be a weak Lewis acid, and the ligand should be a weak base, so that the metal ion forms weak bonds with the six ligands, and hence, the complex gets easily dissociated to five coordinated intermediates, before substitution. In the aquation reaction, the rate decreases in the following sequence of L.

$$[Co(NH_3)_5L] + H_2O \rightleftharpoons [Co(NH_3)_5(H_2O)] + L \qquad (2.17c)$$

$NO_3 > I^- > Br^- > H_2O > C^- > CH_3COO^- > NH_3$, that is, with increasing thermodynamic stability of Co–L bond.

ii. If the six ligands in the complex form σ bonds, there is an increase in electron density around the metal ion. As a result during substitution, an incoming σ bonding seventh ligad cannot be accommodated. The reaction has to proceed through dissociation mechanism.

In the substitution of X by H_2O, in cis- and trans-$Co(en)_2LX$, it has been observed that when L is electron donating group (σ or both σ and π donor) such as OH^-, NH_3, Cl^-, Br^- and NCS^-, the reaction proceeds by SN_1 mechanism.

Similarly in the aquation reaction,

$$[Co(en)_2(xy)(x-py)Cl]^{2+} + H_2O \rightleftharpoons [Co(en)_2(xy)(x-py)H_2O]^{3+} + Cl^- \qquad (2.17d)$$

where $(x - py)$ is substituted pyridine, it is observed that with the increase in the basicity of $x - py$, the rate of aquation reaction increases.

iii. If the ligands in ML_6 are bulky in nature, they produce steric hindrance to each other and prefer to undergo dissociation into five coordinated intermediate, for substitution to take place. For several complexes of the type $[Co(A - A)_2 Cl_2]^+$, where $A - A$ is ethylenediamine or *N*-alkyl ethylenediamine, the rate of substitution of Cl^- by H_2O increases with the increasing bulk of the alkyl group.

Conditions Favouring SN$_2$ Mechanism

i. The metal ion, incoming (X) and outgoing (L) ligands, should be such that the thermodynamic stability of M–L and M–X bonds should be high, so that the metal ion can hold seven ligands and seven coordinated intermediate ML_6X is formed.

ii. If the incoming ligand X is π acceptor in nature, it can accept the electrons from ML_6, where Ls are σ-bonding ligands, to form the intermediate ML_6X.

iii. For the formation of hepta-coordinated intermediate, both the incoming and outgoing ligands should not be bulky in nature.

Different Behaviour of Geometrical Isomers in Substitution Reactions

In the substitution reaction of chloride in an octahedral complex Co(en)$_2$ X Cl, it has been observed that for the π donor ligand X, promoting SN$_1$ reaction, cis isomer reacts faster than the trans isomer, whereas it is reverse in true for the case of π acceptor ligand X, which promotes SN$_2$ mechanism.

NCS$^-$, OH$^-$ and Cl$^-$ act as π donor X group. In the cis isomer of Co(en)$_2$ OH Cl, the Co–Cl bond can be easily dissociated, because the p orbital of OH$^-$ is suitably oriented to give effective overlap with the d^2sp^3 hybrid orbital, vacated by Cl. For the trans isomer, in the penta-coordinated transition state, effective π bonding with OH$^-$ can occur only after rearrangement to trigonal bipyramidal structure. This requires higher activation energy; therefore the SN$_1$ substitution reaction is slow in the trans complex.

When the ligand X is a π acceptor ligand like NO$_2^-$, substitution in trans isomer of Co(en)$_2$NO$_2$Cl is facilitated. This is because, in the back donation of dπ electrons from Co(III) to the vacant π orbital of NO$_2^-$, electron density is lower on both the octahedral faces, adjacent to NO$_2^-$; hence, the incoming group can enter from these sides and hepta-coordinated intermediate is more stabilised. Thus, in the case of the trans isomer, substitution by SN$_2$ mechanism is faster, when ligand X is a π acceptor ligand.

Base Hydrolysis

Substitution of a ligand in octahedral complex by water is called aquation. In the higher pH range, the substitution is by OH$^-$, and this is termed hydrolysis.

$$(ML_6)^{n+} + OH^- \longrightarrow [ML_5OH]^{(n-1)} + L^- \tag{2.18}$$

$$[Co(NH_3)_6]^{3+} + OH^- \longrightarrow [Co(NH_3)_5(OH)]^{2+} + NH_3 \tag{2.19}$$

Aquation proceeds by SN$_1$ mechanism, but base hydrolysis is apparently SN$_2$ type, depending on the concentration of both [Co(NH$_3$)$_6$]$^{3+}$ and OH$^-$.

H$_2$O and OH$^-$ do not differ significantly in nucleophilicity, and therefore if aquation proceeds through SN$_1$, there is no reason, why base hydrolysis should proceed through SN$_2$ mechanism. Hence, alternative mechanism, SN$_1$ conjugate base (SN$_1$CB), has been suggested for base hydrolysis.

Base hydrolysis proceeds in three steps. In the first fast step, one NH$_3$ of [Co(NH$_3$)$_6$] gets deprotonated, and proton so formed combines with the OH$^-$ forming water.

In the second slow rate determining step, the conjugate base (C.B.), [Co(NH$_3$)$_5$NH$_2^-$]$^{2+}$, loses one –NH$_3$ molecule. In the third fast step, [Co(NH$_3$)$_5$NH$_2^-$]$^{2+}$ combines with water molecule to form [Co(NH$_3$)$_5$OH]$^{2+}$.

$$[Co(NH_3)_6]^{3+} + OH^- \xrightarrow{\text{Fast}} \underset{\text{C.B.}}{[Co(NH_3)_5(NH_2)]^{2+}} + H_2O \tag{2.20}$$

$$\underset{\text{C.B.}}{[Co(NH_3)_5(NH_2)]^{2+}} \xrightarrow{\text{Slow}} [Co(NH_3)_4(NH_2)]^{2+} + NH_3 \tag{2.21}$$

$$[Co(NH_3)_4(NH_2)]^{2+} + H_2O \xrightarrow{\text{Fast}} [Co(NH_3)_5(OH_2)]^{2+} \tag{2.22}$$

So, the reaction proceeds through slow SN_1 mechanism (2.21), as in aquation. However, the formation of conjugate base $[Co(NH_3)_5NH_2]^{2+}$, which is a fast reaction, depends on the concentrations of both $Co(NH_3)_6$ and OH^- (2.20), and hence, apparently the reaction is SN_2 type. This is a special case of SN_1 reaction and is termed SN_1CB.

Substitution in Square Planer Complexes and Trans Effect

In the case of the substitution reactions in square planer complexes, $[ML_3 X \xrightarrow{X'} ML_2XX']$, the ligand X may direct the incoming ligand X' to occupy the position trans to it. This labilisation of ligands, trans to the ligand X, is called trans effect.

The ligands can be arranged in the following order of their tendency to bring trans directing effect, $CN \sim CO \sim NO \sim H^- > SR_2 \sim PR_3 > I^- > Br^- > Cl^- > py > NH_3 > OH^- > H_2O$.

By planning the sequence of addition of substituents, the synthesis of a specific isomer of a complex can be attained. For example, cis and trans $[Pt(NH_3)_2Cl_2]$ can be obtained as follows.

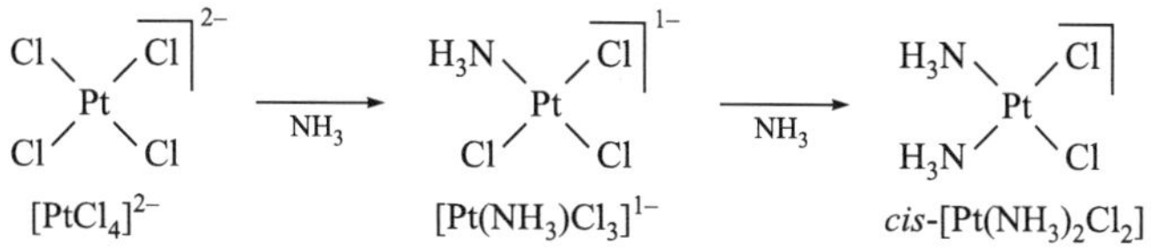

$$
[PtCl_4]^{2-} \xrightarrow{NH_3} [Pt(NH_3)Cl_3]^{1-} \xrightarrow{NH_3} cis\text{-}[Pt(NH_3)_2Cl_2] \tag{2.23}
$$

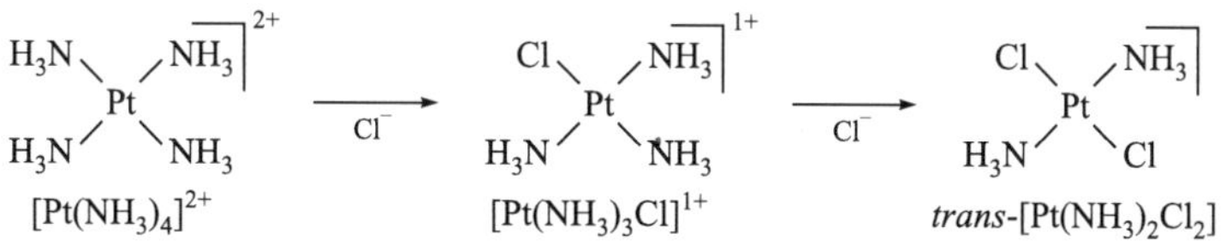

$$
[Pt(NH_3)_4]^{2+} \xrightarrow{Cl^-} [Pt(NH_3)_3Cl]^{1+} \xrightarrow{Cl^-} trans\text{-}[Pt(NH_3)_2Cl_2] \tag{2.24}
$$

Ethylene is also a trans-directing ligand, and the two isomers of the complex $[Pt (NH_3)Cl_2 C_2H_4]$ can be obtained, as follows.

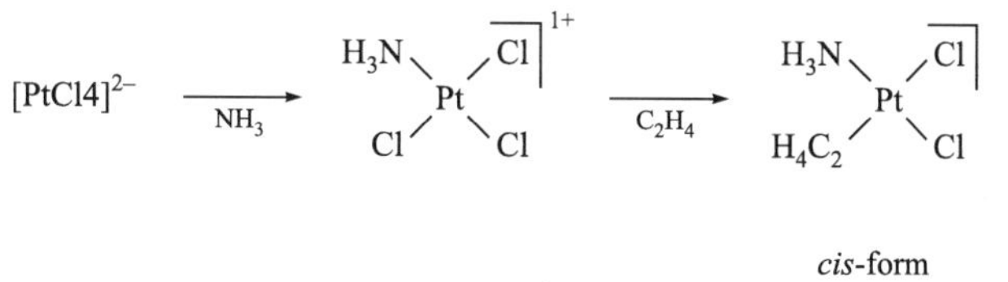

$$
[PtCl4]^{2-} \xrightarrow{NH_3} \quad \xrightarrow{C_2H_4} \quad cis\text{-form} \tag{2.25}
$$

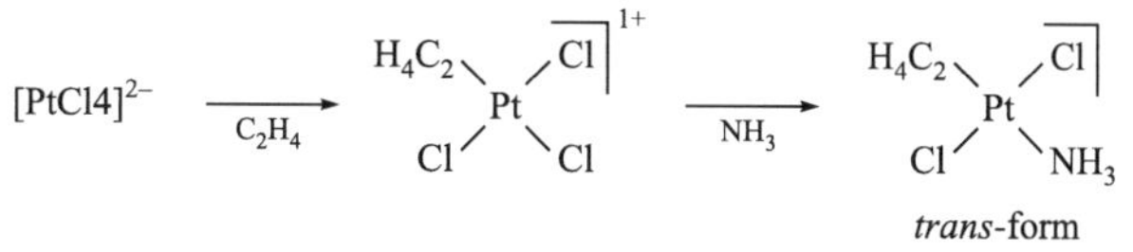

$$
[PtCl4]^{2-} \xrightarrow{C_2H_4} \quad \xrightarrow{NH_3} \quad trans\text{-form} \tag{2.26}
$$

There are two approaches to explain trans effect.

   i. Thermodynamic effect: This involves consideration of thermodynamic effect, as the factor for weakening of the bond in the direction trans to the trans-directing group. Accordingly, trans influence can be defined, as the extent to which the bond, trans to the ligand, is weakened, in the equilibrium state of the complex.

     Two theories have been proposed based on thermodynamic effect.

    a. **Polarisation theory:** It was proposed by Grinberg and is based on Fajan's rule of polarisation. The negatively charged ion or the negative end of the ligand dipole causes polarisation of the metal ion, creating $+\delta$ on the side of the metal ion near to the ligand ion, and the other side gets $-\delta$ charge (Figure 2.16).

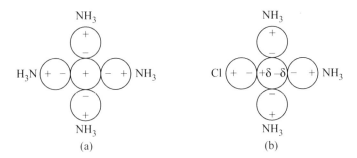

FIGURE 2.16 Polarisation of metal ion electrons by bound ligand dipoles (a) no net polarisation when same ligands in opposite direction (b) net polarisation for different ligands in opposite direction.

If the ligand in the opposite side is also the same, there is a reverse effect. As a result, they cancel each other, and there is no net polarisation of the metal ion (Figure 2.16a). This is true in a completely symmetrical complex $[Pt(NH_3)_4]^{2+}$. However, if one NH_3 molecule is replaced by chloride, in the resultant $[Pt(NH_3)_3Cl]^+$, polarisation due to chloride is more than that due to NH_3; hence, the chloride side of the metal ion will develop $+\delta$, and the position trans to it will develop $-\delta$. This induced $-\delta$ will oppose the $-\delta$ end of the NH_3, which is located trans to the Cl^-, thereby weakening, the $Pt-NH_3$ bond in the trans side. On further substitution the second chloride comes on the trans direction of the existing chloride, resulting in the formation of trans $[Pt(NH_3)_2Cl_2]$.

For the theory to be valid, the trans effect should be dependent on the greater polarisability of the metal ion and the ligand. In keeping with the above expectation, it is observed that the trans effect is in the order of increasing size of the metal ion, $Pt(II) > Pd(II) > Ni(II)$. Similarly, iodide being bigger ion, and ethylene, cyanide and CO being multiply bonded are more polarisable and hence are found to be more trans directing.

b. A second approach, explaining the cause of weakening of the bond, trans to the trans-directing group, is based on the π bonding concept. If a ligand L forms a π bond with the metal ion, involving $M \rightarrow L\ d\pi-d\pi$ or $d\pi-p\pi$ interaction, the electron density shifts from the metal ion to the ligand L. Thus, on the side trans to L, electron density is reduced, and the π bonding of the group L^1 trans to L is weakened (Figure 2.17).

As per this theory, trans $M-L^1$ bond should be weakened only if L^1 is π bonding. However, it is observed that π-bonding ligands like CN^- or PR_3 can labilise an NH_3 molecule at the trans position, though NH_3 is only a σ-bonding ligand. Furthermore, it is observed that π-bonding ligand, ethylene, labilises chloride at trans position, but not the N-coordinating ligands.

Recently, Langford and Gray and Syrkin have extended explanation for the weakening of the trans ligand, due to σ bonding. Langford and Gray considered that the ligands, which lie trans to each other, compete for the p orbital of the metal ion that lies along $L - M - L^1$ axis. Syrkin considered that L and L^1 compete for ns and $(n-1)$ d orbitals of the metal ion. The formation of a strong σ bond by L, using the hybrid orbitals of the metal ion, weakens the overlap of L^1 with the hybrid metal orbitals, with consequent weakening of $M-L^1$ bond.

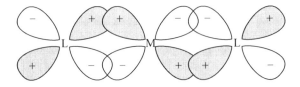

FIGURE 2.17 π bonding concept of trans effect.

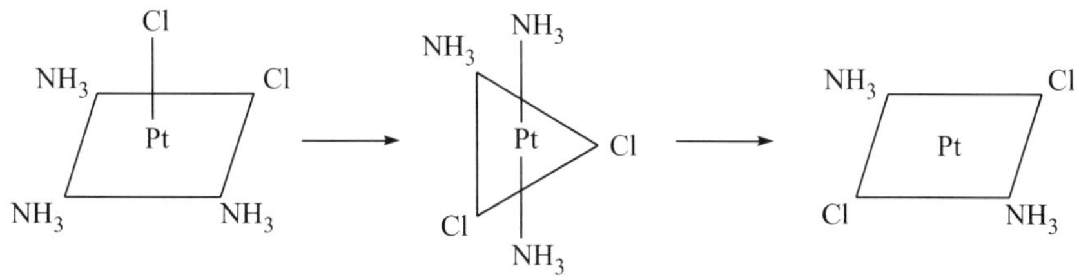

FIGURE 2.18 Formation of trigonal pyramidal intermediate in the substitution reaction of square planar complex.

Both Grinberg's polarisation theory and π bonding theory are based on the consideration of weakening of the bond, trans to the trans-directing group, and presume that the substitution reaction proceeds through dissociation mechanism. However, this is not true in all cases. Another approach to explain trans effect is in terms of the kinetics of the substitution reaction.

ii. **Kinetic effect:** If the substitution reaction in square planar complex is of associative type, first there is a formation of a square pyramidal, penta-coordinated intermediate, which changes to trigonal bipyramidal structure, leading to substitution.

$$ML_3X \xrightarrow{Y} ML_3XY \xrightarrow{-L} ML_2XY$$

In the formation of the trigonal bipyramidal intermediate, it was suggested by Cardwell that due to electrostatic repulsion, the most electron-repelling trans groups of the square planar complex are slightly forced out of the plane. These electron-repelling groups and the incoming ligand Y form the trigonal plane. In the next step, one L from axial direction goes out, resulting in trans complex (Figure 2.18).

However, as per the above theory, in the formation of $[PtCl_2(C_2H_4)_2]$ from $[Pt\,(C_2H_4)Cl_3]^-$, it has to be presumed that C_2H_4 is more electron-repelling than Cl^-. So, an alternative proposal for the stability of the trigonal pyramidal intermediate was extended in terms of π bonding by Chatt and Orgel, independently. As a strong π-bonding ligand accepts π electrons from the metal ion, the electron density around the metal ion is reduced and it can accept the fifth ligand to form the trigonal bipyramidal intermediate. As the d orbitals, involved in π bonding, lie in the trigonal plane, the trans-directing π-bonding ligand occupies position in the trigonal plane of the trigonal bipyramid.

The above kinetic theory requires that the substitution reaction should proceed through the associative mechanism. However, in the substitution of Cl in $[Pt(NH_3)_2Cl_2]$ by various groups, it has been observed that the reaction is insensitive to the concentration of the substituting group. In order to explain this, it was suggested that a square planar complex can be considered as a distorted octahedral complex, if the two axial positions are occupied by solvent molecules. The substitution reaction possibly proceeds through SN_1 mechanism, with a penta-coordinated trigonal bipyramidal or square pyramidal intermediate. The reaction can proceed in two ways (Figure 2.19). In the path I, the first step is a slow reaction and so is rate determining. The substitution of the ligand Y being a fast reaction, the reaction is insensitive to the concentration of Y. In path II, the reaction is dependent on the concentration of both the complex group and the substituting group (Y). Though the rate determining step indicates SN_1 mechanism, the concentration of the intermediate formed, depends on the concentrations of both the complex group and the substituting group.

This mechanism can explain both the cases of substitution reactions, which are dependent or independent of the concentration of the substituting group Y.

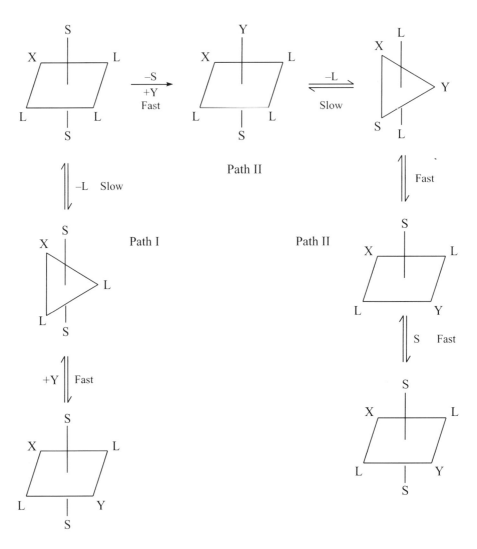

FIGURE 2.19 The SN$_1$ mechanism for substitution reaction in square planar complexes.

Labile and Inert Complexes

If the substitution reaction over a complex is fast, it is said to be labile, while if the substitution is slow, the complex is called inert. However, the boundary between these two types of reactions is not rigid. The reactions can occur in continuously changing rates. According to Taube, if the reaction is complete in less than 1 minute, for 0.1 N concentration of the reactants at room temperature, the system can be termed labile. If the substitution reaction takes more time, the complex is said to be inert.

The lability and inertness are not related to the thermodynamic stability of the system. Thermodynamically stable systems may be kinetically labile or vice versa, because the ease with which substitution reaction takes place does not depend on the strength of M–L bond. For example, the complex $[Mn(CN)_6]^{3-}$ with lower stability constant (log K = 27) has a relatively less and measurable rate of substitution reaction than $[Ni(CN)_6]^{4-}$ with higher log K = 30. Though $[Ni(CN)_6]^{4-}$ has high stability constant, the system is found to undergo substitution. $[Fe(CN)_6]^{4-}$ (log K = 37) is very slow, whereas $[Hg(CN)_4]^{2-}$ with higher log K = 42 is very fast to substitution.

The lability or inertness of a complex has been explained in two ways.

i. **Valence bond (VB) approach:** The substitution reactions proceed through SN_1 or SN_2 mechanism, involving penta-coordinated or hepta-coordinated intermediates, respectively. According to V.B. theory, the penta-coordinated structure will involve dsp^3 hybridisation, and hepta-coordinated structure involves d^3sp^3 hybridisation.

If the octahedral d^2sp^3 hybridisation can easily change to penta-coordinated dsp^3 hybridisation or hepta-coordinated d^3sp^3 hybridisation, substitution reaction can proceed easily through SN_1 or SN_2 mechanism, respectively, and the system is labile.

If, however, change to dsp^3 or d^3sp^3 hybridisation involves higher energy, reaction cannot proceed through SN_1 or SN_2 mechanism, and the system is inert.

In the case of d^0, d^1, d^2 metal ions, there is an octahedral complex formation, due to d^2sp^3 hybridisation. However, one or more low-lying t_{2g} orbitals are still vacant. So, one low-energy d orbital is available for d^3sp^3 hybridisation, and thus, the intermediate hepta-coordinated structure MX_6Y can be easily formed. This will facilitate the substitution reaction can proceed through SN_2 mechanism, and the system is labile.

In the case of octahedral complexes of high spin d^4 to d^6 and d^7 to d^{10} metal ions, there is outer orbital sp^3d^2 hybridisation. The bonds formed will be weak, and the octahedral complex can easily break into penta-coordinate ML_5. Thus, the substitution can proceed through SN_1 mechanism, and the system is labile.

In the octahedral complexes of d^3 and low spin octahedral complexes of d^4, d^5 and d^6 metal ions, there is d^2sp^3 hybridisation. So, the bonds formed are strong, and the octahedral complex does not easily break into penta-coordinated form. That means, the substitution cannot proceed through SN_1 mechanism. Furthermore, in these cases there is no low-energy vacant d orbital available, for d^3sp^3 hybridisation, So, the substitution reaction cannot proceed through SN_2 mechanism, rendering such complexes substitutionally inert.

However, the explanation in terms of the V.B. theory has the following limitations:

a. It predicts about the lability or inertness of the complex, but cannot quantitatively determine the relative order of the lability of the complex.

b. The theory has an inherent presumption that the bonds formed due to outer orbital sp^3d^2 hybridisation are weak. This is not always true.

c. According to VB theory, d^8 octahedral complex should be labile through SN_1 mechanisms, but octahedral Ni(II) complexes are inert.

ii. **Crystal field theory (CFT) approach:** In the case of SN_1 mechanism, the penta-coordinated intermediate is either trigonal bipyramidal or square pyramidal, and in the case of SN_2 mechanism, the hepta-coordinated intermediate is pentagonal bipyramidal.

If the intermediate structure is of lower energy, it can be easily attained and the system is labile. It can be qualitatively seen that in octahedral complexes of d^0, d^1, d^2 metal ions, there is no electron in one of the t_{2g} orbitals, lying in between the axes. Hence, the incoming seventh ligand can enter in that direction and a low-energy hepta-coordinated intermediate can be formed. Such a reaction can easily proceed through SN_2 mechanism.

Similarly, in the cases of the octahedral complexes, where the repulsion between metal d electrons and the six ligands is high, the octahedral structure is destabilised, and the penta-coordinated intermediate can be easily formed, leading to substitution through SN_1 mechanism.

In CFT approach, the CFSE, of trigonal bipyramidal, square pyramidal and pentagonal bipyramidal intermediate structures, have been worked out, for different d electronic configurations of the metal ions. ΔE, the difference in the energies of the original octahedral and the intermediate structures, have also been worked out (Figure 2.20).

ΔE is expected to be always positive; otherwise, the original *Oh* structure should change to intermediate structure, automatically. However, in cases where ΔE is less positive, for either penta-coordinated or hepta-coordinated intermediate or for both, the system is labile.

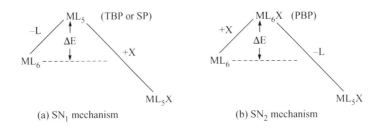

FIGURE 2.20 CFT approach for substitution reactions in octahedral complexes (a) SN_1 and (b) SN_2 mechanism.

If ΔE is more positive for both the intermediate structures, the system is going to be inert. The extent of lability or inertness depends on the magnitude of ΔE.

An observation of the calculated ΔE values for the different intermediate structures shows that it is lower for both types of intermediates in the case of d^0, d^1, d^2, d^5 and d^{10} metal ion complexes. That means, substitution reactions can proceed through both SN_1 and SN_2 mechanism, and the systems are labile. However, as per VB theory, d^0, d^1 and d^2 metal complexes are considered to be labile due to facile SN_2 mechanism and high spin d^4 to d^7 and d^7–d^{10} metal ion complexes are considered to be labile through SN_1 mechanism.

In the case of d^3 and d^8 metal ion complexes, ΔE is high for both penta-coordinated and hepta-coordinated intermediate. As a result, the substitution reaction is slow to proceed through both SN_1 and SN_2 mechanisms, and hence, the system is inert. However, VBT expects d^8 metal complexes to be labile through the SN_1 mechanism.

It is observed, that in case, where there is a disagreement about the lability of the system, as per VBT and CFT, the experimental evidence is found to be in agreement with CFT predictions.

Ni(II) (d^8) complexes are inert. In some cases, substitution reaction on NiL_6 complexes is measurable. In such cases also, it is observed that the Ni(II) complexes are less labile than other more labile complexes. Thus, crystal field theory is able to provide the relative order of labilities of the complexes, rather than the absolute value of the lability.

Stereochemical Changes Involved in Substitution Reactions in Octahedral Complexes

In substitution reactions of octahedral (Oh) complexes, there is a possibility of change in geometrical isomeric form. For example, in SN_1 mechanism of substitution, if the penta-coordinated intermediate is square pyramidal in structure (Figure 2.21), there is no change in geometrical isomeric form.

However, if the intermediate is trigonal bipyramidal (tbp) the stereochemistry may be retained or lost, depending on the direction from which the group Z enters, as shown in Figure 2.22.

In the case of SN_2 mechanism of substitution, there is formation of a hepta-coordinated intermediate. The seventh group may enter from the side close to the leaving ligand. This is termed as cis attack which results in the retention of the geometrical isomeric form (Figure 2.23).

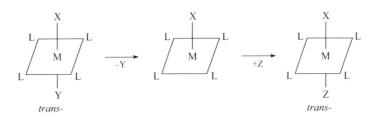

FIGURE 2.21 Stereochemical changes in substitution reaction in Oh complexes following SN_1 mechanism.

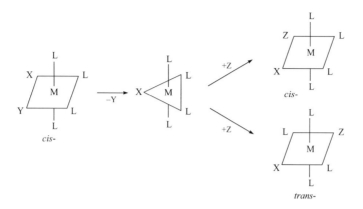

FIGURE 2.22 Stereochemical changes in substitution reaction in Oh complexes following SN$_2$ mechanism.

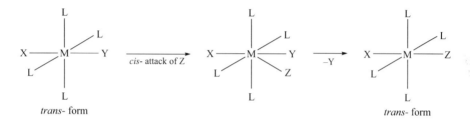

FIGURE 2.23 Retention of geometry of an Oh complex in substitution reaction in cis- attack.

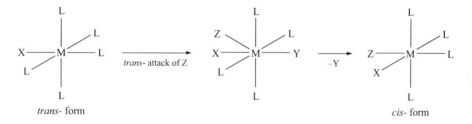

FIGURE 2.24 Change of geometry of an Oh complex in substitution reaction in trans- attack.

If, however, the seventh group enters from the side, opposite to the leaving ligand, (termed trans attack), trans form changes to cis form (Figure 2.24).

Another mechanism of substitution suggested by Brown, Irgold and Nyholm is called edge displacement mechanism. According to this, the incoming ligand shifts the ligand L$_1$ at the other side of the edge, where the group to be substituted is located. The ligand L$_1$ moves along the edge to occupy the position vacated by Y, and Z occupies the position of L$_1$.

With respect to the unsubstituted group X, Y was at trans position, but the incoming ligand Z is at cis position. Thus in this reaction, there is a trans-to-cis conversion (Figure 2.25).

If the complex involves bidentate groups, edge displacement may lead to d − l or l − d conversion.

In the acid hydrolysis reaction of complexes of the type Co(en)$_2$XY, it has been observed that in some cases, cis complexes result in cis products, whereas trans change to cis form. However, in other cases of similar [Co(en)$_2$XY] complexes, both cis and trans forms do not undergo change in geometrical isomeric form during hydrolysis.

$$[Co(en)_2(X)(Y)] \xrightarrow[-Y]{+H_2O} [Co(en)_2 X(H_2O)] \qquad (2.27)$$

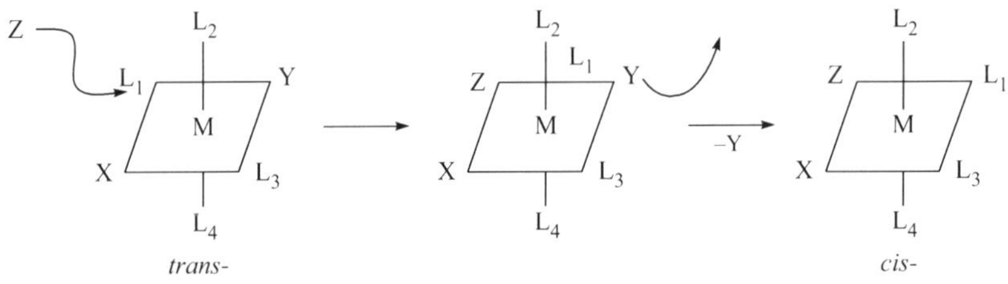

FIGURE 2.25   Edge displacement mechanism for substitution reaction in Oh complexes.

This difference depends on the nature of the unsubstituted group X. The steric course of the reaction depends on how much contribution it can make, to the energy required, for breaking of M–Y bond.

If X is a ligand with π orbitals, like OH^-, halides and NCS^-, the cis form prefers an intermediate, with tetragonal pyramidal structure, and hence, substitution results in retention of the geometrical form.

However, in the case of trans complexes, a trigonal bipyramidal intermediate is preferred and may lead to change to cis form.

In case, where $X = H_2O$ or NO_2^-, without π orbitals, both cis and trans forms undergo acid hydrolysis through tetragonal pyramidal intermediate, without change in geometrical isomeric form.

Isomerisation Reactions

It was observed by Jorgensen that octahedral complexes undergo interconversion from one geometrical isomeric form to another form. Such cis trans conversions are called isomerisation reactions. For example, on boiling for a long time, cis complexes, like $[Co(en)_2NO_2Cl]Cl$ or $[Co(en)_2(NO_2)_2]NO_3$, get converted to the trans form. The green trans form of dichloroethylenediamineCo(III) chloride changes to violet-coloured cis form, on boiling to low volume. There can be a reverse conversion of the violet form to the green form, by evaporation of its hydrochloric acid solution.

The initial suggestion that the isomerisation is an intramolecular process and takes place by opening up of one of the ethylenediamine rings, was ruled out, because the isomerisation process, in presence of labelled chloride (^{36}Cl), shows that there is a random distribution of ^{36}Cl, in the coordination sphere of the isomerised product. Hence, isomerisation must be an intermolecular process. It is further observed that the coordinated chloride is not directly replaced by the chloride, present in the solution. Hence, following mechanism of isomerisation reaction was suggested by Ettle and Jhonson:

$$(I)\quad cis\text{-}[Co(en)_2Cl_2]^+ + Cl^- + H_2O \rightleftharpoons cis\text{-}[Co(en)_2(H_2O)Cl]^{2+} + 2Cl^-$$

$$(2.28)$$

$$(II)\quad trans\text{-}[Co(en)_2Cl_2]^+ + Cl^- + H_2O \rightleftharpoons trans\text{-}[Co(en)_2(H_2O)Cl]^{2+} + 2Cl^-$$

The less soluble cis isomer is obtained from aqueous solution, whereas the still less soluble trans form separates from hydrochloric acid solution as trans $[Co(en)_2Cl_2](H_5O_2)^+Cl_2$. The fact that HCl is precipitant and is not in the coordination sphere was confirmed by precipitating $[Co(en)*Cl_2]^+$ with HCl. On heating the salt at $110°C$, HCl is liberated, but it is not radioactive.

Reaction I is correct, as aquation of cis $[Co(en)_2Cl_2]$ forms cent per cent cis $[Co(en)_2Cl(H_2O)]^{2+}$. However, aquation of trans $[Co(en)_2Cl_2]$ forms directly, a mixture of 35% cis and 65% trans, $[Co(en)_2(H_2O)Cl]^{2+}$. Hence, the reaction in equation (2.28II) may extend further, and there may be formation of a diaquocomplex.

$$\text{trans-}[Co(en)_2(H_2O)Cl]^+ + H_2O \rightleftharpoons \text{trans-}[Co(en)_2(H_2O)_2]^{3+} + Cl^- \qquad (2.29)$$

In the reverse process of the above equilibrium, there may be a formation of both cis and trans forms of $[Co(en)_2(H_2O)Cl]^+$.

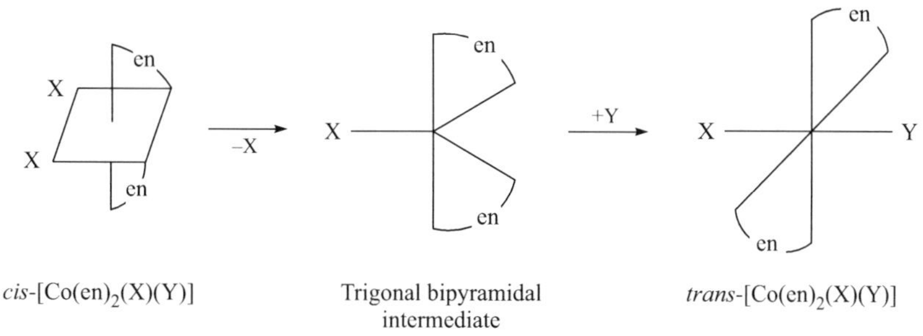

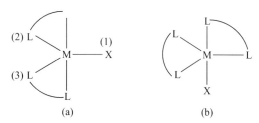

FIGURE 2.26 Isomerisation reaction in Oh complexes by dissociation mechanism.

The isomerisation reaction can be explained by considering dissociation mechanism, involving a trigonal bipyramidal intermediate. As discussed in the case of the substitution reaction, leading to change in the isomeric form, in the present case, when the outgoing ligand Cl⁻ coordinates back, replacing one H₂O, it may occupy another position, leading to cis-to-trans conversion (Figure 2.26).

cis–trans conversion of diaquo-*bis*-ethylenediamine Co(III) complexes has also been studied. The rate of isomerisation reaction is slower in acidic solution than in aqueous solution, showing that the hydroxo complex $[Co(en)_2(OH)(H_2O)]^{2+}$, formed in aqueous solution, undergoes faster isomerisation than $[Co(en)_2(H_2O)_2]^{3+}$. However, the rate of isomerisation decreases at high pH, due to the formation of $[Co(en)_2(OH)_2]^+$, which undergoes slow isomerisation.

Isomerisation reaction has also been studied in Cr(III) complexes. The oxalato complex $K[Cr(C_2O_4)_2(H_2O)_2]$ exists in cis form in solution. On slow evaporation, trans complex is formed, and being insoluble, it gets precipitated.

The reaction proceeds through intermediate trigonal bipyramidal species $[Cr(C_2O_4)_2H_2O]$.

Racemisation Reactions

In the chelates, like $[Co(en)_2Cl_2]^+$ with two bidentate ligands, the trans-form is optically inactive, but the cis form is optically active. When the cis form gets converted into trans form, there is a loss of optical activity. In cis–trans conversion of $[Co(en)_2Cl_2]^+$, it is observed that there is 70% formation of the trans isomer and 30% of racemic mixture of equal amounts of d and l forms of cis-Co(en)₂Cl₂. Similarly, isomerisation of cis-$K[Cr(C_2O_4)_2(H_2O)_2]$ is followed by racemisation.

Hence, not only the loss of optical activity of cis isomer is due to the formation of transform, but there is a change from d to l form of the cis complex. This process of interchange of d and l forms of the complex is called racemisation reaction. It is observed that the rate of racemisation is much faster than that of isomerisation.

Racemisation can be explained, by considering the structure of the trigonal bipyramidal intermediate, formed in the dissociation mechanism of cis–trans conversion of complexes of the type $K[Cr(C_2O_4)_2(H_2O)_2]$, as shown in Figure 2.27.

FIGURE 2.27 Racemisation reaction in Oh complexes of type $K[Cr(C_2O_4)_2(H_2O)_2]$ following dissociation mechanism. (a) and (b) two possible structures of the penta coordinate trigonal pyramidal intermediate.

As discussed earlier, the intermediate with structure shown in Figure 2.27 a can form both cis and trans forms. If the second X is added in the space between 2 and 3, a trans product is formed, whereas addition of X between positions 1 & 2 or 1 & 3 results in the formation of cis product.

However, for the intermediate with structure b in Figure 2.27, there is formation of only cis product. This is followed by racemisation, because the asymmetric structure will leave equal possibility of forming d or l forms of cis $K[Cr(C_2O_4)_2(H_2O)_2]$, leading to racemisation reaction.

Racemisation reaction has also been observed in *tris* chelates $M(A - A)_3$, where there is no possibility of isomerisation reaction. It has been established that such racemisation can proceed through two mechanisms.

i. **Intermolecular mechanism:** The d or l complex $M(L–L)_3$ is considered to dissociate into a *bis* chelate $M(L–L)_2$ and free ligand L–L. The reversible addition of L–L to $M(L–L)_2$ leads to the formation of opposite optical isomeric form of $M(L–L)_3$.

$$M(L-L)_3 \rightleftharpoons M(L-L)_2 + L-L \rightleftharpoons M(L-L)_3 \qquad (2.30)$$
$$\text{Leavo} \qquad\qquad\qquad\qquad\qquad\qquad \text{Dextro}$$

If above is the mechanism of racemisation, its rate should be comparable with the rate of dissociation of $M(L–L)_3$ to $M(L–L)_2$ and also the rate of exchange of the complex $M(L–L)_3$ with the free ligand L–L present in the solution.

It has been observed that only in the case of $[Ni(phen)_3]^{2+}$ complexes (phen = 1,10-phenanthroline) the rate of racemisation is comparable to the rate of dissociation, indicating that the racemisation proceeds exclusively through intermolecular process. In all other *tris* complexes, the rate of racemisation is much faster than that of dissociation, indicating that another intramolecular pathway of racemisation is involved.

ii. **Intramolecular mechanism:** It was suggested by Werner and later confirmed by several workers that *tris* oxalate and *tris* acetylacetonate complexes of Co(III) and Cr(III) undergo racemisation by intramolecular process. This can take place in two ways.

a. Formation of an intermediate of lower or higher coordination number: In this mechanism, suggested by Werner, it is presumed that one or two of the bidentate ligand breaks from the metal ion at one end, resulting in the penta-coordinated or tetra-coordinated intermediate. When they get attached again, there may be reformation of $M(L–L)_3$, with change in the optical isomeric form, leading to racemisation (Figure 2.28).

This mechanism is supported by the observation that in the racemisation of $[Cr(C_2O_4)_3]^{3-}$ or $[Co(C_2O_4)_3]^{3-}$ complexes, there is very slow exchange of free oxalate in solution, ruling out

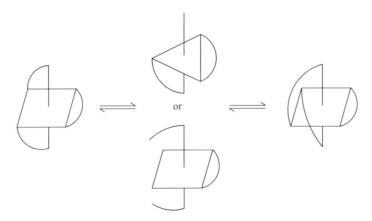

FIGURE 2.28 Racemisation reaction in Oh complexes by intramolecular mechanism with intermediate having lower coordination number.

intermolecular mechanism. However, there is exchange of ^{18}O oxygen of solvent $H_2^{18}O$ with oxalate oxygens, supporting the dissociation of oxalate at one end.

Another possible pathway is the formation of an intermediate with higher coordination number. According to Choronnat, there may be binding of two ligands in the formation of an intermediate cubic structure. The two ligands are lost from the intermediate, reforming the hexa-coordinated structures with a change in optical isomeric form (Figure 2.29). For example, dextro or laevo $K_3Cr(C_2O_4)_3 \cdot 2H_2O$ undergoes loss of H_2O molecule and racemisation, on heating in a sealed tube.

Busch proposed the formation of intermediate hepta-coordinated structure in the racemisation of $Co(EDTA)^-$, catalysed by base. There is probable attachment of OH^- at the seventh position in the intermediate (Figure 2.30).

b. **Formation of an intermediate, with symmetrical structure, due to the distortion of the octahedral structure, without any change in the coordination number:** It was observed by Ray and Dutt that though the complex $[Co(biguanide)_3]^{3+}$ is both thermodynamically and kinetically very stable, it undergoes racemisation reaction. Hence, they argued that the racemisation must be by the distortion of the structure, rather than by the breaking of biguanide at one end. They suggested rhombic or tetragonal twist, leading to an intermediate planar hexadentate structure (Figure 2.31). This can orient back to original octahedral structure with the retention of original optical active form or change to another form, leading to racemisation. This mechanism is popularly known as rhombic or tetragonal twist.

An alternative mechanism was suggested by Bailar and is called triangular twist mechanism. In this process, one of the trigonal planar face of the octahedron undergoes a twist through an angle of 60°, along the C_3 axis, resulting in an intermediate trigonal prismatic structure. When it orients back to the original octahedral form, there may be a change to the opposite optical isomeric form leading to racemisation (Figure 2.32).

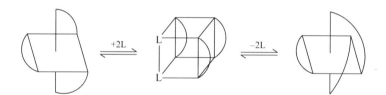

FIGURE 2.29   Racemisation of $K_3Cr(C_2O_4)_3.2H_2O$ through heptcoordinated intermediate.

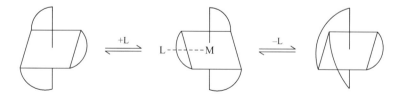

FIGURE 2.30   Base catalyzed racemisation of $Co(EDTA)^-$.

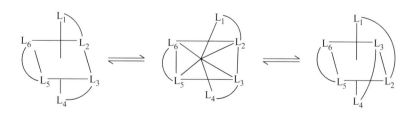

FIGURE 2.31   Racemisation through Rhombic or tetragonal twist in Oh complex.

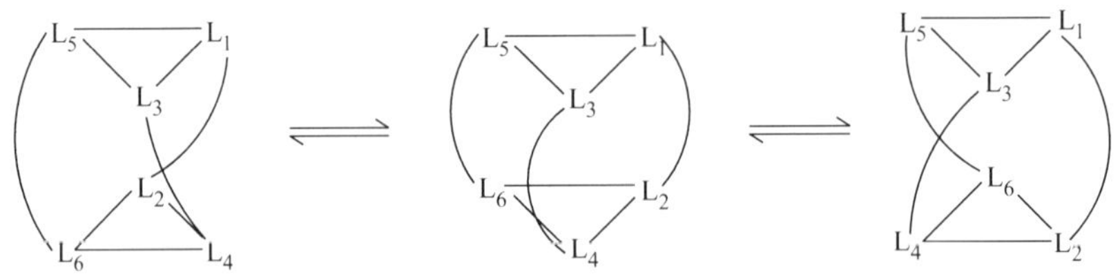

FIGURE 2.32 Racemisation through triangular twist in Oh complex (Bailar's mechanism).

Bailar's mechanism is different from Ray's twist, as well as bond rupture process, suggested by Werner, in the fact that the latter two can give simultaneous racemisation and isomerisation reaction, whereas Bailar's twist does not permit cis–trans isomerisation.

Busch and coworkers observed that the rate of racemisation of Co(EDTA)$^-$ is independent of pH in the range 2–4. As deprotonation of carboxylate group takes place in this pH range, a dissociative mechanism would require dependence on pH. This indicates that the reaction proceeds through intramolecular mechanism. Similarly, racemisation of cis-[M(en)$_2$Cl$_2$]Cl [M=Co(III), or Cr(III)], or [Co(oxalate)$_3$]$^{3-}$ are also through intramolecular twist mechanism.

It is observed that the rate of racemisation of [Fe(phen)$_3$]$^{2+}$ or [Fe(dipy)$_3$]$^{2+}$ is faster than the rate of dissociation of the ligands. This indicates that the process of racemisation is at least partly intramolecular. On the other hand, the analogous Ni(II) complexes undergo racemisation by intermolecular process. This difference in behaviour has been explained, by considering that in the racemisation process, the interatomic distances between the donor atoms of the metal ion increases in the intermediate stage, because of excitation of the low spin state of the metal ion to the high spin state. The racemisation process can be represented by the following equilibria:

$$Fe(L-L)_3^{2+} \xrightleftharpoons{K_2} Fe\cdots(L-L)_3 \xrightleftharpoons{K_3} Fe(L-L)_2 + (L-L)$$

It can be seen that if $K_2 > K_3$, the racemisation will occur by intramolecular process, whereas if $K_2 < K_3$, an intermolecular mechanism, through dissociation, is favoured.

The Fe (II) complexes have diamagnetic low spin d^6 configuration. The expanded intermediate state has high spin t$_2$g^{4}eg^2 configuration. According to crystal field theory, the energy required to attain this state is not very high, because in the high spin state, interelectronic repulsion is released. The intermediate expanded state does not tend to dissociate to [Fe(L–L)$_2$]. Thus, $K_2 > K_3$, and hence, Fe (II) complexes undergo intramolecular racemisation.

Ni(II) complexes are already in high spin state (t$_2$g^{6}eg^2), and hence, expanded excited state is of higher energy. As a result, intermediate expanded state tends to break. Thus, $K_3 > K_2$ and the racemisation proceeds through intermolecular mechanism.

Mechanism of Redox Reactions

Oxidation and reduction reactions are common in biochemical systems. Metalloenzymes, involving metal ions, which can change oxidation states, catalyse the redox reactions. Hence, it is essential to understand the mechanism of such reactions in smaller complex compounds. The redox reaction can be considered to take place in two ways.

1. **Atom or group transfer reactions:** The process can be illustrated by a simpler reaction:

$$SO_3^{2-} + ClO_3^- \longrightarrow SO_4^{2-} + ClO_2^-$$

$$4+ \qquad 5+ \qquad\qquad 6+ \qquad 3+$$

(2.31)

In effect, the reaction involves transfer of one oxygen from ClO_3^- to SO_3^{2-}, leading to oxidation of S and reduction of Cl. The reaction proceeds by transfer of two electrons of S to one of the O of ClO_3^- resulting in the oxidation of sulphur and formation of SO_4^{2-}. The O^{2-} coming out of ClO_3^- leaves two electrons over Cl(V), reducing it to Cl(III). The fact that oxygen is transferred from ClO_3^- is confirmed by starting with ^{18}O labelled ClO_3^-. The resulting SO_4^{2-} is found to have one-fourth of oxygen labelled.

There can be similar transfer of chlorine, resulting in one electron change. The following redox reaction (2.32) is observed to be catalysed, in the presence of chloride ions, by chloride transfer:

$$[Fe(H_2O)_6]^{3+} + [Cr(H_2O)_6]^{2+} \rightleftharpoons [Fe(H_2O)_6]^{2+} + [Cr(H_2O)_6]^{3+} \tag{2.32}$$

There is formation of chloride-substituted $Fe(H_2O)_5Cl$ and subsequent transfer of chloride to $[Cr(H_2O)_6]^{2+}$.

$$[(H_2O)_5Fe(Cl)]^{2+} + [Cr(H_2O)_6]^{2+} \rightleftharpoons [Fe(H_2O)_6]^{2+} + [Cr(H_2O)_5(Cl)]^{2+} \tag{2.33}$$

There may be intermediate formation of chloro-bridged complex.

$$[(H_2O)_5Fe - Cl - Cr(H_2O)_5]^{4+}$$

The chloride transfers one electron to Fe^{3+} reducing it to Fe^{2+} and gets coordinated to chromium as chloride by accepting one electron from Cr^{2+}, oxidising it to Cr^{3+}.

It can be argued that there may be direct one electron transfer from Cr^{2+} to Fe^{3+} and then the Cr^{3+} formed combines with chloride. However, there are two observations, ruling out this possibility. They are:

i. If Cl^- would have got attached with Cr^{3+} after the oxidation reaction, it would not have had catalytic effect.

ii. Cr^{3+} has low tendency to get bound with chloride.

Similar thiocyanate ion transfer can affect one electron redox reaction (2.34).

$$[(NH_3)_5Co(SCN)]^{2+} + [Cr(NH_3)_6]^{2+} \rightleftharpoons [(NH_3)_5Co(SCN)Cr(NH_3)_5]^{4+} + NH_3$$

$$\rightleftharpoons [Co(NH_3)_6]^{2+} + [(NH_3)_5 Cr(SCN)]^{2+} \tag{2.34}$$

However, in the above mentioned chloride and thiocyanate transfer reactions, it is observed that the rate of redox reaction is higher than the rate of atom or group transfer. This shows that the atom or group transfer is only partly responsible for the redox reaction. The reaction is also proceeding partly by direct electron transfer.

2. **Electron transfer mechanism:** It can be illustrated by the mutual oxidation and reduction of two oxidation states of same metal ion.

$$[Fe(H_2O)_6]^{3+} + [Fe(H_2O)_6]^{2+} \rightleftharpoons [Fe(H_2O)_6]^{2+} + [Fe(H_2O)_6]^{3+} \tag{2.35}$$

The fact that such a reaction takes place can be established by using labelled Fe^{3+}. It is observed that as a result of redox reaction, some Fe^{2+} is obtained in labelled form. The reaction is presumed to be proceeding by direct electron transfer from Fe^{2+} to Fe^{3+}.

It is observed that this reaction is also catalysed by the presence of chloride ion. But the reaction does not proceed by chloride transfer, as discussed in the previous type (2.33). This is indicated by the fact that the rate of the redox reaction is very high and within that time transfer of chloride cannot take place. Furthermore, in some reactions chloride is retained with the oxidant after the reaction.

The question that naturally arises is, if there is no group transfer, how does the presence of chloride ion catalyse the reaction.

The explanation is that there is a formation of chloride-bridged intermediate, which facilitates the transfer of electron.

$$(H_2O)_5 Fe^{3+} - Cl - Fe^{2+}(H_2O)_5$$

This is an example of redox reaction, in which there is a transfer of electron from the reductant to the oxidant. Such reactions can involve two mechanisms.

i. **Inner-sphere mechanism:** In this electron transfer process, the reactants share a bridge which assists the electron transfer. It can be illustrated by following reactions:

$$[IrCl_6]^{2-} + [Cr(H_2O)_6]^{2+} \rightleftharpoons [IrCl_6]^{3-} + [Cr(H_2O)_6]^{3+} \qquad (2.36)$$

Unlike the reaction (2.35), which is catalysed by the formation of an external bridge, the above reaction takes place by the formation of an intermediate chloride-bridged precursor binuclear complex, by the replacement of one H_2O of labile $[Cr(H_2O)_6]^{2+}$ by chloride of the inert complex $[IrCl_6]^{2-}$, as follows:

$$Cl_5 Ir^{4+} - Cl - Cr^{2+}(H_2O)_5$$

Up to this point, the reaction is similar to atom transfer process. However, in the next step (2.37), there is electron transfer from Cr^{2+} to Ir^{4+}, and the electron transfer bridge assists the transfer. Thus, a post-cursor complex is formed, which is similar to the precursor, except the oxidation states of the two metal ions. This leads to the formation of reaction products:

$$\underset{\text{Precursor complex}}{Cl_5 Ir^{4+} - Cl - Cr^{2+}(H_2O)_5} \rightleftharpoons \underset{\text{Post cursor complex}}{Cl_5 Ir^{3+} - Cl - Cr^{3+}(H_2O)_5} \rightleftharpoons [IrCl_6]^{3-} + [Cr(H_2O)_6]^{3+} \quad (2.37)$$

Similarly, in the oxidation of $[Cr(H_2O)_6]^{2+}$ by $[Co(NH_3)_5Cl]^{2+}$, there is a formation of the bridged intermediate (Eq. 2.38).

$$\left[Co(NH_3)_5 Cl\right]^{2+} + \left[Cr(H_2O)_6\right]^{2+} \rightleftharpoons \left[(NH_3)_5 Co - Cl - (Cr(H_2O)_5)\right] \rightleftharpoons \left[Co(NH_3)_5(H_2O)\right]^{2+}$$
$$+ \left[Cr(H_2O)_5(Cl)\right]^{2+} \qquad (2.38)$$

The fact that the intermediate bridged complex assists the electron transfer is supported by the observation that the rate of redox reaction between $[Co(H_2O)_6]^{3+}$ and $[Cr(H_2O)_6]^{2+}$ is very slow, whereas it is fast between $[Co(H_2O)_5 - Cl]^{2+}$ and $[Co(H_2O)_6]^{2+}$. This is because in the latter case, there is a possibility of the formation of the intermediate bridged complex $Co^{3+} - Cl - Co^{2+}$, and the electron can be transferred from Co^{2+} to Co^{3+} through the chloride bridge.

The fact that in reaction (2.38), oxidation is not due to chloride transfer is confirmed by the fact that the rate of electron transfer is much faster and within that time chloride transfer cannot take place.

The evidence for the formation of the intermediate bridged complex is indirect in most cases. Only in few cases, it is perceptible. For example, in the oxidation of $[Cr(H_2O)_6]^{2+}$ by $[Ru(NH_3)_4Cl_2]^+$, the formation of the intermediate bridges complex $[(NH_3)_4(Cl)Ru-Cl-Cr(H_2O)_5]$ is indicated by the observation of characteristic peak in the electronic absorption spectrum.

For the inner-sphere mechanism to be effective, one of the reactants should be labile, so that the rate of ligand substitution on it is faster than the rate of electron transfer. The bridging ligand remains with the inert reactants after the redox reaction. Thus, in the inner-sphere electron transfer process, the bridge ligand may or may not be transferred.

For example, in reaction (Eq. 2.37), the chloride stays with iridium even after the reaction. This is because the bridged intermediate $Ir^{4+}-Cl-Cr^{2+}$ changes to $Ir^{3+}-Cl-Cr^{3+}$ after the electron transfer. $Ir^{3+}-Cl$ is less labile than $Cr^{3+}-Cl$, and hence, the bridged complex breaks at Cr(III), after the electron transfer, forming $Cr^{3+}(H_2O)_6$ and $Ir^{3+}Cl_6$, and there is no transfer of chloride bridge.

In reaction [2.38], the bridge intermediate formed is Co(III) – Cl – Cr(II). This changes to Co(II) – Cl – Cr(III), after the electron transfer. However, Co(II) – Cl is more labile than Cr(III) – Cl, and hence, there is formation of [Cr(H$_2$O)$_5$ – Cl] and [Co(NH$_3$)$_5$H$_2$O]. The chloride is transferred from Co(III) to Cr(II).

Thus, the transfer of the bridging atom is not a must in the inner-sphere mechanism of electron transfer. In cases where the bridging atom is transferred, it supports that the formation of the bridge is essential for the redox reaction.

The role of the bridge is to provide a low-energy path for the electron to pass from the reductant to the oxidant. If the bridging ligand is an anion, it also serves to reduce the repulsion between the two metal cations and brings them closer.

Halpern and Orgel observed that the rate of electron transfer depends on the covalency of the bridging atom. For the redox reaction, involving the intermediate (H$_2$O)$_5$Cr(III)–X–Cr(II)(H$_2$O)$_5$, the rate of the redox reaction is in the order X = F$^-$ < Cl$^-$ < Br$^-$ < I$^-$. The covalency of the bond increases, with the increasing polarisability of the bridging atom X.

The hindrance in the formation of the bridge also affects the rate of the reaction. In the case of phthalate bridge (Figure 2.33), the rate of the reaction is in the following order for the isomeric phthalate bridges: para > meta > ortho.

The rate of electron transfer through the bridge is much faster if the bridging ligand has delocalised π electrons, which facilitates the conduction of electrons. For example, maleate bridge, though structurally similar to succinate bridge, favours more electron transfer. The higher rate of redox reaction, in the case of maleate bridge, is because of conjugated double bonds with delocalised π bonds. There is efficient electron transfer through phthalate bridge, also, because of π delocalisation.

There is a very significant enhancement of the rate of redox reaction where pyridine is replaced by pyridine 2 carboxylate, as the bridge (Figure 2.34). In the latter case, there is π delocalisation over the bridge, facilitating electron transfer.

The mechanism, by which the electron is transferred through the bridge, is interesting. Two probable mechanisms have been suggested.

i. **Chemical transfer:** An electron transfer from the reductant is transferred to the bridge, reducing it to a radical anion. This electron is subsequently transferred by hopping process, through the bridge to the oxidant.

   The reduction of the bridge can be confirmed by the ESR spectrum of the system during electron transfer. The ESR signal corresponding to a free radical is observed.

ii. **Resonance transfer:** The electron passes from the reductant to the oxidant by quantum mechanical tunnelling. In effect, it means the leaking of the electrons through the potential barrier of the bridge, which in classical terms should not be penetrable.

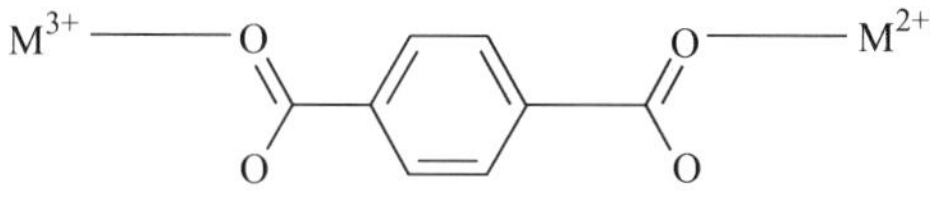

FIGURE 2.33 Inner sphere electron transfer through phthalate bridge.

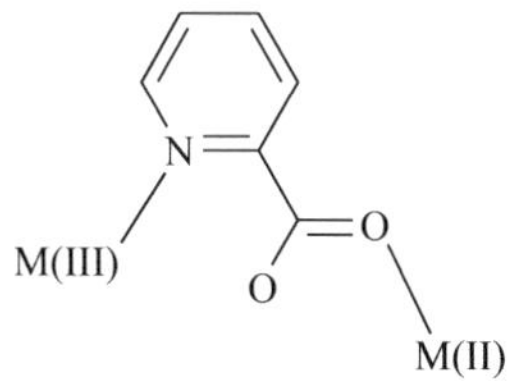

FIGURE 2.34 Inner sphere electron transfer through pyridine-2-carboxylate bridge.

The former mechanism has two advantages over the latter.

a. In a chemical pathway, the electron has not to overcome the activation barrier for both the release of electron from the reductant and the acceptance of the electron by the oxidant simultaneously, as in the resonance transfer, where there is direct transfer of electrons. In the chemical pathway, the barriers are met successively, and hence, total energy required, in the reaction, is relatively less.

b. The reorganisation of the coordination sphere prior to electron loss and electron gain, at the two metal ion centres, is in steps in chemical pathway, whereas it is one step in the resonance pathway. Hence, electron transfer by chemical pathway is easier.

iii. **Outer-sphere mechanism:** If the formation of a bridged precursor complex is not possible, the redox reaction has to proceed through outer-sphere mechanism. This can happen under three conditions:

a. If both the oxidant and reductant complexes are inert in nature, bridge formation cannot take place.

b. If none of the reactant complexes have bridging ligands or H-bonding ligands, or if one of the complex has bridging ligand but the other complex, to which the bridging ligand is to be attached, is inert, bridge formation cannot take place.

c. Even in cases, where bridge formation is possible, but the rate of electron transfer is much faster than bridge formation by ligand substitution, electron transfer cannot be considered to take place by inner-sphere mechanism.

In such cases, electron transfer proceeds through outer-sphere mechanism. All the three types can be illustrated by following examples. In the redox systems $IrCl_6^{2-}$ $IrCl_6^{3-}$, $[Fe(CN)_6]^{3-}$ $[Fe(CN)_6]^{4-}$ and $[Fe(bipy)_3]^{2-}$ $[Fe(bipy)_3]^{3+}$, the self-exchange electron transfer occurs through outer-sphere mechanism, because both the reactants are coordinatively saturated and inert.

In the redox reaction between $[Co(III)(H_2O)_6]$ and $[Co(II)(H_2O)_6]$, the electron transfer takes place by outer-sphere mechanism, though Co(II) is substantially labile. This is because the inert reactant $[Co(III)(H_2O)]$ does not have a bridge-forming ligand.

In the reaction between $[Fe(CN)_6]^{3-}$ and $Co(EDTA)]^{2-}$, there is a formation of the bridge intermediate $[(CN)_5Fe^{3+}-CN-Co^{2+}(EDTA)]^{5-}$. However, the detailed study of the kinetics of the reaction shows that the electron transfer proceeds through outer-sphere mechanism. This is because the bridged binuclear complex is a dead-end species, which does not react further to give the redox products. Hence, the electron is transferred from Co^{2+} to Fe^{3+} through outer-sphere mechanism.

In the reaction of $[V(H_2O)_6]^{2+}$ and $[Co(N_3)_2(en)_2]^+$, the oxidant has a bridging ligand, but the reductant $[V(H_2O)_6]^{2+}$ is only border line labile. The rate of electron transfer is faster than the rate of substitution, forming the bridge precursor complex. Hence, the oxidant complex prefers to receive the electron mainly by mechanistically simpler, outer-sphere pathway.

In the outer-sphere mechanism of electron transfer, there is no bond making or breaking between the reactants. The electron transfer is considered to take place by electron tunnelling, that is, the possibility of electron leaking through a potential energy barrier. Though it appears to be impossible as per classical theory, it is possible as per quantum mechanical concept.

In physical terms "tunnelling effect" means extension of electronic orbitals of the reactants in space, so that the electron can be transferred at distances, considerably greater than that would correspond to actual collision of the reactants.

A theoretical treatment of the outer-sphere electron transfer was suggested by Marcus and Hush, considering the electron transfer to be an adiabatic process, without involving any change in energy. According to them, the reactants have to be brought together from an infinite distance to form an outer-sphere complex, leading to electron transfer and formation of products.

Suppose φ_1 and φ_2 are the wave functions of the initial and the final states, at distance of the reactants, such that there is no interaction between them. The state of transferring electron (outer-sphere complex) when the reactants are brought closer can be represented by a single wave equation:

$$\Psi = (1-m)^{1/2}\varphi_1 + m^{1/2}\varphi_2 \qquad (2.39)$$

The mixing parameter **m** is the average fraction of the electron, which has been transferred. It changes from 0 to 1 during the reaction.

The basic assumption is that there is small overlap between φ_1 and φ_2, so that the interaction is minimum and is just sufficient to lower the barrier energy, so that the electron can freely move through it.

The formation of the outer-sphere complex involves three factors:

i. The thermodynamic driving force of the reaction

ii. The work involved in bringing together the two reactants and also separating them after electron transfer. The forces which hold the reactants in the outer-sphere complex are the electrostatic interaction, van der Waals forces and H bonding between the reactants.

iii. The reorganisation of the coordination and solvation spheres of the two reactants during the reaction.

The outer-sphere complex formation is shown in the potential energy diagram (Figure 2.35). It is evident that ΔG^* is the difference in the free energies of the reactants and the transition state, that is, the free energy change to bring the reactants together from infinite distance to form the outer-sphere complex. ΔG^* depends on various factors, which are shown in the following equation:

$$\Delta G^* = \frac{W^r + W^p}{2} + \frac{\lambda_0 + \lambda_i}{4} + \frac{\Delta G^0}{2}\frac{[\Delta G^0 + W^p - W^r]}{4(\lambda_0 + \lambda_i)} \tag{2.40}$$

where ΔG^0 = the difference in the energies of the reactants and the products (Figure 2.35), that is, the standard free energy change of the reaction, corresponding to the thermodynamic driving force of the reaction.

W^r and W^p = the work necessary to bring the reactants together and to separate them, respectively.

λ_1 = the free energy change involved, in organising the coordination sphere of the reactants, that is, the change in bond length and bond angle of the reactants during the electron transfer reaction.

λ_0 = the free energy change involved, in organising the solvation spheres of the two reactants. It depends on the changes in the polarisation of the solvent polarity.

The rate constant for electron transfer K_{et} is related to the free energy change ΔG^* as follows:

$$K_{et} = \underline{K}\, A_r^2 \exp\left(-\Delta G^*/RT\right) \tag{2.41}$$

where $\underline{K}$ is the transmission coefficient and is equal to 1 for adiabatic processes. A_r^2 is the effective collision frequency, equal to $Z = 10^{11}$ dm^3/mol/s.

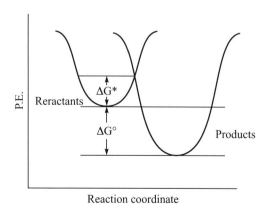

FIGURE 2.35 Potential energy diagram for formation of outer sphere complex formation.

By using the above Marcus–Hush equation, k_{11}, the self-exchange constant for a complex system can be determined. The outer-sphere electron transfer rate constant k_{12} for a system, involving two different complex ions, $[M(H_2O)_6]^{3+}$ and $[M^{-1}(H_2O)_6]^{2+}$, can be determined, by using the following equation:

$$k_{12} = \sqrt{k_{11} \cdot k_{22} \cdot k_{12}} \cdot f_{12} \qquad (2.42)$$

where k_{11} and k_{22} are the self-electron exchange rate constant of each reactant and can be determined. k_{12} is the equilibrium constant of the following reaction:

$$[M(H_2O)_6]^{3+} + [M'(H_2O)_6]^{2+} \rightleftharpoons [M(H_2O)_6]^{2+} + [M'(H_2O)_6]^{3+} \qquad (2.43)$$

The value of k_{12} is approximately $1 \cdot f_{12}$, which is the ratio of the collision frequencies associated with each species, involved in the reaction system, and is equal to $Z_{12}^2 / Z_{11} \cdot Z_{12}$.

The value of electron transfer constant k_{12} thus calculated can be compared with the experimental value. The agreement between them supports that the reaction proceeds through outer-sphere mechanism.

As the value of electron transfer rate constant depends on the value of ΔG^* (Eq 2.41). Hence K12 is also dependent on the value of ΔG^*, that is, on all the energy changes involved in going from reactants to products (Eq 2.40). Equation (2.40) also affects k_{12}. Following are some illustrations.

If the donor and acceptor ions are oppositely charged, the electrostatic attraction in the formation of the outer-sphere complex is higher, for example, in the reaction between $[Fe(H_2O)_6]^{2+}$ and $[Fe(CN)_6]^{3-}$, and hence, the energy change involved is lower, and the outer-sphere electron transfer reaction between them is faster.

The electron transfer rate constant for $Fe^{3+}(H_2O)_6 - Fe^{2+}(H_2O)_6$ reaction is found to be lower than $Fe(CN)_6^{3-} - [Fe(CN)_6]^{4-}$ and $[Fe(bipy)_3]^{3+} - [Fe(bipy)_3]^{2+}$. This is because the former is an outer orbital complex, whereas latter two are inner orbital. The Fe–O bond distance changes significantly in the two oxidation states in the outer orbital complex. Hence, the transfer of electron from Fe^{2+} to Fe^{3+} involves more energy in the outer orbital complex, and electron transfer reaction becomes slower.

The redox reaction of $[Co(NH_3)_6]^{3+}$ with $[V(H_2O)_6]^{2+}$ is faster than that with $[Cr(H_2O)_6]^{2+}$. This is because the oxidation of Cr(II) to Cr(III) involves more reorganisation of bond length, involving higher energy, than the oxidation of V^{2+} to V^{3+}.

Furthermore, the rate of electron transfer is slower if the change of oxidation state involves change in electron spin. For example, k_{12} for $Co(H_2O)_6^{3+} - Co(H_2O)_6^{2+}$ is low, because Co^{2+} is high spin and Co^{3+} is low spin. Thus, for electron transfer to take place from Co^{2+}, it has to be raised to excited spin state. This requires higher energy, making the reaction slower.

However, if the ligand is strong field, as in $[Co(phen)_3]^{3+}$ $[Co(phen)_3]^{2+}$ system, the ligand field stabilisation energy is much higher when Co^{2+} is oxidised to low spin Co^{3+} state. This compensates for the extra energy required for spin pairing, and hence, electron transfer rate constant remains high.

Reactions on Coordinated Ligands

Besides the redox reactions of the metal ion centre, or the substitution reactions of the ligands in the metal complexes, reactions can be carried out over the ligand, without breaking of the M–L bond. The coordination of the ligand with the metal ion results in change in electron density over the ligand, and this may change the reactivity of the ligand molecule, compared to the free ligand molecule. The change in electron density, in the coordinated ligand, depends on the nature of metal ligand bond and can be of two types.

1. If the metal ion is of class A and the ligand is a σ base, there is a shift of electron density from the ligand to the metal ion, through the $L \rightarrow M$ σ bond. In certain cases, there may be ligand-to-metal electron transfer due to π bond formation also. This creates a lower electron density site (+d) over the ligand molecule, and hence, a nucleophilic attack over the coordinated ligand is facilitated, compared to that on the free ligand. This can be illustrated by following examples.

a. **Hydrolysis of Amino Acid Esters:** On coordination of the amino acid ester, with a metal ion, like Co(II) or Cu(II), there is formation of a five-membered chelate ring. There is drainage of electron density from carboxylate oxygen to the metal ion, and this lowers the electron density over the ester group thereby creating $+\delta$ over the carboxylate carbon atom. This facilitates the nucleophilic attack of water or OH$^-$ is facilitated, leading to hydrolysis (Eq. 2.44).

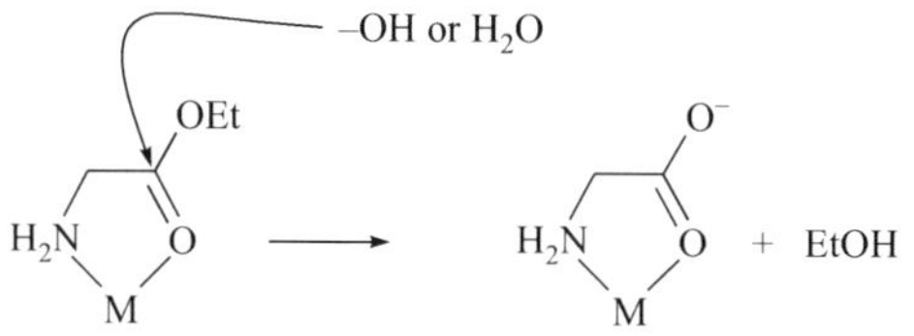

$$\text{(2.44)}$$

Similarly, the hydrolysis of the amino acid amide is facilitated, on coordination with the hard acid metal ion.

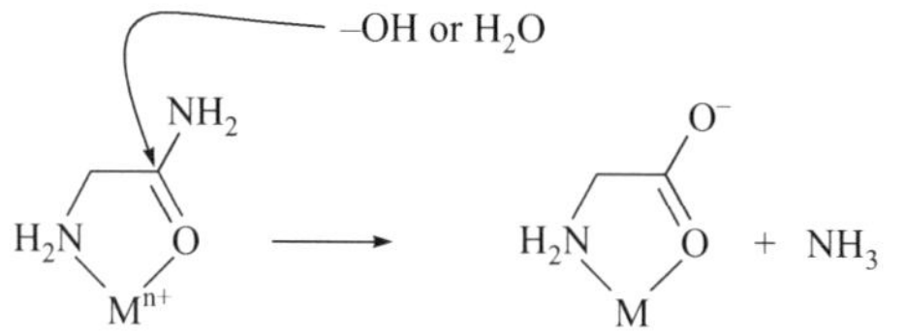

$$\text{(2.45)}$$

The breakdown of the dipeptides or polypeptides into amino acid, by nucleophilic attack of H_2O molecules, is also accelerated by coordination with the metal ion.

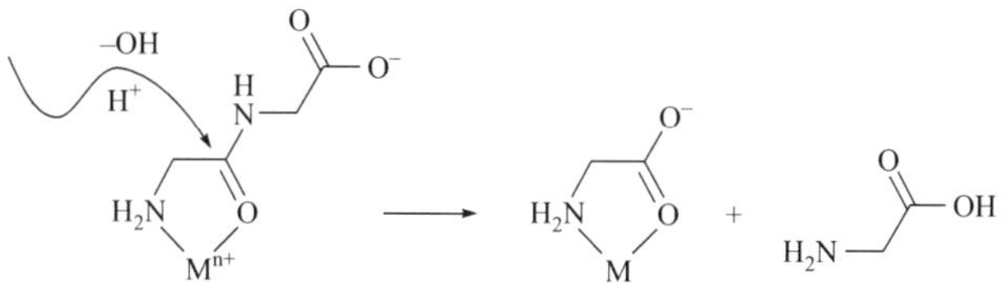

$$\text{(2.46)}$$

In the above studies of the activation of the ligand on coordination with the metal ion, the complexes formed are labile. That means, there is uncertainty about the mode of binding of OH$^-$ or H_2O with the complexes. To resolve this ambiguity, the activation of hydrolysis of the above ligands on coordination with inert metal ions like Co (III) has been studied. Instead of using aquated Co(III) ion, Co(III) complex of the quadridentate ligand, hexa methylene tetra amine ($H_2N–CH_2\ CH_2\ NH–CH_2\ CH_2\ NH–CH_2–CH_2–NH_2$) has been used. The tetra amine ligand remains strongly bound to the metal ion and does not get dissociated during the reaction. It occupies four coordination sites and leaves two vacant sites for the ester or amide of the amino acid, whose hydrolysis is catalysed by the complex. This inert complex makes it possible to investigate the two possible alternative pathways of nucleophilic attack by OH$^-$.

i. The OH$^-$ remains non-coordinated with the metal ion, and thus, there is an intermolecular nucleophilic attack of the OH$^-$ on the bidentate chelated ester or amide, causing its hydrolysis, as in (2.47).

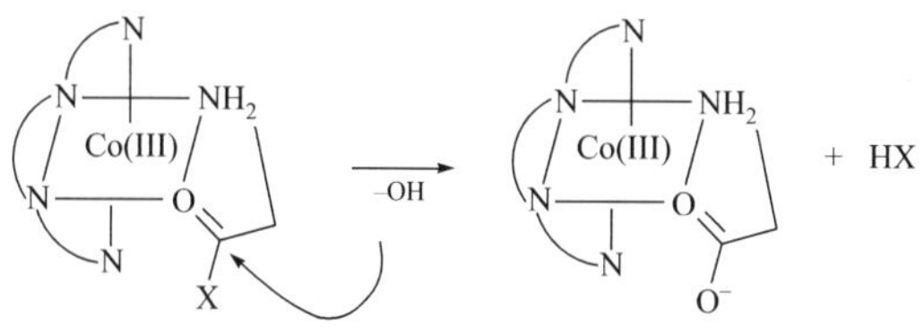

$$\text{(2.47)}$$

ii. The amino acid ester or amide coordinates with the Co (III) complex only from $-NH_2$ end as monodentate ligand, leaving the sixth position free for the coordination of OH^-. The coordinated $-OH^-$ has an intramolecular attack on the amino acid derivative and causes hydrolysis, as shown in (2.48).

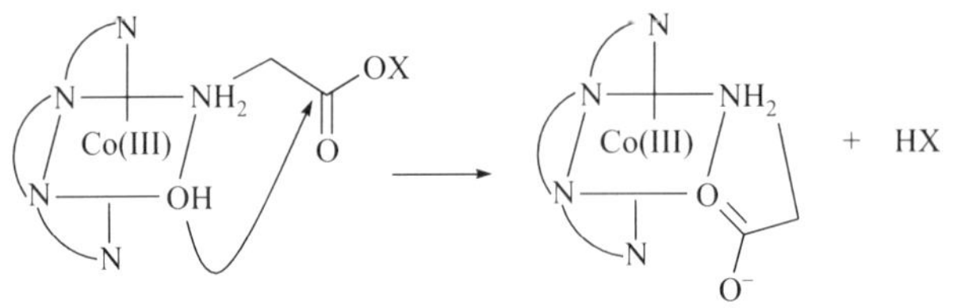

(2.48)

It has been confirmed by studies using ^{18}O-labelled water that the product after hydrolysis has labelled OH^- coordinated to Co(III), supporting the monodentate coordination of the amino acid derivative and the intramolecular nucleophilic attack by the coordinated OH^-. This pathway is preferred because the coordinated OH^-, though depleted in σ base character, due to binding with the metal ion, is suitably oriented to attack the coordinated amino acid derivative to bring about the hydrolysis. Thus, the metal ion not only polarises the C=O group for easy nucleophilic attack, but also helps the two reactants to come closer and also facilitates their reaction.

The knowledge of the mechanism of the metal complex catalysed hydrolysis of carboxyl compounds, provides a basis for understanding the action of Zn-containing enzymes peptidase, catalysing the hydrolysis of peptides and of zinc-containing carbonic anhydrase enzyme, which catalyses the reaction between carbon dioxide and water (to be discussed in Chapter 4).

b. Schiff base formation: Another type of reaction, catalysed by class A metal ions, is Schiff base formation. The Schiff base formation is due to nucleophilic attack of the amine over the carbon atom of aldehyde or keto group. In the case of an aromatic aldehyde, with a coordinating group at ortho position, like salicylaldehyde, it has been observed that the Schiff base formation is catalysed in the presence of a metal ion. This is because the O^- of the OH group and the oxygen of the aldehyde or keto group get bound with the metal ion, forming a chelate. This results in the shift of electron density from O → M and in turn creates a region of low electron density at the keto carbon atom. Hence, the nucleophilic attack by the amine is facilitated.

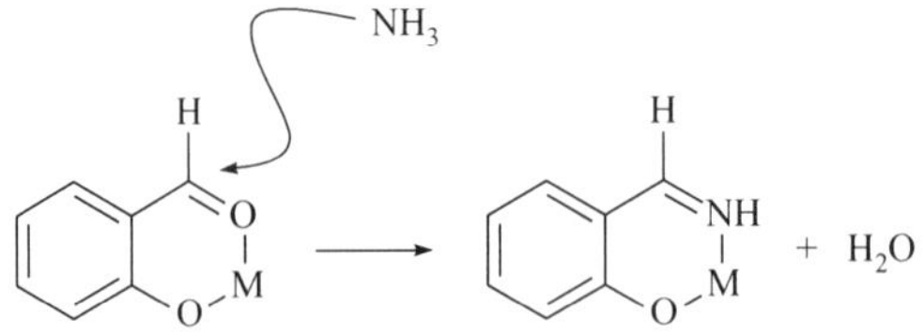

(2.49)

In this reaction also, it has been suggested that nucleophilic attack of the amine may be intramolecular. The amine gets coordinated with the metal ion and thus is suitably oriented to have a nucleophilic attack over the carbon atom of the coordinated aldehyde. In the Schiff base formation of salicylaldehyde with amino acids, it has been suggested that there is formation of a mixed-ligand complex by the simultaneous coordination of the aldehyde and the amino acid with the metal ion. This is followed by the attack of amino group of the acid on the aldehyde group, leading to the formation of the tri dentate Schiff base as in (2.50).

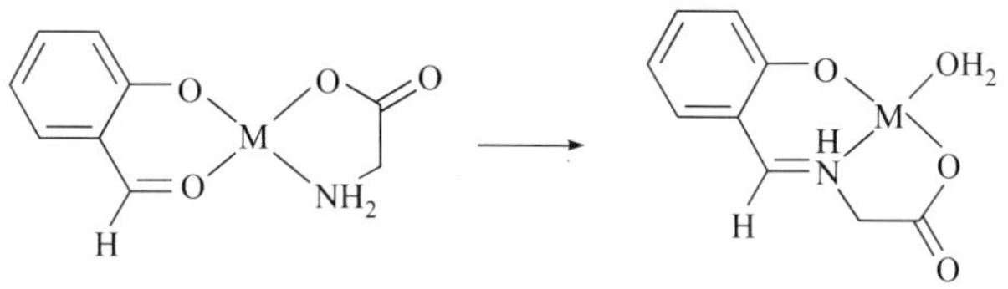

(2.50)

This indicates that the presence of the metal ion converts a bimolecular Schiff base formation reaction into an unimolecular reaction and accelerates it.

Amine exchange reaction over Schiff base also takes place by a nucleophilic attack of the second amine on the azomethine C atom of the Schiff base. This is also catalysed by the coordination of the azomethine nitrogen of the Schiff base with the metal ion. Due to coordination with the metal ion, the electron density over the azomethine C of the Schiff' base is lowered, and there is an increase in $+\delta$ over it. Hence, the nucleophilic attack of the amine, leading to transamination, is facilitated, as shown in Eq. (2.51a).

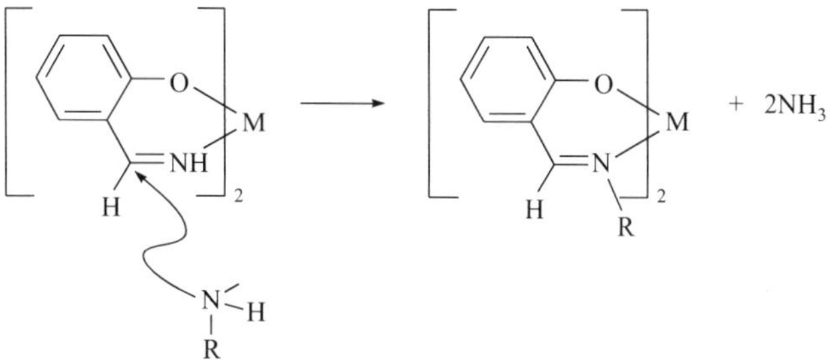

(2.51a)

If the amine exchange is affected by a diamine, it leads to the formation of tetra dentate Schiff base, as in (2.51b).

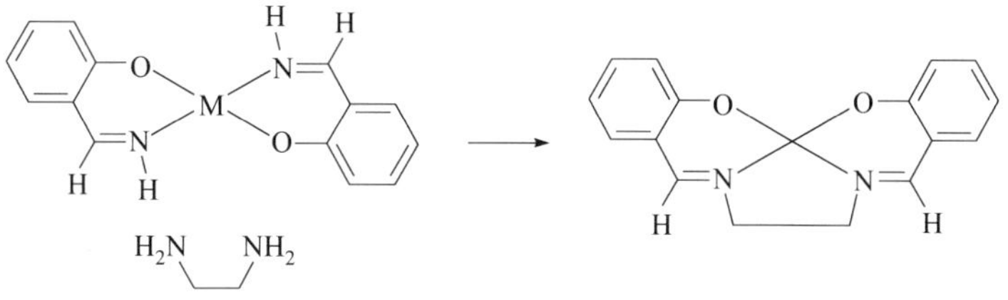

(2.51b)

Such metal catalysed transamination reactions allude to the possibility of role of metal ion in amino transferase reaction as in Eq. (2.52), and isomerisation reaction Eq. (2.53), affected by pyridoxal phosphate (a vitamin B_6 derivative) as coenzyme.

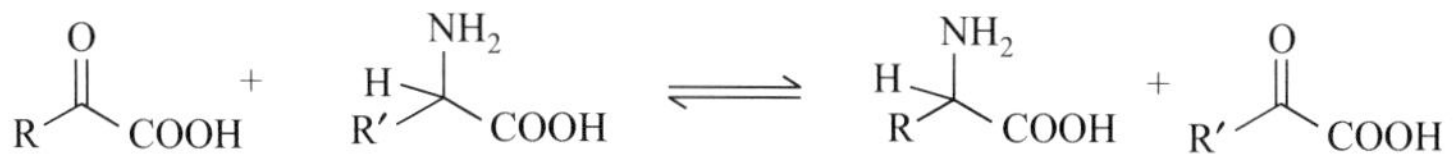

(2.52)

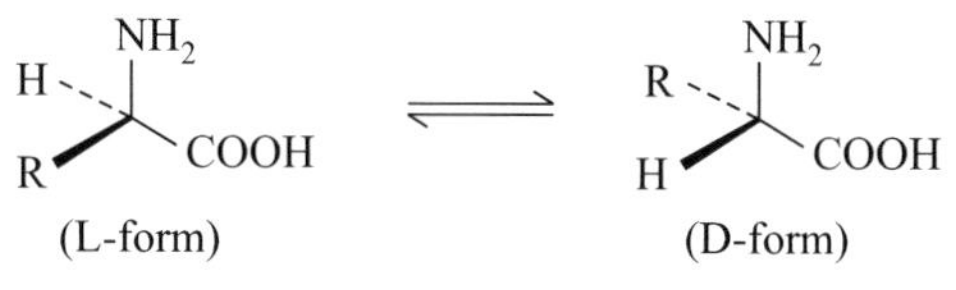

(2.53)

Model studies of reactions of above types with amino acids have been found to be catalysed by metal ions such as Cu(II), Al(III) and Fe(III).

It has been proposed that there is intermediate formation of a metal Schiff base complex, by reaction of pyridoxal and amino acid, catalysed by the metal ion (Eq. 2.54). The metal Schiff base complex rearranges from aldimine to ketimine form, which gets hydrolysed to yield pyridoxalamine and the ketoacid. Thus, the amino acid is converted to ketoacid.

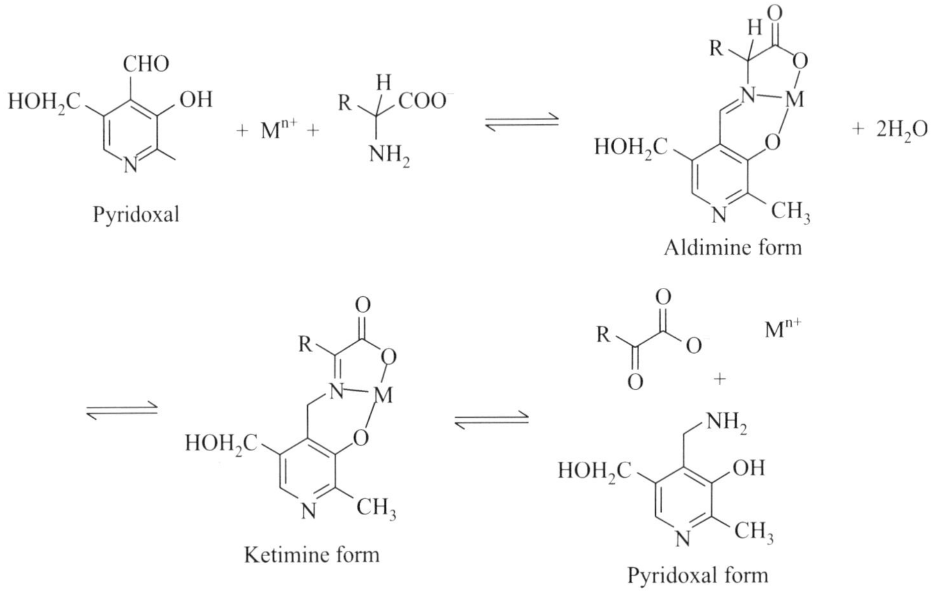

(2.54)

In the aminotransferase reaction (2.52), the pyridoxal amine formed can react with another keto acid and convert it to amino acid form, and in the process pyridoxal is reformed. However, if the reverse reaction of ketimine-to-aldimine conversion takes place, followed by hydrolysis, the original amino acid and pyridoxal are regenerated, and the amino acid may change to another optically active form leading to racemisation (Eq. 2.53).

Though the above reactions, mimicking the role of vitamin B_6 in biochemical reactions, are catalysed in the presence of metal ion, direct involvement of metal ion, in vitamin B_6 catalysed reaction, is not definitely known.

c. **Polyphosphate hydrolysis:** ATP breaks into ADP and orthophosphate by the nucleophilic attack of H_2O, at the phosphorous atom. This reaction leads to liberation of energy and is a phenomenon of great importance in biochemical reactions.

It has been shown that the hydrolysis of ATP is catalysed in the presence of hard acid metal ions. The metal ion gets bound to the α-and β-phosphate oxygens; thus, the nucleophilic attack of H_2O over the terminal g phosphorous is facilitated, and ATP breaks into ADP and monophosphates (2.55).

On coordination with the metal ion, there is a drift of electron density from O → M, leading to lowering of electron density and increase of $+\delta$ at the terminal phosphorous. This facilitates the nucleophilic attack of water or OH⁻. Furthermore, it is likely that the OH⁻ gets coordinated with the metal ion, at a position other than those occupied by the triphosphate. Thus, the reaction between OH⁻ and triphosphate is facilitated.

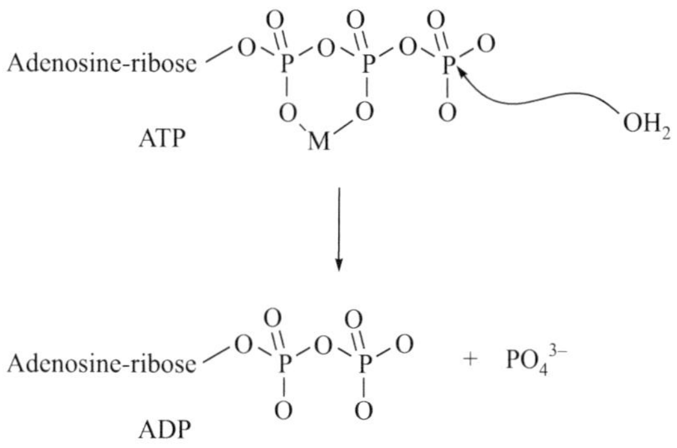

$$(2.55)$$

The above model systems throw light on the mechanism of the reactions of the enzymes ATPase and kinases (to be discussed in Chapter 3).

Metal ions also catalyse the hydrolysis of phosphate esters. In this also, the coordination of the oxygen of the phosphate ester with metal ion can be considered to polarise the P–O bond, facilitating nucleophilic attack of water on the phosphorous, leading to easy hydrolysis. However, an alternative explanation has been proposed. The phosphate ester coordinates with the metal ion through the ester oxygen. This results in the shift of electron density from RO $\rightarrow$ M, and the P–OR bond is polarised, thereby the ester group gets dissociated. Water reacts with the products, forming phosphate and alcohol (2.56).

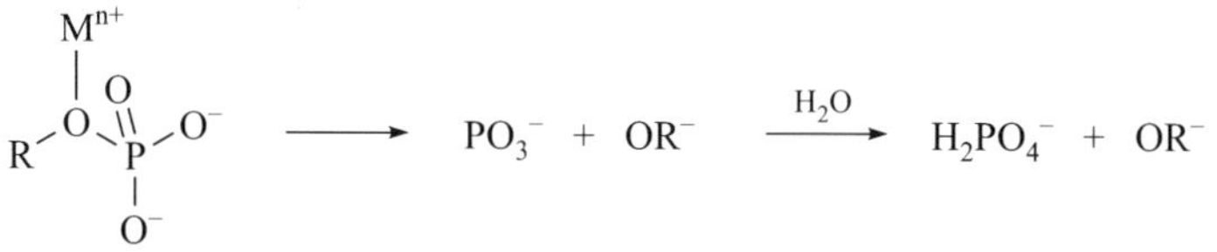

$$(2.56)$$

The above reaction can help to understand the role of zinc in the enzyme alkaline phosphatase, catalysing the hydrolysis of natural phosphate esters (to be discussed in Chapter 4).

d. **Electrophilic substitution on coordinated ligands:** If the ligand is bound to a class A metal ion, through σ bond, the electrophilic substitution reaction over it is retarded. This is because on coordination with the metal ion, there is a drainage of electrons from the ligand to the metal ion, and hence, electron density over the electrophilic substitution sites is reduced. As a result, the electrophilic attack is inhibited.

For example, nitration and bromination reactions on salicylaldehyde are retarded, on its binding with the metal ion like Cu(II) (2.57).

In free salicylaldehyde, the electrophiles NO_2^+ or Br^- attack at 3 and 5 positions. On coordination of salicylaldehyde with the metal ion, the sites of substitution remain unchanged, but the rate of substitution is reduced, because of shift of electron density to the metal ion.

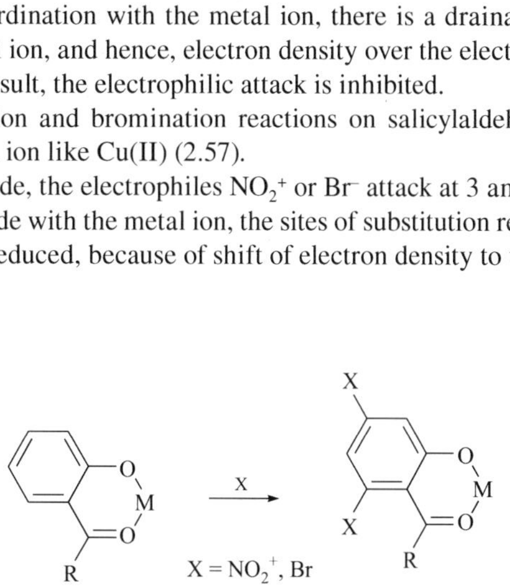

$$(2.57)$$

2. Besides the σ or π bonding from the L → M, there can be M → L back bonding due to interaction between the filled dπ orbitals of the metal ion and the vacant pπ or dπ orbitals over the ligand. This happens in the case of complexes of class B metal ions with π acidic ligands. The relative strength of L → M σ or π bond and M → L π bond determines the residual electron density over the ligand, after coordination with the metal ion.

If M → L π bonding is predominant, there is an increase in the electron density over the ligand, leading to its greater electrophilicity. Thus, the coordinated ligand becomes more susceptible to attack by the electrophiles and less susceptible to nucleophilic attack. Following are two illustrative examples.

a. In β-diketonate complexes, there is a ligand-to-metal σ bond and also M → L π interaction, due to the interaction between metal dπ orbitals and the π orbitals, delocalised over the β-diketonate ion. Thus, the electron density over the metal ion is transferred on the coordinated β-diketonate ion, with π delocalisation over the whole ring, constituted of the ligand and the metal. The two CO groups become equivalent with $V_{co} \sim 1,600\,cm^{-1}$ in the IR spectrum, intermediate between double- and single-bonded CO. This means, the keto character is lost and the coordinated β-diketonate does not undergo the nucleophilic Schiff formation reaction.

The π delocalised ring has pseudoaromatic character, as supported by the appearance of the signal of γ carbon proton at ~3.7 τ, comparable with the proton signal of aromatic compounds. The carbon atom of coordinated β-diketonate undergoes electrophilic nitration and bromination reactions, like the aromatic compounds, whereas such reactions are not possible on the free ligand (2.58).

$$X = NO_2^+, Br \qquad (2.58)$$

b. π Bonding of small molecule with electron-rich molecules leads to the transfer of electron density on the small molecule, and this makes it a stronger nucleophile. For example, when N_2 molecule gets coordinated to $M(R_3P)_4$, where M = Mo, W or Fe, there is an increase in electron density over N_2 and the electron-rich N_2 can bring about nucleophilic attack on proton. Thus, coordinated nitrogen can get protonated.

If the metal ion in the complex $M(R_3P)_4$ is molybdenum, the coordinated nitrogen gets only monoprotonated. However, if M = W, diprotonation of nitrogen occurs. This is because there is more π back donation possible from Tungsten, and so, the electron density on the coordinated N_2 is higher enough, to bind with two protons.

This reaction throws light on the role of iron and molybdenum in the nitrogen binding sites of the enzyme nitrogenase, catalysing conversion of N_2 to ammonia (to be discussed in Chapter 8).

Model Systems versus Metalloenzymes

It is evident from the above discussions that the metal ion-catalysed reactions throw light on the mechanism of metalloenzyme reactions. However, the metalloenzymes are not bare metal ions, but metal ions bound to apoenzymes, therefore metal complexes with vacant coordination sites can serve as better models for the metalloenzyme reactions. For this reason, the catalytic roles of Co(III) tetra amine complexes in amide hydrolysis and of trialkyl phosphine complexes of metal ions in activation of N_2 molecules to protonation have been discussed in this chapter.

The ligand bound to the metal ion is supposed to behave like the apoenzyme in natural systems. The σ or π character of the ligand affects the electron density over the metal ion and thus controls the hard or soft acid behaviour of the metal ion and in turn its catalytic activity. In these model systems, the ligand is considered to be a spectator ligand, supposedly not taking part in catalysis. However, this may not be true in the biochemical metalloenzymes with apoenzyme and substrate molecules with large side groups. In most cases, the apoenzyme has the catalytic role, and the metal ion only orients the apoenzyme and the substrate suitably to react most effectively. Apparently, synthesised metal complexes of small ligands may not be ideal models for the metalloenzymes and may not be able to compete with the metalloenzymes in efficiency.

However, working with model systems is easier than with metalloenzymes. Furthermore, variation in the metal ion and the ligands can be made in the model catalytic complex to design the best catalyst. Such variations in the metalloenzymes may not be possible, because any change, made in metalloenzyme, may deactivate it. So, the better option will be to use the knowledge of the effect of various factors on catalytic activity, as gained in models systems and extrapolate it to understand the role of metalloenzymes. This emphasises the importance of Inorganic Chemistry in understanding biochemical processes.

QUESTIONS

1. What does the overall formation constant depict?
2. List the factors that affect stability of a metal complex.
3. The heat evolved in the formation of a gram mole of the complex is called _____.
4. Why does chelate formation result into more stable metal complex?
5. What is the Irving–Williams order? What is its importance?
6. Explain the anomalous positions of Cr (II) and Cu (II) in the Irving–Williams order.
7. How does the size of the metal ion affect the stability of metal complexes of macrocyclic ligands?
8. What is "multiple juxtapositional fixedness" in metal complexes?
9. Explain the role of π-acidic ligands in stabilising the ternary complexes.
10. Why is the effect more when the second bidentate ligand is O^-—O^- coordinating?
11. How is this effect important in biochemical systems?
12. What are the other factors in ligands affecting $\Delta \log K$ in mixed-ligand complexes?
13. What is intramolecular interligand interaction?
14. What are the different types of inter-ligand interactions?
15. What is the order of strength of electrostatic, hydrophobic and stacking interligand interactions?
16. Explain how ligand side group–metal interaction stabilises metal complexes.
17. Why the ligand substitution reactions in metal complexes are considered as nucleophilic substitution reactions?
18. Which conditions favour SN_1 reactions in metal complexes?
19. Trans effect is related to which reactions?
20. How does Taube identify a labile metal complex?
21. Substitution reactions for labile metal complexes depend on their thermodynamic stability. Do you agree with this statement? Why?
22. Explain how valence bond theory tells that in the case of metal complexes of d^0, d^1, d^2 metal ions, substitution reaction is expected to proceed through SN_2 mechanism. Will such complexes be labile or inert?
23. On boiling for a long time, some cis complexes, like $[Co(en)_2NO_2Cl]Cl$ or $[Co(en)_2(NO_2)_2]NO_3$, get converted to the trans form. What is this reaction called?
24. What is rhombic or tetragonal twist?
25. What was the alternate method of rhombic twist? Explain.
26. How is Bailar's mechanism different from Rays twist?
27. Which factors are responsible for formation of outer-sphere complex?

3

Alkali and Alkaline Earth Metal Ions in Biochemical Systems

The four cations Na^+, K^+, Mg^{2+} and Ca^{2+} are widely distributed in all living organisms. Mainly, they have role as constituents of the body parts or body fluids. They also play other roles like transmission of nerve impulses and muscle contraction, through transmembrane concentration gradient. They stabilise the structures of proteins, cellular membranes and skeletal mass. They also act as enzyme activators.

Alkali Metal Ions

Both Na^+ and K^+ act as cofactors for ATPase enzyme. However, roles of Na^+ and K^+ are, mainly, antagonistic (opposite to each other), and this is a controlling factor in metabolism. K^+ activates respiration in muscles, kidney, adipose and erythrocytes, and helps protein synthesis and acetyl choline synthesis, whereas Na^+ opposes all these functions.

A large number of enzymes require K^+ for showing their maximum activity. For example, pyruvate kinase, an enzyme in glycolytic cycle, in respiration process, is activated by K^+. As K^+ has low charge density, it cannot play the role of Lewis acid, in activating the enzyme. It activates pyruvate kinase by bridging the enzyme and the substrate, thus bringing them closer and facilitating the phosphorylation of pyruvic acid.

The extra- and intracellular concentrations of Na^+, K^+, Mg^{2+} and Ca^{2+} are different. In the case of mammalian blood cells, the extracellular blood plasma has higher concentration of Na^+ and Ca^{2+} and lower concentration of K^+ and Mg^{2+}, whereas inside the red blood cells, situation is opposite.

The concentrations in mmole/L of the alkali and alkaline earth metals, in intra- and extracellular fluids, are shown in Table 3.1.

The normal osmotic transport of metal ions, through semi-permeable membrane, is from higher concentration to lower concentration, leading to equal concentrations on both sides. This is called passive transport. However, as stated above, in living cells a concentration gradient (difference in concentration of ions) is maintained, across the semi-permeable membrane, defying passive transport.

The transmembrane concentration gradient of Na^+ and K^+ plays an important role in a variety of transport mechanism, used by cells to accumulate and expel nutrients and toxic ions.

Furthermore, the concentration gradient, across the membrane, causes the generation of an electrical potential of about 60 mV in the membranes of the excitable tissues. This transmembrane potential inhibits the flow of ions, from one side of the membrane to another.

When there is an excitement of the tissue, the permeability of the alkali metal ion increases. There is diffusion of Na^+ inside the cell and of K^+ outside the cell, and there is lowering in the transmembrane potential. Thus, the flow of ions in the channels, in the membrane, is voltage gated (regulated by voltage). Electrical signals are transmitted between nerve cells by a mechanism involving ion channels.

Table 3.1

Metal Ion	Intracellular Concentration	Extracellular Concentration
Na^+	10	145
K^+	140	5
Mg^{2+}	30	1
Ca^{2+}	1	4

This process of ion transport in the direction of concentration gradient is a passive transport, governed by the osmotic phenomenon. However, after few milliseconds of the transmembrane potential change, the original intra- and extracellular concentration levels of the alkali metal ions and transmembrane potential are restored, by the back diffusion of K^+ inside the cell and of Na^+ outside the cell. This process, of transport of potassium inside the membrane, is against concentration gradient and is called active transport. The mechanism of both passive and active transports, through the cell membrane, needs to be explained.

The cell membrane is constituted of phospholipids and proteins, as discussed in Chapter 1. The membrane allows restricted transport of metal ions and nutrients. The plasma is a region of high dielectric constant, whereas the membrane has low dielectric constant. Hence, the membrane should inhibit the passive transport of the charged alkali metal ions across it. Experimental studies show that the cell membrane is almost impermeable to Na^+ and K^+. However, on addition of valinomycin, there is a significant increase in the permeability of K^+, whereas there is a slight increase in the permeability of sodium ion. Some of the synthetic crown ethers (cyclic polyethers) (Figure 3.1a) and cryptands (bicyclic polyether diamines) (Figure 3.1b) also assist in the transport of the charged metal ion, across the hydrophobic neutral lipid membrane.

The ionophores bind with the metal ion through the polar ($-\delta$) oxygen site and show selectivity in binding with the alkali metal ions. The selectivity is primarily based on the charge and size of the metal ion. The cyclic ligands form most stable complexes with the metal ions which fit in the cavity of the ligand ring, optimally. For example, Na^+ binds most selectively with dicyclohexyl-16-crown-5 and [2,2,1] cryptand, whereas bigger K^+ is bound selectively to [18-crown-6] and [2,2,2]-cryptand.

As stated above, the neutral molecule valinomycin, a cyclic dodecadepsipeptide is more selective of K^+. However, in this case the selectivity is not based on the best fit in the valinomycin, because both Na^+ and K^+ can be accommodated comfortably in the central core of this ligand. The selectivity is due to the difference in the hydration energies of the ions. The macrocyclic ligand can coordinate with the metal ion only if it liberates sufficient energy to overcome the loss of hydration energy of the alkali metal ion. The Na^+, with a small size, has higher hydration energy than K^+, and hence, valinomycin binds favourably with K^+.

The cyclic ligand encapsulates the charged metal ion. After binding a metal ion, the polar functional groups fold inwards and thus provide a neutral hydrophobic organic cover. This promotes the solubility of the metal–ionophore system in the lipid membrane and helps in the transport of the alkali metal ions,

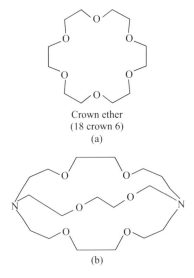

Crown ether
(18 crown 6)
(a)

(b)

FIGURE 3.1 Synthetic macrocycles (a) crown ethers (b) cryptands.

through the hydrophobic hydrocarbon tail regions of the lipids, in the bilayer. The rate of transport of the metal ion through the membrane depends on the lipophilicity of the alkali metal–ionophore complex.

The transport of ionophore encapsulated metal ion, through hydrophobic medium, can be illustrated by the following experiment. It is observed in Figure 3.2a that the Na^+ nitrophenolate cannot pass through hydrophobic chloroform. However, chloroform has crown ether dissolved in it (Figure 3.2b). Na^+ is taken up by the crown ether from the NaCl solution in the left column and is transferred to the aqueous solution in the right column, as seen by the colour in the right-hand side. Thus, Na^+ passes through the neutral hydrophobic chloroform.

Na^+ passes through the chloroform being encapsulated, in the holes of the dissolved crown ether. The presence of ionophores, in the cell membranes of the microorganisms, has been shown since the early 1950s. Their molecular weights range from 500 to 2,000 daltons. The ionophores present are usually monobasic carboxylic derivatives of ethers. These ionophores form channels for the positively charged ion to pass through. The negatively charged ionophore forms a neutral complex with the metal ion. However, if the ionophore is a neutral species, it forms a positively charged complex with the alkali metal ion, and it needs a permeant anion, to move through the membrane.

Thus, there is carrier ionophore-assisted, passive transport of the metal ion from the region of higher concentration to the lower concentration region across the cell membrane, and this explains the phenomenon of passive transport.

The question, how there is higher concentration of Na^+ outside the cell membrane, whereas there is higher concentration of K^+ inside the cell, that is, the existence of concentration gradient across the cell membrane, has to be answered. As per normal osmosis, this concentration gradient cannot be maintained. In dead cells, the concentration gradient is not retained. This shows that in the living cells, there is some special mechanism of the transport of the alkali metal ions, against the concentration gradient. This is called "active transport".

The energy required for the transport of metal ions against the concentration gradient is released due to Na^+- and Mg^{2+}-promoted, enzyme ATPase, catalysed, hydrolysis of ATP to adenosine diphosphate

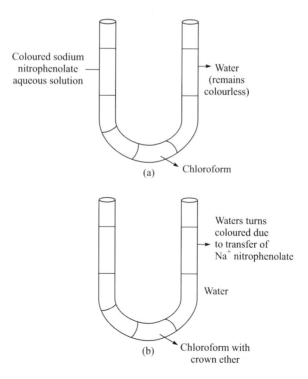

FIGURE 3.2 Experiment showing transport of ionophore encapsulated metal ion, through hydrophobic medium (a) no transport in absence of crown ether (b) transport in presence of crown ether.

(ADP) and inorganic phosphate. Hence, this phenomenon of active transport is called "sodium pump mechanism".

The liberated energy helps in pushing Na^+ outside the membrane. K^+ outside the cell membrane binds with the enzyme – phosphate, and enters inside the cell. Three Na^+ ions and two K^+ are involved in the reaction, as shown by the following equation:

$$3Na^+ \left(in\right) + 2K^+ \left(out\right) + Mg^{2+} \left(in\right) + E - ATP + H_2O \longrightarrow$$

$$3Na^+ \left(out\right) + 2K^+ \left(in\right) + Mg^{2+} - ADP + HPO_4^{2-} + H^+ + E$$

It is illustrated in Figure 3.3.

A cyclic representation of the process is shown in Figure 3.4.

The figure shows that the phosphorylated enzyme E_1PO_4 undergoes conformational change (eversion) to E_2PO_4 outside the cell membrane. As the binding of Na^+ with E_2 form of the enzyme is weak, Na^+ is released outside the cell membrane. It is like a motion though a revolving door that brings bound Na^+ ion to the outside of the cell membrane. Three Na^+ ions are replaced by two K^+ in E_2PO_4 because the K^+ forms strong bond with the E_2 form of the enzyme and the resulting $2K^+-E_2-PO_4$ is pushed inside the cell. Here, the enzyme in $2K^+ E_2-PO_4$ gets dephosphorylated and changes back to E_1 form. As K^+ has

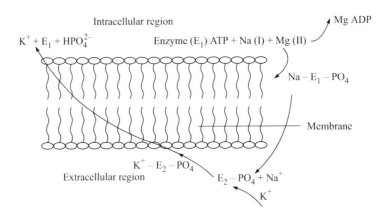

FIGURE 3.3   Active transport of sodium through cell membrane by sodium pump.

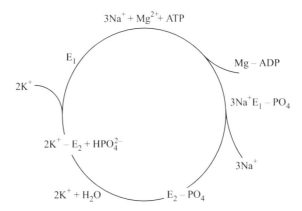

FIGURE 3.4   Cyclic presentation of sodium pump.

weak binding tendency to the E_1 conformation of the enzyme, it gets dissociated from K–E_1. Thus, K^+ and E_1 are released, inside the cell, and the cycle continues.

The ionophores present in biological membrane have much greater degree of selectivity, but in alkali metal transport, they have to discriminate only between Na^+ and K^+, as other alkali metal ions, such as Li^+, Rb^+ and Cs^+, are in negligible amounts. As the difference in the sizes of Na^+ and K^+ is large (38 Å), natural ionophores, in the cell membrane, can discriminate them very efficiently.

Role of Na⁺/K⁺ Pump in Brain Activity

Brain is the centre of central nervous system in vertebrates. About 14–16 billion neurons (nerve cells) are present in the cerebral cortex in human brain and about 40–60 billion are in cerebellum. The neurons in brain are considered basic working units designed to transmit information to other nerve cells, muscles or gland cells. Neurons communicate with each other through synapses. Information is transmitted in the form of electrical signals.

Signal transmission in neurons in brain occurs by flow of Na^+ and K^+ ions aided by Na^+/K^+ pump. The flow produces an imbalance of electric charge – an electrical spike – which is called an action potential (AP). Occurrence of AP is also called "firing" of a neuron. An AP is thus a message, in the form of an electric impulse, caused by a rapid change in the cell's membrane potential from its resting potential, in the event of the cell getting a stimulus. The sodium/potassium pump maintains the resting potential of the cell (which is $-70\,mV$). After an AP is produced, the Na^+/K^+ pump restores back the original state of the Na^+ and K^+ ions, making the neuron ready for next AP when needed.

Thus, till the end of the 20th century, the role of Na^+/K^+ pump in brain was considered to be only an indirect one, helping signal transmission. Research in last 20 years, however, has pointed towards more direct role of Na^+/K^+ pump in functioning of brain neurons. It has been suggested that the Na^+/K^+ pump plays an important role in brain coding and computation. The pump controls the mode of activity of the Purkinje neurons in cerebellum, which are responsible for motor control. Purkinje neurons are large cells found within the Purkinje layer in cerebellum and have fundamental role in motor control. Recent studies have shown that the activity of these neurons is dictated by Na^+/K^+ pump. The pump controls the firing pattern of these neurons. The activity of the pump depends on the concentration of the intracellular Na^+ and extracellular K^+ ions. The relative change in these concentrations and the rate of these changes decides the kind of firing exhibited by the neurons. The identified firing modes of the neurons are

- **Quiescent:** relatively quiet or inactive mode or silent mode
- **Tonic:** slow firing
- **Burst:** fast firing
- **Bimodal (tonic and quiescent):** combination of the two modes
- **Trimodal (tonic, burst, quiescent):** combination of the three modes
- **Bimodal (burst and quiescent) repeat:** the two modes repeat for certain time duration

In the bimodal and trimodal repeat modes, the pump sets length of each constituent mode, that is, the time duration for a particular mode of firing. The repeat length of trimodal pattern is fixed for a single Purkinje cell but varies among different Purkinje cells from 20 seconds to 20 minutes. This duration and transition from one mode to other is dependent on the intra- and extracellular concentration of the ions, controlled by the Na^+/K^+ pump.

Forrest's numerical model proposes that intracellular Na^+ concentration is a memory element, which records firing history. Furthermore, the Na^+/K^+ pump "reads" this memory setting to dictate the timing and duration of long quiescent periods. These long quiescent periods, the longer time scales, may permit storage and short-term processing of sensory information in the cerebellar cortex, which is connected to a network of these neurons. Several animal studies have shown this kind of memory effect and information processing dictated by the Na^+/K^+ pump, though more studies are needed in humans to completely understand this role.

Sodium and Blood Pressure

The blood pressure (BP) in the human circulatory system, largely due to the work done by the heart, is often measured for diagnosis. It is usually expressed as systolic pressure (maximum, during one heart-beat) and diastolic pressure (minimum, in between two heart beats) and is measured in millimetres of mercury above the atmospheric pressure. The values largely accepted as normal are 80–120 mm, though the accepted values differ with age and health. In case the systolic pressure consistently remains higher, the condition may result into hypertension and/or cardiovascular diseases. Regulation of BP is a highly complex process, depending on relation between renal, neural, cardiac, vascular and endocrine activities under the influence of genetic and environmental factors.

Relation between dietary sodium intake and BP has been an area of intense research. Huge amount of data has been acquired confirming role of sodium, its regulation by kidney in humans and its relation to BP, hypertension and cardiovascular diseases. It is in general accepted that excess of sodium consumption leads to an increase in BP, and WHO suggests that sodium consumption should not exceed 5 g/day for healthy adult humans. USFDA has suggested a daily intake of not more than 2.3 g/day. Sodium is mainly consumed as salt (NaCl) in food, though body receives it through water and other food sources also.

In general, it is considered good to reduce daily sodium intake slightly as a preventive measure. Though bulk of the research data indicates direct relationship between high sodium intake and BP and hypertension, high sodium intake and increased cardiovascular risk and mortality, the recent research does indicate that the relation is not linear, as thought of earlier. In fact, it describes a J-shaped curve. Thus, increased intake of sodium increases risks, and taking very low amount of sodium also increases the risks and diseases.

The correlation between dietary sodium consumption and hypertension was reported by Dahl in 1972. This work and subsequent studies resulted into development of model for BP regulation by Guyton. This model suggested that kidney maintains the balance between sodium intake, extracellular volume and BP. The concept of *natriuresis* was introduced as mechanism through which kidney preserves a normal BP by regulating volume homeostasis and sodium reabsorption. Natriuresis is the process of sodium excretion in urine by kidney. The process is promoted by natriuretic peptides. It lowers the concentration of sodium in blood and decreases volume of blood. Natriuresis is similar to diuresis (the excretion of an unusually large quantity of urine), except that in natriuresis, the urine is exceptionally salty. This model of BP regulation is considered as a "two-compartment" model in which the extracellular fluid volume within the intravascular space is in equilibrium with the interstitial space volume. Since sodium is the major cation in the extracellular fluid, any change in removal of sodium in urine will cause increase in the intravascular fluid volume, thereby increasing BP. Recent studies have suggested modification of this model to a "three-compartment" model. Two important factors that have contributed to newer understanding are as follows:

- Studies indicating that body can accumulate excess of sodium without being osmotically active, the salt being stored in a third body compartment, like tissues of skin and muscles.
- Possibility of measurement of tissue sodium content in muscles and skin using ^{23}Na – magnetic resonance imaging (MRI). This was not known earlier; hence, this part of sodium in body was overlooked. These data show that sodium can get accumulated in skin and muscles without concomitant water accumulation. These studies have also shown that the concentration of sodium in these tissues increases with age.

New studies have also added immune system as a regulator of sodium content in body, along with kidney. A further set of studies have shown that a diet rich in sodium affects intestinal microbiota, causing inflammation in the gut. Hypertensive patients also show similar effect in gut microbiota.

Temporary Memory Loss

Transient memory loss (amnesia) is a memory loss, which cannot be said to be due to common neurological condition, such as epilepsy or stroke. In this, the memory of recent event is lost so that patient cannot remember where one is or how he arrived there.

In aged persons, sudden loss of sodium levels due to medications, or heart problems, can cause memory loss. Electrolytes, Na^+ and K^+, should have proper balance in the body, otherwise there can be effect on vital body systems. Severe imbalance of electrolytes can cause serious problems in brain, and this can cause temporary memory loss. Na^+/K^+ imbalance affects the production and activity of neurons. Hence, proper consumption of food and fluids should be maintained for balanced electrolytes in the body.

Calcium

Calcium has, mainly, the role as the constituent of extracellular structure of the body. Calcium is essential for the development of extracellular structures such as egg shells, bones and teeth. It is stored in the form of small crystals in the tissues, so that deposition and reabsorption are rapid and calcium can be mobilised and deposited, for the formation of extracellular structure.

Bone is mainly basic calcium phosphate, $Ca_{10}(PO_4)_6(OH)_2$, like mineral apatite. Apart from Ca^{2+} and PO_4^{3-}, there are other inorganic ions such as Mg^{2+}, Na^+, K^+, CO_3^{2-}, F^- and Cl^-. X-Ray structure reveals that the inorganic material is in the form of imperfect small crystals. There is hexagonal arrangement of Ca^{2+} and PO_4^{3-} around columns of OH^-. Besides this, bone contains large amount of the fibrous protein, called collagen. These provide a framework, around which there is nucleation and growth of calcium hydroxy apatite, for the formation of bone. The mode of deposition determines the shape of the bone. A number of biochemicals, such as parathyroid hormone, vitamin D, calcitonin, calsequestrin and ostocalcium, control the mobilisation and deposition of calcium. The pattern of deposition differs in bone, teeth and shell, proving that there is no single mechanism of calcification (calcium deposition).

Two types of cells that regulate bone formation are called osteoblast and osteoclast. The former helps bone formation, whereas the latter causes bone erosion, leading to the formation of deep tunnels in the bone matrix, causing decalcification. This results in the serious disease, called osteoporosis, specially in women, after menopause.

Teeth are also made of basic calcium phosphate. However, crystals in the teeth enamel are larger and thinner than those in bones. It is advised to add fluoride in the tooth paste, as it occupies some of the sites of OH^- in the hydroxyapatite of teeth structure, forming fluorapatite, which is less sensitive to acid degradation. However, excess of fluoride in water reacts with the hydroxyapatite and causes degeneration of bone and teeth. This leads to the disease called fluorosis.

Calcium is present in other organisms also. Egg shells and shells of smaller organisms are made of calcium carbonate. Crystals of inorganic salts are used by the shells in navigational devices, due to their gravity sensor property. In the vertebrates, the balance organ is present in the inner ear and is constituted of calcium carbonate. The exact mechanism by which the calcium mineral works in body balance is not known.

Calcium carbonate is also present in sea corals and molluscs. Calcium sulphate is stored as gravity devices in jelly fish. Stone formation in different organs of vertebrates is due to calcium oxalate deposition.

Other Biochemical Roles of Calcium

Besides the above extracellular roles of calcium, it is also an important constituent of the intracellular material and has an important role in muscle contraction, hormone secretion, glycolysis (glucose breakdown), gluconeogenesis (glucose synthesis) and cell growth. The changes in the concentration of intra- and extracellular calcium also govern the metabolic regulations, known before, as the calcium messenger system.

There are three important aspects of calcium biochemistry as follows:

    i. **Distribution of calcium across the semipermeable cell membrane:** The concentrations of the intra- and extracellular Ca^{2+} are 10^{-6} and 10^{-3} mol/dm^3, respectively. Though the intracellular concentration of Ca^{2+} is less, its concentration in certain intracellular organelles, such as endoplasmic reticulum (ER) and mitochondria, may be considerably high. The concentration gradient across the cell membrane is important for the biological activity of Ca^{2+}. Whenever the concentration of intracellular calcium increases, it is expelled through the semi-permeable cell membrane, against concentration gradient. This is controlled by an active transport phenomenon, similar to "sodium pump" and is called "calcium pump". This process can occur in two ways:

a. The transfer of Ca^{2+} from lower concentration, inside the cell, to the higher concentration outside the cell, across the semipermeable cell membrane, is associated with the hydrolysis of ATP, assisted by (Ca^{2+}, Mg^{2+}) ATPase enzyme (E_1) ATP break down and the product ADP and $(Ca^{2+}, Mg^{2+}, E_1, PO_4)$ are pushed out of the cell membrane by the energy liberated due to the hydrolysis of ATP. In the extracellular region, there is a change in the conformation of the enzyme. Ca Mg E_1P form changes to $CaMgE_2PE_2$ form. E_2 form of the enzyme has much lower affinity for Ca(II), and hence, Ca^{2+} gets dissociated and is released in the extracellular solution. Mg^{2+} remains bound to E_2P and is pushed back inside the cell. The enzyme comes back to the original E_1 form, and Mg^{2+} is liberated. E_1 is again available for the second cycle.

Calcium pump differs from sodium pump in the fact that for every Na^+ purged out of the cell membrane, there is an ion of K^+ transferred from the extracellular region. However, in the purging out of Ca^{2+}, there is no counter transport of Mg^{2+} from the extracellular region to the intracellular region. Mg^{2+} which acts as a coenzyme in ATPase, along with calcium, comes out of the intracellular region and finally is pumped in from the extracellular region. There is no change in Mg^{2+} concentration on either side (Figure 3.5).

b. Transport of Ca^{2+} ions against activity gradient across plasma membrane may also be accomplished by coupled transport of other ions, like Na^+, in opposite direction (antiport). This has been observed in the large cells of giant squid, axon and plasma membrane vesicles from other tissues. Details of the exchange process is not yet evident, but from thermodynamic calculation of the exchange process, based on the concentrations of ions inside and outside the cells, it is known that for every Ca^{2+} to be pumped out of a cell, against the concentration gradient, three Na^+ ions should pass in the opposite direction, inside the cell, along the concentration gradient.

c. Besides the above two types of Ca^{2+} transport, there are channels in the plasma membrane for Ca^{2+} transport, which can be opened or closed by the action of biochemicals or changes in the membrane potential. For example, Ca^{2+} channels can be opened by amino acid N-methyl-D-aspartate.

ii. **Ca^{2+} messenger system:** The transient increase in Ca^{2+} in the cytoplasm, which triggers various biochemical processes, is controlled by some hormones, called the first messenger. The entity which controls the action of the hormone is termed the second messenger. In some cases,

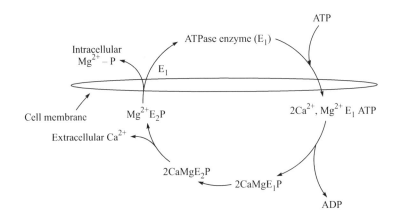

FIGURE 3.5 Active transport of calcium through cell membrane by calcium pump.

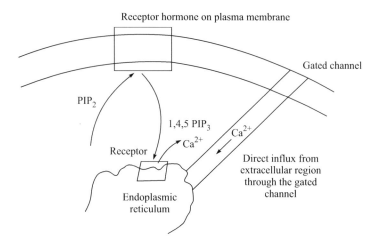

FIGURE 3.6   Calcium messenger system.

Ca^{2+} itself may act as second messenger. Some extracellular Ca^{2+} may be released in the cytoplasm by normal osmotic process, and this triggers the release of larger amount of Ca^{2+}. Thus, it is a case of calcium-induced calcium release. This mechanism is effective, in the case of cells having gated plasma membrane Ca^{2+} channels (of the type c, discussed above), for entry of extracellular Ca^{2+} inside the cytoplasm.

However, in most of the other cases, Ca^{2+} is released in response to intracellular messenger, inositol 1,4,5-triphosphate, (1,4,5 IP3). This is produced upon binding of phosphotidylinositol 4,5-biphosphate (PIP2) with a hormonal receptor on the plasma membrane, and getting hydrolysed to diglycerol and (1,4,5 IP3). The latter reacts with the receptor on the ER, which triggers the release of Ca^{2+}. However, original concentration of Ca^{2+} is soon revived. Excess Ca^{2+} in the intracellular region may be returned to the extracellular region or ER by Ca – ATPase-directed calcium pump, against concentration gradient. Ca^{2+} concentration in ER may alternatively be replenished, by direct influx of Ca^{2+} from extracellular medium, through the gated channel of type (c) (Figure 3.6).

iii. **Effect of change of calcium concentration:** As stated earlier, the change in the concentration of cytoplasmic Ca^{2+} influences the cellular machinery. This is because the excess calcium binds with certain proteins. These proteins should have selectivity for Ca^{2+}, compared to other ions. The protein should have a binding constant for Ca^{2+}, such that it should combine with Ca^{2+}, significantly, only when the concentration of calcium in the cytoplasm is high, due to the stimulus, which triggers the messenger system, for calcium release. Furthermore, the calcium protein should be kinetically labile and the protein should undergo change in conformation, on binding with calcium, so that there is a change in its interaction with other molecules.

The intra- and extracellular proteins which bind with calcium and assist some biochemical processes are as follows.

Intracellular Calcium-Binding Proteins

Troponin

Calcium, in millimole concentration, is necessary for muscle contraction. Muscle contraction is governed by the release of calcium from sarcoplasmic reticulum and the Ca^{2+}-binding proteins, present on the muscle fibre. When an impulse comes in the nerve ending in the muscle fibre, there is a release of Ca^{2+} from sarcoplasmic reticulum. This Ca^{2+} interacts with the regulatory proteins present in the muscle. In the higher vertebrates, the skeletal muscles contain the regulatory protein, troponin.

Troponin consists of three different components, troponin I, troponin T and troponin C. Troponin C has molecular weight 18,000 daltons and consists of 150 amino acid residues. It is the Ca^{2+}-binding subunit of troponin. It has two globular parts, each with two potential Ca^{2+}-binding sites, separated by nine turns in the protein structure. All the four calcium-binding sites have similar structures. In each one, there are two α-helics, separated by 12 amino acid loops, wrapping around the calcium. The ligand atoms of the protein binding with Ca^{2+} are oxygen atoms (carboxylate, hydroxide), located at the corners of a pentagonal bipyramid. This structural arrangement is termed "EF" or hand type (Figure 3.7). This is because the helix loop helix structure of the protein and the calcium-binding site is comparable with a hand with five fingers (Figure 3.7b). The fore finger, in the plane, represents the E helix of Figure 3.7a. The thumb, directed perpendicular to the plane, represents the F helix. The remaining fingers make the calcium-binding loop.

Out of the four calcium-binding sites, two have high affinity, and two have low affinity to bind with Ca^{2+}.

The high-affinity Ca^{2+} sites bind with Mg^{2+} competitively, though with much less affinity, and are called Ca^{2+} and Mg^{2+} sites. The low-affinity Ca^{2+}-binding sites bind with Mg^{2+} weakly and hence are called calcium-specific sites. Binding of calcium with low-affinity site is the crucial step in the muscle contraction and hence is called the regulatory site. Skeletal muscle troponin C can bind four calcium, whereas cardiac muscle, troponin C, binds with only three calcium.

As the magnesium concentration inside the muscle cell is high, the Ca^{2+}–Mg^{2+} sites are fully occupied by Ca^{2+} and Mg^{2+}, in the resting state of troponin. The release of calcium in the cytoplasm leads to Ca^{2+}-binding to Ca^{2+}-specific sites. Troponin C undergoes change in conformation, with consequent contraction of the muscle. It has been shown by NMR studies that the conformation change, induced due to Ca^{2+} binding with troponin C, is mainly confined to the region where Ca^{2+} ions are bound, and it cannot be propagated to other domains.

Calmodulin

In eukaryotes, there is a calcium-binding protein, calmodulin, consisting of 148 amino acid residues. It resembles troponin C closely. X-Ray study of bovine brain calmodulin shows that it has a dumb bell shape, with two globular parts, connected by an eight-turn helix each with two Ca^{2+}-binding sites and having no direct contact.

Each calcium-binding site has helix loop helix "EF" hand-type structure, as in troponin C. The two calcium-binding sites of calmodulin also have higher binding constant with Ca^{2+} (2×10^{-5} M^{-1}), whereas the other two have lower binding constant (3×10^4 M^{-1}).

The Nuclear Magnetic Resonance (NMR) spectral study shows that calmodulin undergoes change in structure on binding with Ca^{2+}. One probable change suggested is that the hydrophobic parts of the two globular portions of the proteins get exposed, on binding with Ca^{2+}, and calmodulin protein can better

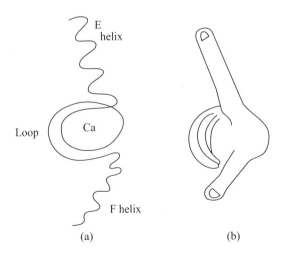

FIGURE 3.7 Troponin C (a) the EF arrangement (b) comparison with hand.

bind with other hydrophobic molecules such as drugs, small peptides and the target proteins, on which it brings catalytic reactions.

This Ca protein acts as a phosphatase, causing dephosphorylation of target protein. It also acts as a kinase, causing phosphorylation of target protein.

Troponin C and calmodulin are closely related intracellular Ca^{2+}-binding proteins, with similar "EF" hand-type structure and similar regulatory roles. They are considered to belong to the "calmodulin super family". The other members of this family having characteristic EF hand-type structures are as follows.

Parvalbumin

It is a lower molecular weight (12,000 daltons) water-soluble, calcium-binding protein, found in the white muscles of fishes, amphibians and chickens. It also occurs in different mammalian tissues, including the nervous system. It has 103 amino acid residues. It has two Ca^{2+}-binding sites per mole. The protein has two helix oriented in an EF hand-type structure, and the Ca^{2+} are octahedrally coordinated. The structure of the complex was determined by the substitution of Ca^{2+} by Cd^{2+}. Ca^{2+} with quadrupolar nucleus shows broadened NMR signals, whereas Cd (I = 1/2) shows sharp signals in NMR spectrum, and the spectrum supports the octahedral structure, of the substituted Ca^{2+} site.

Parvalbumin is similar to calmodulin and troponin, but does not appear to exert a direct regulatory function, like the other two Ca^{2+} proteins. It is involved in muscle contraction process in the fish, amphibian and chicken. It has been suggested that it has the role of taking up the Ca^{2+}, released from calcium troponin complex in the muscle cells; thus, it keeps the Ca^{2+} concentration low, in the cytoplasm. Parvalbumin also occurs in non-muscle tissues, indicating that it has roles, other than muscle contraction.

Calbindin

It is an intracellular Ca^{2+}-binding protein with no definitely known function. It is considered to be involved in transport of Ca^{2+} in intestine and placenta.

This protein occurs in various tissues of mammals in D9k form, and in birds in D28K form. However, the two forms of the proteins do not have closely related structure. Only similarity is that their synthesis depends on vitamin D.

The protein D9k has two Ca^{2+}-binding sites, both Ca^{2+} being roughly octahedrally coordinated. Though the overall structure of each unit is EF hand type, as in calmodulin, troponin C and parvalbumin, there is difference in structures of the peptides at the two calcium-binding sites.

There is only one carboxylate at site I, whereas there are three carboxylates at site II. However, strength of binding of both the sites with Ca^{2+} is alike.

Sarcoplasmic Calcium-Binding Proteins (SCP)

This is another member of the calmodulin super family. It is found in the muscles of vertebrates and invertebrates. It has four domains. However, only one half of the proteins, containing domains III and IV, have two calcium-binding EF hands, similar to calbindin D9k and globular domains of troponin C and calmodulin. The other half does not have loop structure. The function of this protein is not definitely known.

Calpin

It is a calcium-activated intracellular protease. It has four domains. The first and third domains have no distinct sequence in the protein structure. The second domain has structure, similar to proteolytic enzyme papain, whereas the fourth domain has structure similar to calmodulin. Hence, it is called calpin. This is an unusual combination of a regulatory protein (calmodulin) with a target enzyme (papain).

Protein Kinase C (PKC)

It is a calcium protein, with EF hand-type calcium-binding site as in calmodulin. The calcium loaded, active form of protein kinase C (PKC) causes phosphorylation of the cytoplasmic proteins.

Annexins

This is a class of intracellular proteins, which interact with the membrane or membrane cytoskeleton protein, depending on the calcium ion concentration. The protein has three calcium-binding sites, which are located on the membrane binding side of the protein. The calcium-binding sites do not have "EF" hand-type structure, and hence, it does not belong to calmodulin super family. When the calcium concentration is above $200\,\mu M$, these proteins mediate fusion or aggregation of vesicles. They are also involved in cell signalling pathway. However, this class of proteins have no known intracellular role, though they are intracellular proteins.

Role of Extracellular Calcium-Binding Protein

Following are the significant biochemical roles of the extracellular Ca^{2+}-binding proteins.

i. **Enzyme action:** Ca acts as a cofactor for enzymes like amylase and microbial proteinase. Staphylococcal nuclease, which shows hydrolite activity towards DNA or RNA, also has Ca(II) as cofactor. The apoprotein is bound to only one Ca(II), which has higher coordination number (6, 7 or 8), and can orient the apoenzyme to the conformation which is catalytically most suitable for binding with the substrate.

ii. Mammary glands produce a calcium-binding enzyme protein activator, α-lacto albumin. The Ca-binding constant of the enzyme protein is $107\ M^{-1}$. It is involved in the conversion of glucose into lactose. It has a molecular weight 15,000 daltons. The Ca is bound to the protein through seven oxygens, at the corners of a distorted pentagonal bipyramid, two oxygens are from carbonyl group, three from carboxylate of the aspartyl residues and two from water molecules. The structure has superficial similarity to calmodulin structure.

iii. In the serine proteases like trypsin and chymotrypsin, there is one Ca^{2+} bound to the four oxygens of the amino acid residues of the protein and two water molecules, resulting in octahedral structure. The protein is produced inside the cell but is not bound to Ca there. It gets bound to Ca in the extracellular region and is converted to active enzyme. Thus, the unwanted activity of the enzyme inside the cell is avoided.

iv. Phospholipase A2 is a Ca-dependent enzyme, located in the outer membrane, and catalyses the hydrolysis of 1,2-diacylglycero-3-phospholipids, releasing free fatty acids, which are digested in the liver, liberating energy. There is Ca^{2+} bound to four amino acid residues of the protein and two water molecules. NMR studies show that on Ca binding, the protein undergoes change only in the region of Ca binding, and the overall structure of the protein remains unchanged.

v. **Clotting of blood:** In the case of bleeding due to injury, coagulation of blood occurs to prevent excessive blood loss. This process proceeds in several steps like a cascade and involves many enzyme proteins. These proteins contain the amino acid residues γ-carboxy glutamate (gla) and β-hydroxy aspartate (Hya). Gla is formed by a vitamin K-assisted process. The best known gla-containing protein is prothrombin, which has a major role in blood clotting. It is a Ca(II) protein bound to up to ten Ca(II). In the presence of Ca^{2+} ions, prothrombin and other vitamin K-dependent proteins bind to the membrane of the platelets, containing phospholipids, leading to blood coagulation.

vi. **Stimulus-controlled secretion:** The secretion of biochemicals, involved in control and defence mechanism, is stored in vesicles and granules. When there is a stimulus, the granules migrate to the periphery of the cell and are extruded, a process called exocytosis. The process needs energy and is assisted at some stage by Ca(II) proteins.

Ca (II) also helps nerve impulse transmission. As a result of excitation (electronic in nature), there is ejection of neurotransmitters adrenaline, dopamine or norepinephrine. This is controlled by calcium proteins. Similarly, Ca controls the secretion of other hormones like histamine and insulin.

Role of Magnesium

Magnesium

Magnesium is fourth in abundance in biological systems. Half of it is present in the skeleton, and the remaining is present in the enzymes.

The concentration of Mg(II) is high inside the cell, whereas it is low in the plasma. As the intracellular concentrations of K(I) and Mg(II) are high, ribosomal RNA and proteins in the cells are bound to these metal ions, more to Mg(II), with higher Lewis acidity. Mg(II) binds at the phosphate end of the nucleic acid and reduces the electrostatic repulsion between the phosphates and thus stabilises the interaction between the base pairs in RNA. This results in the increase in the melting point of RNA. It has been observed that some RNA molecules show catalytic activity to promote hydrolysis of RNA – phosphodiesters. These RNA enzymes are called ribozymes. They are bound to five $[Mg(II)(H_2O)_6]^{2+}$, which retain their solvent environment and are H-bonded to O and N atoms on the bases, phosphates and sugar rings of RNA. These five Mg^{2+} help to stabilise the structure of RNA. One more Mg(II) is bound to the RNA and has catalytic role in phosphodiester hydrolysis. Mg(II) is bound to the RNA at three sides and to the phosphate oxygen of the substrate phosphodiester. Thus, there is formation of the intermediate RNA phosphodiester–Mg–RNA complex, leading to the hydrolysis of the phosphodiester (Figure 3.8).

Mg^{2+}-containing proteins are involved in several other enzymatic reactions. Mg^{2+} may participate in the enzymatic reaction in two ways. Mg^{2+} may bind with the substrate, and the enzyme protein gets bound with the substrate part of the Mg(II)–substrate complex (Figure 3.9a). On binding with Mg(II), the substrate can be activated by the enzyme favourably. Alternatively, Mg(II) may directly bind to the enzyme protein (Figure 3.9b). However, the Mg(II) binds weakly with the proteins, and the Mg(II)-activated proteins are not isolated in metal bound form. Probably, in this case, on binding with Mg(II) the enzyme gets suitably oriented to interact with the substrate.

Mg(II) can be replaced in vivo, in most of the Mg(II)-activated enzymes, by Mn(II). This may be, because both have spherical electronic distribution with electronic configuration $d^0(Mg^{2+})$ and $d^5(Mn^{2+})$.

Following are some of the enzymes containing Mg(II) or Mn(II). The discussion is based on the various roles of Mg(II)- and Mn(II)-containing enzymes in the glycolysis and the Krebs cycle in glucose oxidation.

1. **Kinases:** Metabolic energy liberated by oxidative degradation of glucose is stored in biological systems as P–O bond energy of the nucleotide ATP. Whenever energy is required, ATP molecules undergo hydrolysis, under the influence of kinases, to produce ADP, inorganic phosphate and 7.3 Kcal of energy per mol. The inorganic phosphate is transferred to some other molecule, resulting into its phosphorylation. The phosphate from the phosphorylated molecule can be transferred to ADP, regenerating ATP. This reverse reaction is also catalysed by kinases.

    Following are some kinases:

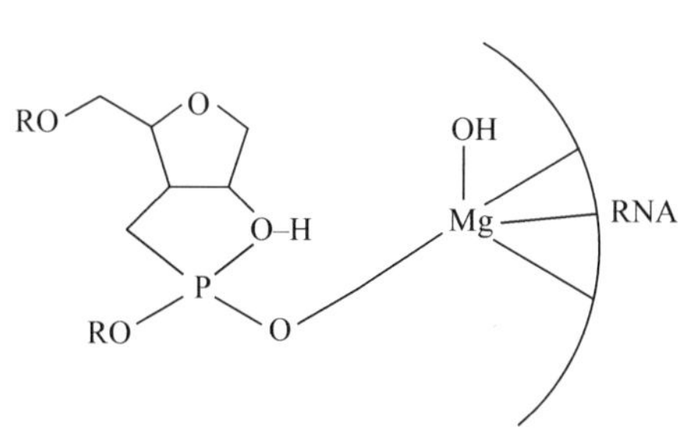

FIGURE 3.8 RNA phosphodiester – Mg – RNA complex.

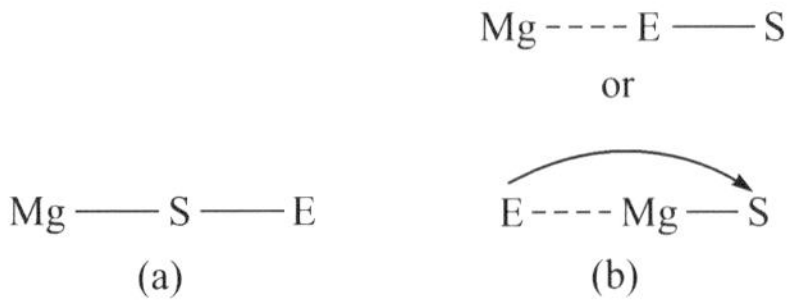

$$Mg ---- E —— S$$
$$or$$
$$Mg —— S —— E \qquad E ---- Mg —— S$$
$$(a) \qquad\qquad (b)$$

FIGURE 3.9 Binding modes of Mg-protein in enzyme action (a) Mg-Substrate-Enzyme (b) Mg-Enzyme-Substrate.

a. **Creatine kinase:** Creatine is an amino acid present in muscles. Its concentration in the blood is normally low, but shows increase in the case of renal dysfunction. Creatine kinase is a Mg(II) protein which assists transfer of phosphate from ATP to creatine (Eq. 3.1). It has a molecular weight of 82,000 daltons and occurs in heart and skeletal muscles. This enzyme has a dimeric structure and has one –SH group in each subunit, to maintain suitable conformation.

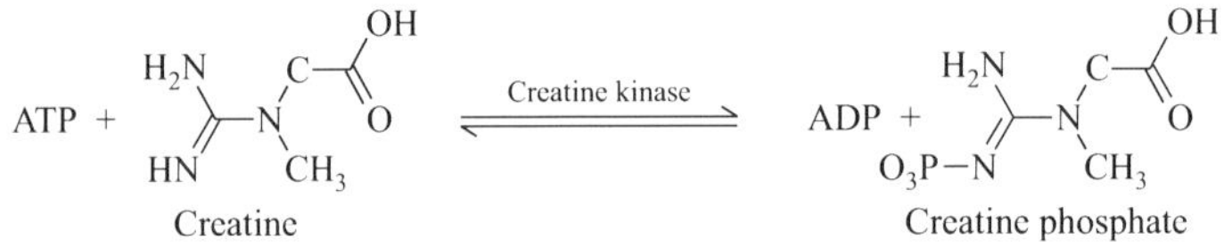

$$ATP + \text{Creatine} \;\xrightleftharpoons{\text{Creatine kinase}}\; ADP + \text{Creatine phosphate} \qquad (3.1)$$

It has been shown that there is formation of the intermediate ternary complex E–Mg–ATP. In the reverse process of Eq. (3.1), creatine phosphate catalyses the phosphorylation of ADP by phosphocreatine through the formation of the intermediate E–Mg–ADP. Thus, creatine kinase helps in storing the chemical energy of surplus ATP, by catalysing the formation of phosphocreatine. When there is an injury to skeletal or heart muscle, creatine kinase level in blood serum increases. Thus, the detection of creatine kinase level in blood serum is an indication of heart infarction.

b. **Pyruvate kinase:** The energy liberated by the metabolic oxidative degradation of glucose is the sustainer of all life activities. It proceeds in two stages. In the first stage (glycolysis), the six carbon skeleton of glucose breaks to three carbon pyruvic acid. In the second stage, called citric acid cycle or Krebs cycle, pyruvic acid is completely oxidised to carbon dioxide and water. This proceeds through the phosphorylation of pyruvic acid.

Pyruvate kinases are Mg(II)- or Mn(II)-containing enzyme proteins which assist the transfer of phosphate from ATP to pyruvate (Eq. 3.2). They also need K^+ to bring the necessary conformational charge and to keep the enzyme in the active form.

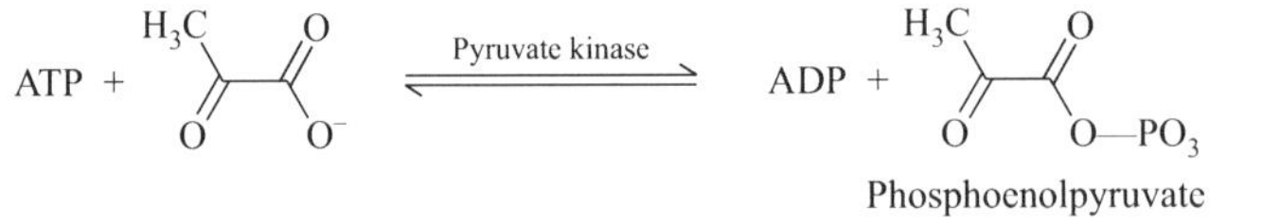

$$ATP + \text{(pyruvate)} \;\xrightleftharpoons{\text{Pyruvate kinase}}\; ADP + \text{Phosphoenolpyruvate} \qquad (3.2)$$

In the intermediate stage (Figure 3.10), the metal ion (Mg^{2+} or Mn^{2+}) gets bound to the phosphate end of substrate ATP and helps the transfer of phosphate to pyruvate.

The above intermediate was suggested by earlier proton relaxation studies. However, more detailed studies show a low binding of Mg with the enzyme. The distance between Mg and ATP is 1.5 nm which is large for any interaction. Mg(II) may only be orienting the enzyme protein suitably to interact with ATP, for its breakdown to ADP and phosphate and transfer of phosphate to pyruvate. This indicates that the protein has the enzymatic role and that Mg(II) acts as a coenzyme.

In the reverse reaction of (3.2), the phosphoenolpyruvate can transfer the phosphate group to ADP, regenerating ATP. This reverse reaction is also catalysed by pyruvate kinase.

c. **Hexokinase:** In the first stage of glycolysis, a glucose molecule is activated through phosphorylation by ATP and is converted to glucose-6-phosphate. This reaction is catalysed by a Mg(II)-activated enzyme hexokinase, Eq. (3.3).

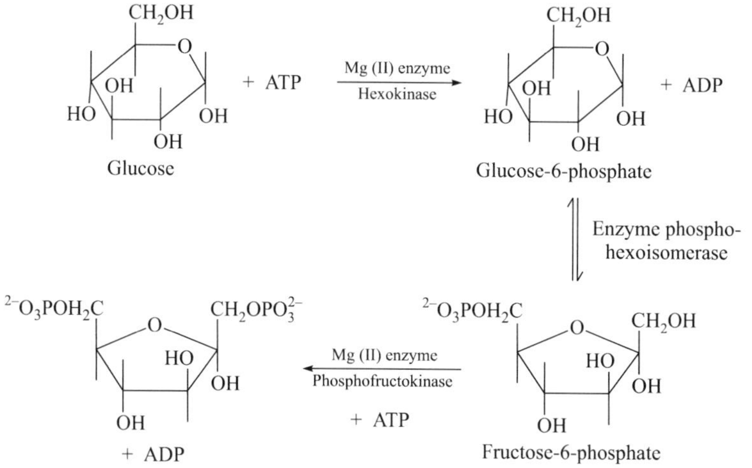

$$(3.3)$$

Glucose-6-phosphate isomerises to fructose-6-phosphate, catalysed by a non-magnesium enzyme phosphohexoisomerase (Eq. 3.3).

d. **Phosphofructokinase:** Fructose-6-phosphate undergoes further phosphorylation by ATP to form fructose-1,6-diphosphate (Eq. 3.3). This reaction is catalysed by Mg(II)-containing enzyme phosphofructokinase. Fructose-1,6-diphosphate splits into dihydroxy acetone phosphate and isomeric 3-phosphoglyceraldehyde, catalysed by the enzyme fructose diphosphate aldolase (Eq. 3.4). In subsequent step, dihydroxy acetone phosphate converts to 3-phosphoglyceraldehyde, catalysed by the enzyme triosephosphate isomerase. The two molecules of 3-phosphoglyceraldehyde, generated from fructose-1,6-phosphate, undergo oxidation to 3-phosphoglycerate (Eq. 3.4). The enzymes involved in the later reactions, as shown in Eq. (3.4), do not involve Mg(II).

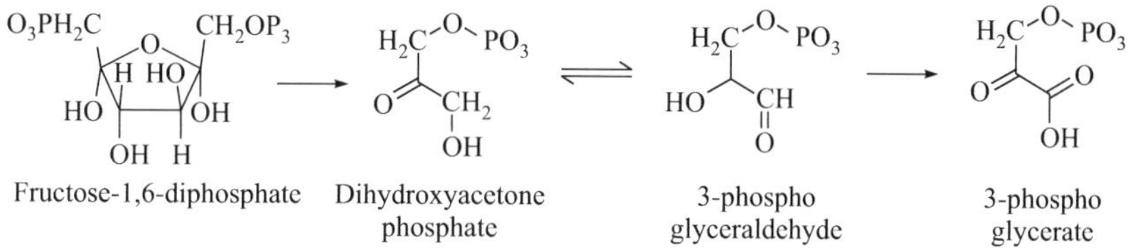

$$(3.4)$$

2. **Non-kinase Mg(II)-containing enzymes:** These are involved in further breakdown of 3-phosphoglycerate and are as follows.

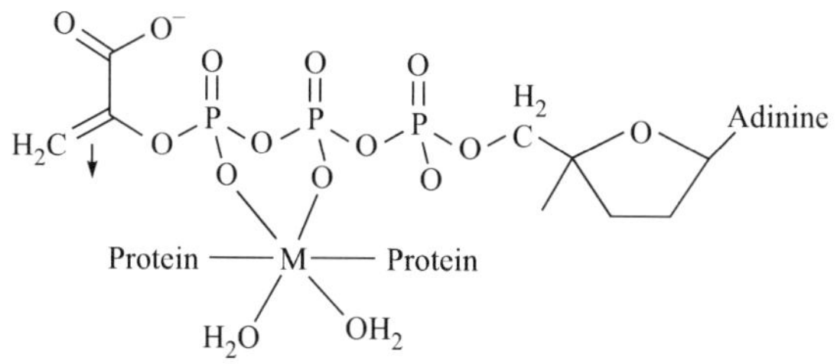

FIGURE 3.10 Intermediate formed in the transfer of phosphate from ATP to pyruvate by enzyme Pyruvate kinase.

a. **Phosphoglyceromutase:** This is a Mg(II)-activated protein enzyme, which catalyses the transfer of phosphate from 3 to 2 position in phosphoglycerate, leading to the formation of 2-phosphoglycerate (Equation 3.5).

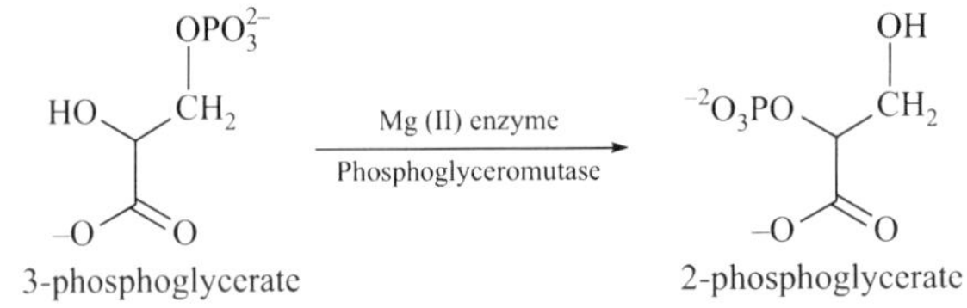

$$\tag{3.5}$$

b. **Enolase:** This is a Mg(II)- or Mn(II)-containing protein, which catalyses the loss of a molecule of water from 2-phosphoglycerate, to form the high energy compound phospho-enolpyruvate (Eq. 3.6).

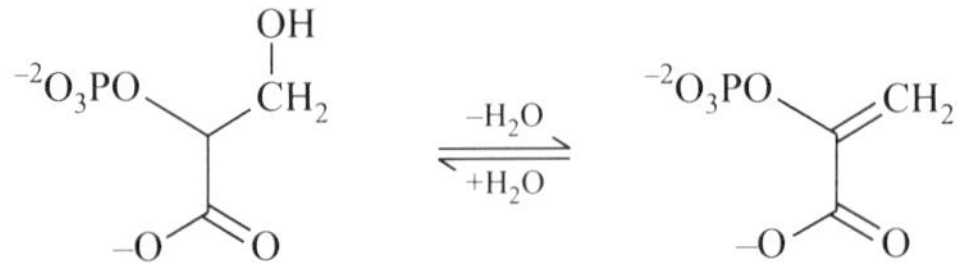

$$\tag{3.6}$$

Phosphoenolpyruvate transfers phosphate to ADP forming ATP, and pyruvic acid is generated. As stated earlier (Eq. 3.2), this reaction is catalysed by Mg(II)-containing enzyme, pyruvate kinase.

c. **Isocitrate dehydrogenase:** The pyruvic acid, produced in glycolysis, is converted in the Krebs cycle to citric acid. There is oxidative decarboxylation of pyruvic acid and subsequent combination with oxaloacetate, resulting in the formation of citric acid. Hence, the Krebs cycle is also called citric acid cycle or tricarboxylic acid cycle.

The enzyme pyruvate dehydrogenase, which catalyses the oxidative decarboxylation of pyruvic acid, involves Mg(II) as a cofactor.

In the next stage of the Krebs cycle, citric acid isomerises to isocitrate. Furthermore, isocitrate is oxidised to a ketoglutarate. This reaction is catalysed by the NAD+-linked enzyme, isocitrate dehydrogenase. This enzyme is activated by Mg(II) and Mn(II) (Eq. 3.7).

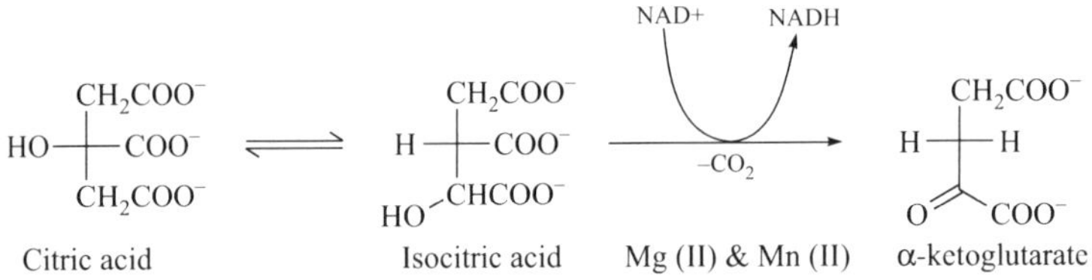

$$\tag{3.7}$$

d. **α-Ketoglutarate dehydrogenase:** α-Ketoglutaric acid undergoes an oxidative decarboxylation reaction through some intermediate stages to form succinate, catalysed by the enzyme α-ketoglutarate dehydrogenase. This is a complex enzyme which needs Mg(II), thiamine pyrophosphate, lipoic acid and coenzyme A, as co factor (Eq. 3.8).

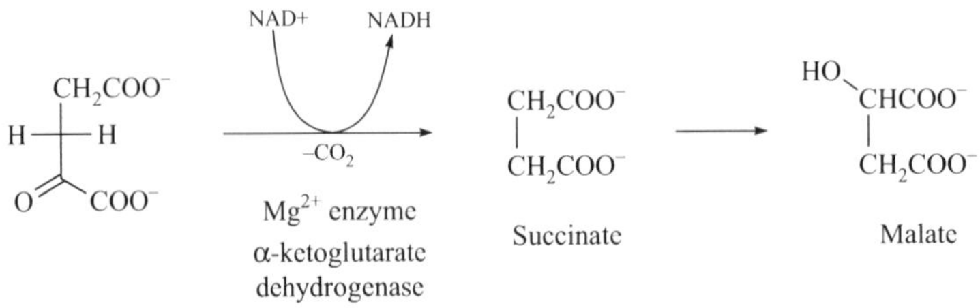

$$\tag{3.8}$$

Succinate is oxidised to fumarate, which undergoes reversible hydration to malate. The last reaction in the Krebs cycle is the oxidation of L-malate to oxaloacetate. The Krebs cycle is completed with the regeneration of oxaloacetate, which was used in the formation of citric acid.

e. **Pyruvate carboxylase:** Many cells have an alternative pathway of conversion of pyruvic acid to oxaloacetate. It is the carboxylation of pyruvate. Pyruvate carboxylase is the Mg^{2+}- or Mn^{2+}-containing enzyme [molecular weight 65,000 daltons], which assists the carboxylation of pyruvate. There are four molecules of biotin (vitamin H coenzyme) associated with the enzyme. Biotin acts as the CO_2 carrier, which is transferred to pyruvate to form oxaloacetate (Eq. 3.9).

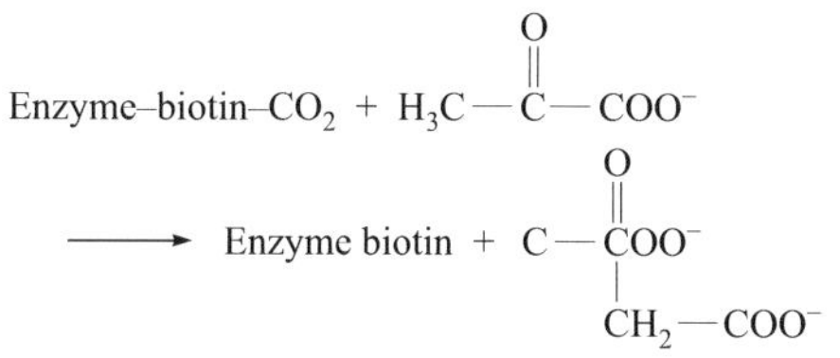

$$\text{Enzyme–biotin–CO}_2 \ + \ \text{H}_3\text{C}-\overset{\overset{\text{O}}{\|}}{\text{C}}-\text{COO}^-$$

$$\longrightarrow \ \text{Enzyme biotin} \ + \ \text{C}-\overset{\overset{\text{O}}{\|}}{\underset{\underset{\text{CH}_2-\text{COO}^-}{|}}{\text{COO}^-}} \tag{3.9}$$

f. **Phosphoglucomutase:** When glucose is not used for immediate combustion, glucose-6-phosphate is converted to glycogen in muscle tissues. For this, glucose-6-phosphate first rearranges to glucose-1-phosphate. This rearrangement is assisted by the enzyme phosphoglucomutase, which is having Mg^{2+} as a cofactor. The protein has a serine residue, the –OH of which gets phosphorylated and assists transfer of phosphate to 1 position.

The role of Mg(II) in glycolysis and citric acid cycle, leading to the oxidation of glucose, is shown in Figure 3.11a and b.

Following the above pathway, glucose undergoes oxidation, using oxygen, to produce CO_2 and H_2O, and liberating 671 Kcal/mole of energy. In the various stages, ADP is converted to ATP. Glycolysis involves use of ATP. Overall calculation shows that eight molecules of ATP are produced in the glycolysis process. In the Krebs cycle, 14 molecules of ATP and one GTP (= ATP), that is, 15 molecules of ATP are produced per pyruvate (shown in brackets in Figure 3.11a and b). As two pyruvates are produced from glucose, the total number of ATP synthesised in the Krebs cycle is 30. Thus, total number of ATP produced, utilising the energy liberated in the oxidation of one molecule of glucose, is 38. Formation of each molecule of ATP from ADP requires 7.3 Kcals. Hence, the formation of 38 molecules of ATP stores 277.4 Kcals, out of 671 Kcals of energy, liberated by the oxidation of one molecule of glucose.

The present discussion of the glycolysis and the Krebs cycle is only a background for understanding the various roles of Mg(II) enzymes. The readers, interested in the details of glucose metabolism, are advised to refer to the text books in biochemistry, to understand the sequence of the reactions, more exhaustively.

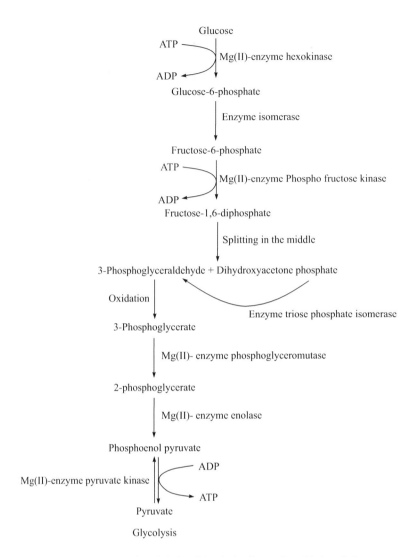

FIGURE 3.11a Role of Mg in the glycolysis and citric acid cycle, leading to the oxidation of glucose.

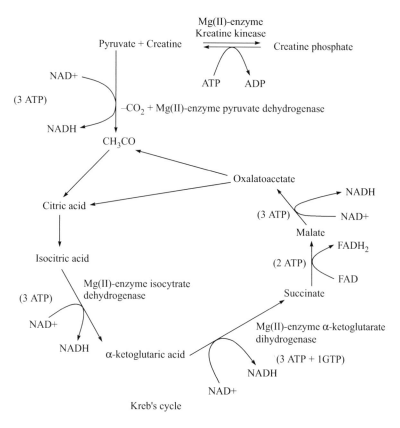

FIGURE 3.11b   Role of Mg in Krebs cycle.

QUESTIONS

1. Write broadly the important functions of Na^+, K^+, Mg^{2+} and Ca^{2+} cations in living systems.
2. What are alkali and alkaline earth metals? Why are they named so?
3. List some functions of Na^+ and K^+ in biological systems.
4. What is passive transport of ions? How is this related to phospholipid membranes in biological systems?
5. What are ionophores?
6. What do you understand by concentration gradient across cell membrane in living organisms?
7. What is sodium pump mechanism?
8. List some important functions of calcium ion in biological system.
9. Give an account of calcium pump.
10. How does magnesium ion stabilise the interaction between base pairs in RNA?
11. How many magnesium ions are present in ribozymes?
12. What is a phosphodiester bond?
13. Which metal ion is known to replace Mg^{2+} ion in vivo?
14. What are kinases? Give some examples of kinases.
15. Explain role of Ca^{2+} in muscle contraction.

4

Zinc in Biochemical Systems

Zinc is present in biological systems as a trace metal. However, it is indispensable for the growth and development of plants and animals. It is necessary for the synthesis of nucleic acids and for the maintenance of their conformation. Deficiency of zinc affects cell growth, division and differentiation, and thus leads to abnormalities in the composition and functions of cells.

It is thus apparent that Zn(II) is an essential trace element and should be adequately supplied in the diet. Zn(II) is absorbed in the small intestine, the absorption being minimum in the stomach and long intestine.

Zn(II) absorption is dependent on the biochemical ligands present in the diet, which form complex with Zn(II) and increase or decrease its uptake. The biochemicals, with cysteine and histidine coordinating sites, increase Zn(II) uptake.

Zn(II) is more efficiently absorbed from food from animal sources than from plant sources. Phytate, present in plants, forms strong chelate with Zn(II) and makes its absorption by the membrane difficult. However, in a balanced food, where other more effective Zn(II) binding ligands are present, along with phytate, its effect may be nullified and Zn(II) is easily absorbed by the membrane. Infants can absorb Zn(II) better from breast milk than from cow's milk.

The presence of minerals, like calcium, lowers down the zinc absorption. Similarly, a large excess of Cu(II) in the diet reduces Zn(II) absorption. This antagonism is due to various factors, involved in the absorption of trace metals. Another metal ion of nutritional significance, reducing zinc absorption, is iron.

This raises the question that food or vitamins, fortified with iron, may reduce the absorption of essential trace metal ion Zn(II) and adversely affect the health. However, it has been observed that iron given with milk does not affect Zn(II) absorption, emphasising that various factors play important role in trace metal absorption.

The human body contains 2–3 g of Zn(II) bound to proteins. Half of the Zn(II) is in the muscles, and 30% is in the bones. Zinc is also present in brain. In the human beings, normal Zn level in blood plasma is 0.7 to 1.3 mg/L.

Zn(II) occurs in biochemical systems mainly in the form of metalloenzymes. Main Zn(II)-containing enzymes are discussed below.

In the enzymes, Zn(II) may have four different roles:

i. The first type of Zn(II) is directly involved as a Lewis acid catalyst.

ii. In some enzymes, there may be two Zn(II) centres present. One is of the first type and has catalytic activity. The second type of Zn(II) is not itself catalytic, but regulates the enzymatic activity of the first type of Zn(II). It may act as an activator or inhibitor.

iii. The third type of Zn(II) stabilises the overall conformation of the enzyme protein, mainly stabilising its quaternary structure, so that the protein acts as an efficient catalyst. Thus, this type of Zn(II) acts as a coenzyme.

iv. There is a fourth type of Zn(II) in the enzymes, with none of the above three types of roles. This is called a non-catalytic type of Zn(II).

In the enzymes, Zn(II) mainly acts as a Lewis acid catalyst, because, this metal ion with d^{10} configuration has no redox chemistry. As a Lewis acid catalyst, it assists the reactions, in which there is a nucleophilic attack over the substrate.

Zn(II) is preferred by the nature as a Lewis acid catalyst, because of the following reasons:

i. Though Zn(II) ion has a bigger size than H^+, it has double positive charge. Hence, the positive charge density on Zn(II) is more, and it attracts the electrons strongly, and is a stronger Lewis acid. Further, Zn^{2+} does not hydrolyse till pH 8. So, it acts as a super acid; that is, the metal ion remains unhydrolysed and can act as Lewis acid at higher pH, where H^+ concentration is very much reduced.

ii. There is no ligand field stabilisation in d^{10} metal ions, and hence, there is no preferred coordination number in Zn (II) complexes. It can take up coordination numbers 4, 5 or 6. The stability of Zn(II) complexes is determined by the Zn–L distance and the repulsion between the ligands. In tetrahedral complexes, Zn–L distance is shorter, but the interligand repulsion is more. With increasing coordination number, M–L distance becomes longer, but the interligand repulsion becomes less. Thus, due to the cumulative effect of the two factors, the 4, 5 and 6 coordinated complexes have small difference in stabilisation energies and can change from one form to the other, easily. In enzymes, the number of ligands bound to Zn(II) is less than 6. There is a vacant coordination site available for the substrate binding. This implies that there is no significant energy requirement for the binding of the substrate with the Zn(II) of the enzyme.

iii. Zn (II) complexes are labile, and hence, exchange of ligands can take place rapidly. The lability of Zn(II) complexes is because Zn(II) has no preference for coordination number. Substitution reaction can take place at the tetrahedral complex site by the addition mechanism, through the formation of penta-coordinated intermediate, involving small energy barrier. Similarly, the hexa-coordinated complex site can undergo substitution, involving dissociation mechanism. The formation of the penta-coordinated intermediate needs small energy change, and thus, substitution is facilitated.

The following are the Zn(II)-containing metalloenzymes, where Zn(II) may have a catalytic or structure stabilising roles.

Peptidases

Carboxypeptidase

These are exo peptidases, which catalyse the hydrolysis of the peptides or proteins at the C-terminal amino acid residue. These are of two types:

i. **Carboxypeptidase A:** This catalyses the hydrolysis of those proteins which have a hydrophobic aliphatic or preferably aromatic side chain at the C-terminal amino acid residue.

ii. **Carboxypeptidase B:** This requires the presence of a positively charged side chain at the C-terminal of the protein, over which hydrolysis is to be catalysed.

Both carboxypeptidases A and B are formed by the hydrolytic action of trypsin, on the respective procarboxypeptidase precursors, in the pancreas.

Carboxypeptidases A and B, obtained from bovine, are Zn (II)-containing proteins (molecular weight 34,600 daltons), with 307 amino acid residues in A and 308 amino acids in B. The amino acid sequence in carboxypeptidases A and B are similar, up to 49%. Both types have only one Zn (II) with catalytic activity.

The Zn(II) ion is situated at the centre of the protein and is bound to its 69 histidine, 72 glutamate, 196 histidine residues and a water molecule, in a distorted tetrahedral geometry.

On demetallation, the enzyme loses the catalytic activity, but retains the overall structure. The activity is revived on the addition of Zn(II) or Co(II) or other bivalent metal ions, of similar size like Ni(II) or Mn(II).

The substrates on which hydrolysis is catalysed by carboxypeptidase are dipeptides, like glycyl tyrosine. The main features of the binding of the substrate, with the apoprotein and the zinc centre, are as follows (Figure 4.1):

i. The C-terminal carboxylate group of the substrate interacts electrostatically with the guanidine group of 145 arginine of the apoprotein. It has been suggested that the interaction could first be with arginine 71 and then finally with arginine 145.

ii. As a result of the positioning of the carboxylate at arginine 145, the C-terminal hydrophobic aromatic group (hydroxy phenyl) of glycyl tyrosine inserts itself in the hydrophobic pocket of the apoprotein and displaces several water molecules.

iii. It had been suggested earlier that the substrate dipeptide gets bound to the Zn (II) from the peptide O, displacing water molecules. This is by analogy with the model studies, where metal ion is known to act as a catalyst. Binding with the metal ion causes polarisation of the peptide $-C^{+\delta}=O^{-\delta}$. This facilitates the nucleophilic attack of water molecules on the $+\delta$ carbon atoms, leading to the acceleration of the hydrolysis of the dipeptide. However, now it has been established that the coordination of amide CO with Zn^{2+} is not necessary. At the same time, this has not been completely ruled out. It has been suggested that since some flexibility of the coordination type of Zn^{2+} is allowed, along with H_2O coordination, peptide CO coordination, or its electrostatic interaction with Zn(II) centre can also take place, resulting in penta-coordination around Zn (II) in the intermediate stage. Thus, –CO may be activated for nucleophilic attack by H_2O or OH^-.

However, in either case, the main role of the metal ion is to activate the deprotonation of the coordinated H_2O to form a coordinated OH^- ion. The H_2O molecule is also H-bonded to the glutamate 270. This also assists the formation of OH^-, by transferring the proton to the carboxylate group of glutamate 270, forming glutamic acid. This OH^-, a more potential nucleophile than H_2O, affects a nucleophilic attack on the peptide CO. The peptide CO may be activated to nucleophilic attack by OH^-, possibly but not necessarily due to the coordination of –CO with Zn (II). Mainly, the activation of –CO to nucleophilic attack is due to the H bonding of the –CO group with the arginine 127 residue and the protonation of the amide nitrogen due to H bonding with tyrosine 248. Both the processes make the carboxyl C electron deficient and more prone to attack by coordinated $–OH^-$.

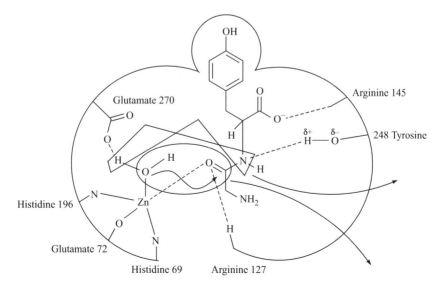

FIGURE 4.1 Binding of dipeptide substrate with Zn(II) carboxy peptidase

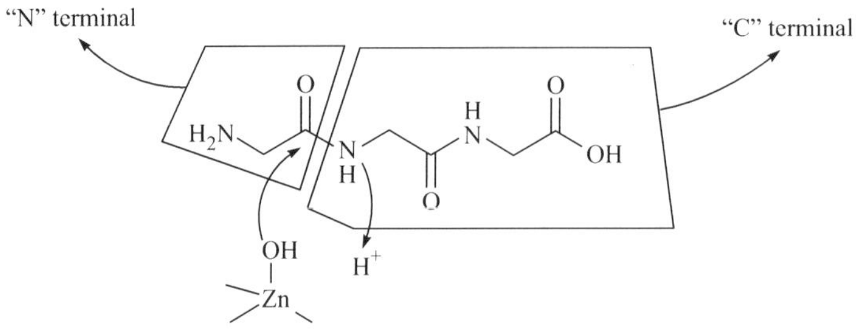

FIGURE 4.2   Hydrolysis of N-terminal amino acid site by Zn(II) amino peptidase

iv. As a result of the nucleophilic attack by –OH on the carbon of the peptide –CO, the C–N bond breaks, and peptide NH gets converted into $-NH_2$ by getting the proton from glutamic acid at 270 (proton previously transferred from coordinated H_2O). The C-terminal amino acid, tyrosine, is released. Glutamic acid at 270 gets converted back to glutamate form and temporarily coordinates with Zn(II).

v. The N-terminal amino acid (glycine) formed due to the hydrolysis of the dipeptide leaves the coordination site of the metal ion, and in its place one water molecule gets bound to the metal ion. The Zn(II) glutamate bond breaks, and the coordinated H_2O forms the H bond with the glutamate 270. Thus, the enzyme is regenerated for the fresh cycle.

Aminopeptidase

These Zinc (II) protein enzymes catalyse hydrolysis of the N-terminal amino acid sites of polypeptides and proteins, as shown in Figure 4.2.

The mammalian zinc aminopeptidases are oligomeric (molecular weight 200,000 daltons), whereas corresponding microbial enzyme is monomeric (molecular weight 40,000 daltons).

The former have two Zn(II) per mole, whereas latter may have one or two. Zn(II) can be removed from the enzyme by 1,10-phenanthroline, without affecting the hexameric structure. Zn(II) can be replaced by Co(II), retaining the enzyme activity. In the mammalian amino peptidase, replacement of one Zn(II) by Mg(II) or Mn(II) enhances the catalytic activity, but the replacement of both the Zn(II) ions, by these metal ions, does not retain the catalytic activity. This shows that at least one Zn(II) is essential for the catalytic activity. The other one may have regulating role, which can be maintained by the substituted Mg(II) or Mn(II). Hence, the substitution of the second Zn(II), with catalytic role, by other metal ions, makes the enzymes catalytically inactive.

Alkaline Phosphatase

These are Zn(II)-containing proteins which cause hydrolysis of monophosphate esters of primary alcohols, secondary alcohols, phenols and mono-substituted phenols (Eq. 4.1). They also catalyse trans phosphorylation reaction. Besides Zn(II), the enzyme also has Mg(II) in its structure (Figure 4.3).

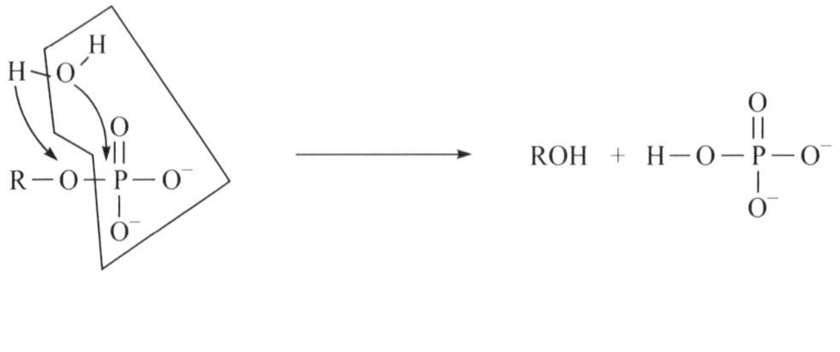

$$R = CH_3, (CH_3)_2CH,$$

$$\text{or} \qquad (4.1)$$

They are called alkaline phosphatase because of their catalytic activity at around pH 8. They are obtained in the bacteria *Escherichia coli* and in the mucosa, placenta and bones of the mammals. Increase in the alkaline phosphate level in the serum is an indication of bone tumour, hepatitis and gall bladder diseases.

The enzyme occurs as a dimer of m.w 94 K daltons, in bacteria and of molecular weight 240–280 K daltons in animals and human beings. Each monomer of the enzyme contains two Zn(II) at a distance of 4 Å and a Mg(II) situated at a distance of 5–7 Å. Both the Zn(II) act as the catalytic sites. The activity of the enzyme is inhibited on addition of EDTA, showing that the presence of metal ion is essential for the catalytic activity. Though the enzyme needs Mg(II) to exhibit full activity, its role is not definitely known.

The first Zn (Zn_1) is bound to the apoprotein through histidines 331, 372 and 412 or histidines 331, 412 and aspartate 327. There are also water molecules bound to Zn_1 (Figure 4.3).

The second Zn (Zn_2) is bridged to Zn_1 through aspartate 369 and histidine 370. Zn_2 is bridged to Mg(II) through aspartate 51. Mg(II) is also bound to aspartate 153, Thr 155 and glutamate 322. Zn_2 is also situated close to the hydroxyl group of the serine 102 residue. The crystal structure reveals that it is not at direct bonding distance. The structure is shown in Figure 4.3.

A possible pathway for the catalytic reaction could be as follows (Figure 4.4):

i. The phosphate group of the phosphate ester substrate gets bound to Zn_1, by displacing a water molecule. The binding of phosphate with Zn_1 is facilitated by interaction with arginine 166 residue. As a result of binding, the phosphorus of the phosphate is activated for nucleophilic attack. The binding of phosphate with Zn_1 causes steric change in the enzyme, and hence, serine 102 moves closer to the Zn_2. This results in the lowering of pKa of the –OH of the serine, and it gets deprotonated, and the alkoxo O^- of the serine gets bound to Zn_2.

ii. There is a nucleophilic attack of the Zn_2 bound serine alkoxo, on the activated phosphorus of the substrate phosphate ester, bound to Zn_1. This leads to polarisation of the P–OR bond, thereby RO^- gets liberated from the ester.

iii. The phosphate Zn_1 bond breaks, with the formation of phosphoseryl intermediate bound to Zn_2. The delinking of phosphate from Zn_1 results in strong binding of the water molecule on Zn_1 leading to its deprotonation. This proton is taken by the leaving RO^- to form alcohol.

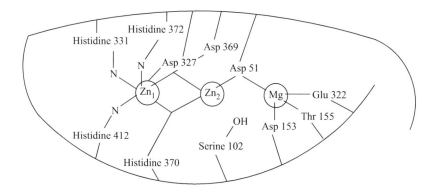

FIGURE 4.3 Alkaline phosphatase

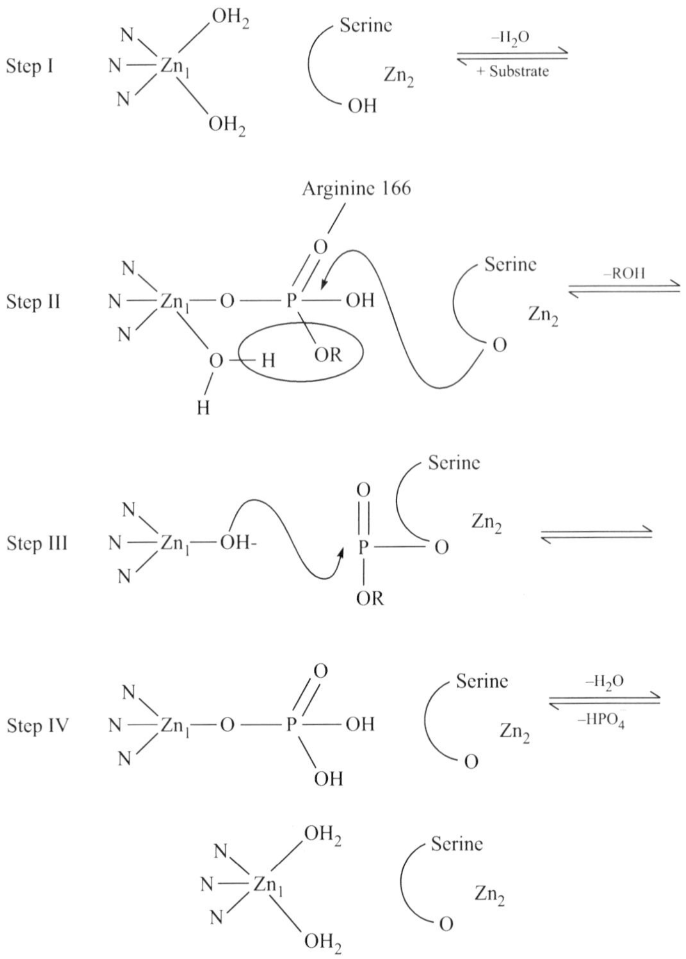

FIGURE 4.4 Mechanism of the enzymatic reaction of alkaline phosphatase.

The deprotonation of the water leaves OH^- coordinated to Zn_1. This is a strong nucleophile and attacks on phosphate seryl derivative on Zn_2 and causes displacement of seryl group, which takes back its position close to Zn_2. After the displacement of OH^-, the phosphate comes back to Zn_1. There is further reaction of one water molecule liberating phosphate. The seryl OR^- takes up one proton to reform the serine residue. Thus, the enzyme gets back into its native form for the second cycle.

There has been a suggestion that before the attack of OH^- of Zn_1 on the phosphoseryl group on Zn_2, there is formation of a binuclear intermediate, by the linking of phosphate to both the Zn ions. However, this is doubtful, because there is no direct evidence of bonding of phosphate with Zn_2.

It is evident that the deprotonation of H_2O, forming $-OH^-$, is an important step, in the reactivity of the enzymes. The deprotonation of water bound to Zn(II) occurs only at a high pH, and hence, the formation of $-OH^-$ takes place at a high pH only. This explains why alkaline phosphatase is active at high pH only. The mechanism of the enzyme action is shown in Figure 4.4.

In the presence of another alcohol, the phosphoenzyme at step IV may transfer the phosphoryl group to that alcohol, instead of liberation of phosphoric acid. This results in the formation of the phosphoester of the new alcohol. Thus, alkaline phosphatase assists trans phosphorylation reaction.

Alkaline phosphatase also has an important role in the biomineralisation process. It causes hydrolysis of organic phosphates, resulting in higher concentration of free phosphate ion. As a result, the ionic product $[Ca^{2+}][PO_4^{3-}]$ exceeds the solubility product of calcium phosphate, leading to its precipitation.

Another class of enzymes, called acid phosphatase, are active at low pH. They are exactly like alkaline phosphatase, causing hydrolysis of phosphate ester, but they are binuclear Fe(III) proteins. The reaction proceeds in the same way as in alkaline phosphatase. Fe(III) being a strong Lewis acid, the water coordinated to Fe(III) is more polarised, and has a lower pKa value and gets deprotonated at low pH. Thus, the active OH^- is made available in the lower acidic pH range, making the enzyme active at low pH.

Some of the plant phosphatases contain mixed Fe(III) Zn(II) binuclear sites. These Fe(III)-containing phosphatase are purple in colour due to Fe(III) bound to the tyrosinate residue. The colour is due to phenolate O^- to Fe(III) charge transfer.

Carbonic Anhydrase

Carbonic anhydrase is a Zn(II) protein, which catalyses the reversible hydration of carbon dioxide. It is present in animals, plants and certain microorganisms. It was discovered in 1930, and the presence of Zn (II) in the enzyme was established in 1939. The hydration reaction, catalysed by the enzyme, can be shown as follows:

$$CO_2 + H_2O \rightleftharpoons HCO_3^- + H^+ \qquad (4.2)$$

The hydration reaction is important, because CO_2 is generated by the oxidation of carbohydrate and it has to be converted into carbonic acid in the blood, to be carried back by deoxyhaemoglobin to the lungs. In the lungs, carbonic anhydrase (CA) catalyses the dehydration of HCO_3^-, so that CO_2 can get exhaled. It may be noted that the nomenclature of the enzyme is based on this reverse reaction of removal of water from carbonic acid.

Carbonic anhydrases (CAs) are of three types, A, B and C, occurring in different concentrations. They have small difference in their structure. The structure of carbonic anhydrase B, occurring in human and bovine, has been determined, most exhaustively. It is monomeric and contains 1 atom of Zn(II) bound to protein with molecular weight 30,000 daltons. The protein has 256–265 amino acid residues. The protein has an extensive structure of polypeptide chain and has an ellipsoidal structure like a rugby ball, divided into two halves. The Zn(II) is present at the bottom of the large hydrophobic cavity and is almost at the centre of the ellipse, approximately 1,800 pm from the surface. The Zn(II) is bound to three histidine residues (His 117, His 94 and His 96) in a distorted tetrahedral geometry, with the fourth coordination position being occupied by water molecule (Figure 4.5). The largest departure from tetrahedral structure is of 20°. There is also a suggestion of binding of a second water molecule leading to penta-coordination. This inference is drawn from the observation that on binding of the substrate, one water molecule is retained and thus a penta-coordinated structure is attained. The substrate probably displaces one coordinated water molecule.

The water in enzyme is unusually ordered, due to H bonding. The coordinated water molecule is H-bonded to other non-coordinated water molecules and also to the threonine 199 of the protein, which in turn is H-bonded to glutamate 106. These amino acids are about 4 Å away from Zn(II) and form a hydrophilic zone around the metal ion. Water molecule and OH^- ion (formed by deprotonation of H_2O) are bound at sites Y and Z of the hydrophilic zone (Figure 4.5).

On the back side of the cavity, there is a hydrophobic zone, created by valine – 143, leucine – 198 and tryptophan – 209. The ligands, with hydrophobic groups, get coordinated to the metal ion, at site X. In this hydrophobic region also, there are water molecules, which are H-bonded to the coordinated H_2O.

The unusual ordering of the water molecules may have some role in the coordination of water with the tetrahedral Zn(II) centre and its reaction with substrate CO_2.

Zn(II) can be removed from the enzyme by adding 1,10-phenanthroline or 8-hydroxyquinoline. The structure of the apoenzyme is same as that of the native enzyme, showing that there is no change in the tertiary structure of the apoprotein on demetallation. On addition of Zn(II), the catalytic activity is fully revived. However, Zn(II) binds with the apoprotein at a much slower rate than with smaller ligands, showing that the active site of the apoprotein has highly ordered structure and is deep seated. The strong binding of the metal ion, with the highly structured protein, is reflected in the fact that the log K for the

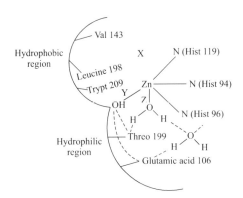

FIGURE 4.5   Structure of carbonic anhydrase B.

Zn–apoprotein complex is 12.5, which is much higher than the stability constant of Zn(II) complexes with analogous smaller polypeptide ligands.

The activity of the demetallated apoenzyme is regenerated, by adding other bivalent metal ions, such as Co(II), Mg(II) or Mn(II). However, the activity is less than the Zn(II) enzyme. This may be because there is a change in the coordination geometry, on the substitution of another metal ion, and this in turn affects the ordered water structure in the enzyme.

It is thus evident that in the enzyme, the active site and the water molecules have ordered structure, and there is a fine balance between the hydrophobic and hydrophilic interactions. These factors play an important role in the binding of the inhibitors. The positively charged complex namely $[Zn(N)_3 H_2O]^{2+}$ attracts the negatively charged inhibitor ion. They get bound strongly at the hydrophilic sites Y and Z, removing water and maintaining four coordination. This inhibits the activity of the enzyme. Inhibitor neutral ligands, like imidazole, occupy the hydrophobic site X, retaining water molecule, and result in penta- or hexa-coordinated structure. Thus, further binding of the substrate CO_2 with Zn(II) is hindered, and the activity of the enzyme is inhibited.

Mechanism of Enzymatic Reaction

The water bound to Zn(II) undergoes deprotonation at the physiological pH and is in the form of coordinated OH^-. The carbon dioxide is attracted in the cavity by binding to the hydrophobic part or to the metal ion. The coordinated OH^- is an efficient nucleophile and can affect a nucleophilic attack over CO_2. Kenner and coworkers have proposed that the H bonding with threonine 199 places OH^- suitably, for the nucleophilic attack over CO_2. The CO_2 is located between the positively charged Zn (II) and –NH of threonine. The interaction of CO_2 with the positive charges may shift electron density from it and activate it for nucleophilic attack by OH^-. The suggested activation of CO_2 is a possibility and is yet to be confirmed.

There is a formation of HCO_3^- by the combination of OH^- and CO_2. At this stage, one more water molecule gets bound to Zn(II), resulting in a penta-coordinated structure. HCO_3^- is subsequently liberated, and the enzyme is regenerated for the second cycle. The turn over number for carbon dioxide hydration is 10^6 mol/s; that is, one molecule of the enzyme can hydrate 10^6 molecules of CO_2 per second. Uncatalysed hydration rate of CO_2 is 7×10^{-4} mol/s.

Alcohol Dehydrogenase

This is a Zn(II)-containing protein which catalyses the reversible oxidation of primary or secondary alcohols to aldehydes or ketones, respectively. It is dependent on pyridine nucleotide NAD^+ for the catalytic activity. The system $NAD^+/NADH$ acts as a coenzyme. Since NAD^+ gets converted into NADH, during the reaction, it can also be considered as a co substrate (Eq. 4.3).

$$(4.3)$$

In the alcohol dehydrogenase, obtained from yeast or mammal, each molecule contains four Zn(II), and these perform both catalytic and structural role. Horse liver alcohol dehydrogenase (LADH) is dimeric. There are two identical subunits of weight 40,000 daltons each. Each monomeric unit contains two Zn(II) ions. X-Ray diffraction shows that the two Zn(II) are of different types.

The first type of Zn(II) is close to the surface of the subunit. It is bound to the S of four cystenyl residues, cys 97, 100, 103 and 111.

The second type of Zn (II) is located deep inside the protein and is bound to two cystenyl S atoms (cys 46 and cys 174), an imidazole N atom of histidine 67 and a water molecule. Thus, the Zn(II) has a distorted tetrahedral structure. The second type of Zn(II) is essential for catalytic activity, whereas the role of the first type is not definitely known. The dimeric structure of the enzyme is independent of both the Zn(II). Hence, the first type of Zn is neither involved in catalytic or structural role and is classified as non-catalytic Zn(II).

The X-ray study shows that each monomeric subunit has two domains of protein, which can rotate independently. Normally, the two protein domains remain in an orientation, called the open form. During the intermediate stage of the catalytic reaction, the protein domains get into closed form, and this regulates the reaction. The sequence in which the enzymatic reaction proceeds is as follows (Figure 4.6):

i. In the open form of the protein, a molecule of NAD$^+$ is held inside it by electrostatic or H bonding interaction with the protein.

ii. The substrate alcohol molecule enters through the crevice between the two domains and gets coordinated to the Zn(II), displacing the water molecule.

iii. At this stage when NAD$^+$ is bound with the protein and the alcohol is bound to Zn(II), there is rotation of one domain of the protein leading to the closed form of the monomer. The binding of alcohol with the positively charged metal ion, and also the vicinity of the positively charged NAD$^+$, lowers down the deprotonation constant of the alcohol. It gets dissociated, and the proton is released outside the cavity of the enzyme monomer.

iv. The closure of the cavity results in the binding of Zn with the alkoxide in such a way that the latter is suitably oriented to the nicotinamides (NAD$^+$). The C–O bond of the alkoxide gets polarised, in turn polarising the C–H bond, so that its proton is transferred as hydride, to the nicotinamide, forming NADH. And the alkoxide is converted to aldehydes by the loss of H$^-$. Though the dehydrogenation of alcohol is an oxidation process, but Zn(II) ion has only a Lewis acid role in polarising the C–O bond, leading to hydride transfer.

v. From the resulting Zn–aldehyde complexes, the aldehyde gets detached and is replaced by water molecule. Additional water molecules also enter the cavity, leading to the opening of the monomer subunit. NADH, so formed, is released, and the enzyme gets back to the native form, ready for the next cycle (Figure 4.6).

The pictorial representation of the closed and open forms of the protein, as shown in Figure 4.6, are only illustrative simplified forms, for understanding the steps of the enzymatic reaction in the natural system.

Exchangeability of Zn(II) and Co(II) in the Above Enzymes

The activity of the demetallated apoproteins of the above Zn(II) enzymes are revived on addition of Zn(II) or Co(II). The Co(II)-substituted enzyme exhibits as much activity, as the Zn(II)-containing enzyme.

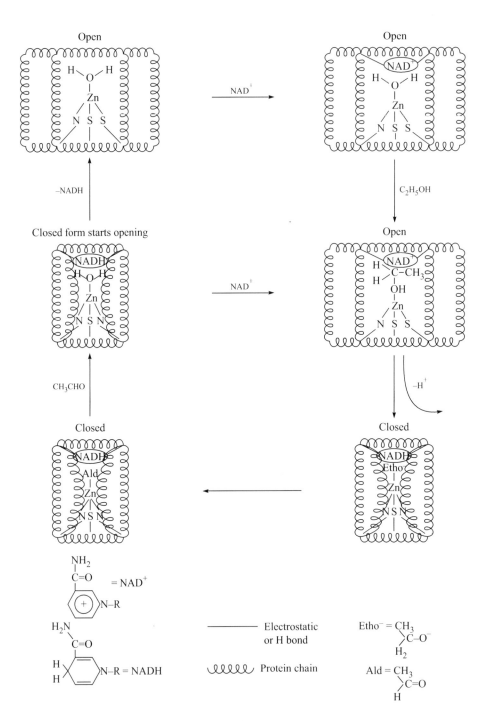

FIGURE 4.6 Mechanism of enzymatic action of alcohol dehydrogenase.

This shows that conformation of apoprotein remains intact upon removal of Zn(II), and Co(II) occupies the same protein site.

Yeast, grown in the presence of excess supplemental Co(II), produces alcohol dehydrogenase in which Zn(II) is partially displaced by Co(II).

When *E. coli* is grown in the presence of relatively high concentration of Co(II), an active Co(II) alkaline phosphatase is synthesised.

The exchangeability of Zn (II) and Co (II) may be due to the following reasons:

i. Since there is no CFSE in d^{10} Zn(II) ion, structure of its complex is solely dictated by electrostatic forces. In case of high spin Co(II) with d^7 configuration, there is a small difference in stabilisation energies in tetrahedral (−8 Dq) and octahedral (−12Dq) forms, and hence, in Co(II) complexes, one form can change into other easily, as in Zn(II) complexes.

ii. In high spin tetrahedral complex of Co (II), the electronic configuration is $e^4 t2^3$, and it behaves like a spherical ion, as d^{10} Zn(II).

iii. Furthermore, Co(II) and Zn(II) have similar ionic radius. Hence, Co(II) is the best substituent for Zn(II) in the enzymes.

The advantage of Co(II) substitution is that it is paramagnetic (S = 3/2). The electron spin relaxation is fast at high temperature, and hence, NMR spectra can be recorded and at low temperature, electron spin relaxation is slow, and hence, ESR can be obtained. Thus, Co(II) substitution can help in monitoring the structure and activity of Zn(II) enzymes.

Aspartate Transcarbamoylase

This is a Zn(II)-containing protein which catalyses the condensation of carbamoyl phosphate (I) with *l*-aspartate (II) leading to the formation of carbamoyl-*l*-aspartate (III), which is a precursor for the synthesis of pyrimidines in the biochemical systems (Eq. 4.4).

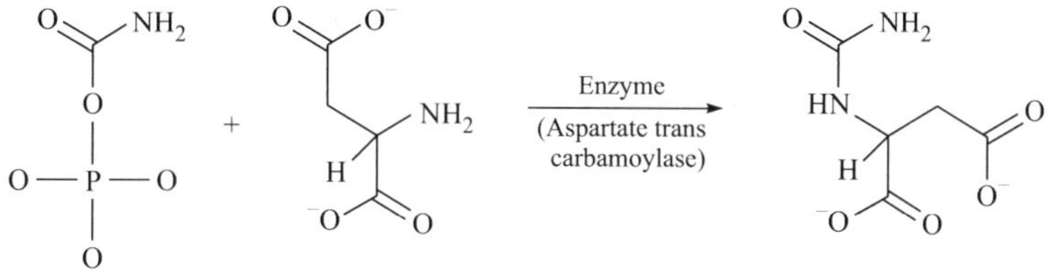

(4.4)

The enzyme protein, obtained from *E. coli*, has a molecular weight of 310 K daltons and has six Zn(II) per molecule. The enzyme has four subunits, which are of two types: (i) catalytic or C type and (ii) regulatory or R type. Zn(II) are bound to the regulatory subunits of the enzyme, which is not required for catalytic activity. One Zn(II) is bound to the four cysteinyl residues at the C-terminal region of each chain of the R subunit. Thus, the Zn(II) has a tetrahedral geometry. The R chains interact with the C chain in the region where Zn(II) is bound. The polypeptide loops, between the cysteinyl ligands, binding with Zn(II), are in close contact with the polar regions of the C chain, which is postulated to be the active catalytic site. Thus, the Zn(II) is responsible for the interaction between the R and C chains. It has also been recognised that Zn(II) has a structural role in maintaining quaternary structure of the protein, necessary for catalytic activity. If Zn(II) is removed, the subunits become unstable, the quaternary structure breaks, and the catalytic activity is lost. On addition of Zn(II), Cd(II), Mn(II), Co(II) or Cu(II), the dimeric structure of R subunits is restored and the enzyme is activated. Thus, Zn(II) has structural role in this enzyme and assists the catalytic activity of the protein, by maintaining the required quaternary structure.

DNA and RNA Polymerases

As stated in the introductory part of this chapter, Zn(II) has an important role in the biochemistry of nucleic acids. It has been recognised that Zn(II) is essential for the transfer of genetic information. It is now known that the enzymes DNA and RNA polymerases, which catalyse the replication, and transcription of nucleic acids in prokaryote and eukaryote are nucleotide transferases, with stoichiometric amount of intrinsic Zn(II). The molecular weight of DNA polymerase from different sources varies, from 110 to 150 K daltons, with 1 or 2 g atom of Zn(II) per mole. In case of RNA polymerase, the molecular weight varies from 380 to 700 K dalton, and there are 2 g atoms of Zn(II) per mol. The native enzyme is Zn–Zn RNA polymerase. The absorption spectrum of Co(II)-substituted RNA polymerase shows two absorption bands at 584 and 730 nm, indicating that Co(II) centres are tetrahedral and the two centres are non-equivalent. There are three probable roles of Zn(II) in the enzyme.

i. **Catalytic effect**: The fact that 1,10-phenanthroline inhibits activity of DNA or RNA polymerase indicates that Zn(II) has a catalytic role in the DNA or RNA synthesis. o-phenanthroline is considered to bind with zinc and thus makes it unavailable for the DNA or RNA synthesis. Zn(II) is presumed to act as a Lewis acid and binds with the 3′ OH primer terminus of DNA or RNA. The OH gets deprotonated, and the O⁻ affects a nucleophilic attack over the α-phosphorous atom of the incoming ATP to propagate polymerisation. The reaction can be shown as follows:

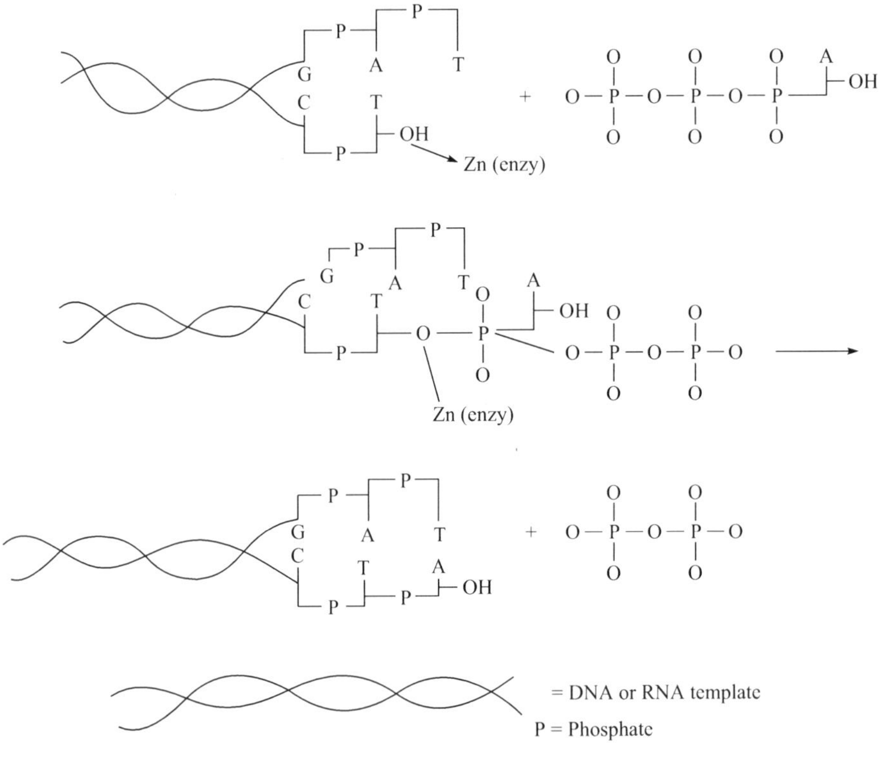

$$(4.5)$$

= DNA or RNA template

P = Phosphate

However, inhibition of enzymatic activity is not a definite evidence for the catalytic activity of the Zn (II) centres. Zn(II) may have structural role only. Inhibition can be caused, because 1,10-phenanthroline binding may interfere with the structural role of Zn (II) in the enzymatic activity.

ii. Regulatory role – The Zn (II) metal ion may have an important role in stabilizing the DNA – enzyme complex. The binding with the Zn(II) metal ion may also bring a conformational change in DNA, which accelerates its reaction with the nucleotide, thus leading to faster transcription process.

iii. Structural Role – The Zn (II) present in the enzyme may have a structural role in maintaining a specific conformation of the apo enzyme, suitable for catalysing the polymerization reaction.

Other Roles of Zinc

1. **Ca (II) antagonism:** Zinc ion has a strong interaction with membrane of the cells of skeletal muscles, brain, intestine, liver and sperms and also with blood cells. Binding with Zn(II) gives structural stability to the cell membrane.

   In case intracellular calcium content is high, there are alterations in the cell membrane, causing damage to it. This damage can be diminished or inhibited by supply of zinc. The mechanism of the reversal of the effect of calcium is that zinc inhibits the calcium stimulated activity of ATPase. Thus, breakdown of ATP is retarded, and there is an increase in intracellular ATP level. Zinc mainly deactivates the calcium protein calmodulin and inhibits the calmodulin-stimulated activity of Ca–Mg ATPase. Calmodulin is mainly responsible for the calcium-induced damage of cell membrane. Hence, shrinkage of the cell membrane and also various other enzyme and cellular functions, stimulated by calmodulin, are inhibited by zinc.

2. **Treatment of sickle-cell anaemia:** In cases of sickle-cell anaemia, there is modification in haemoglobin structure, which results in the polymerisation of oxygenated haemoglobin. As a result, the blood cells get deformed in shape of sickle. The deformation causes damage to the membrane of the blood cell. The damage is irreversible, and the cell does not get back to original shape even after deoxygenation. The deformed blood cells get occluded in the small blood vessels, and thus, oxygen transport is retarded.

   If zinc is administered to the patient of sickle-cell anaemia, there is a significant decrease in the number of sickle cells. As discussed earlier, Zn(II) can inhibit the damage to the cell membrane. In case of sickle cells also, Zn(II) interacts with the sickle-cell membrane and revives the cell to normalisation. Zn(II) does not interact with the modified oxyhaemoglobin, whose polymerisation is supposed to cause the formation of sickle cell. Thus, Zn(II) cannot inhibit the polymerisation process, though it inhibits formation of sickle cells. Hence, it can be argued that besides polymerisation of oxyhaemoglobin, the deformation of the cell membrane may be an important factor contributing to the formation of sickle cell.

   It is known that calmodulin (Ca–protein)-assisted enzyme activities can cause cell membrane damage (discussed earlier). It can, therefore, also cause the deformation of haemoglobin. Hence, initiation of sickle cell formation may be due to calmodulin-stimulated enzyme activities, and it can be argued that Zn(II)-assisted recovery of sickle cell is due to deactivation of calmodulin.

3. **Increase in oxygenation of haemoglobin:** It has been observed that Zn(II)-bound haemoglobin shows increasing tendency to bind with oxygen. Zn(II) has greater affinity for oxyhaemoglobin than haemoglobin, showing that oxygen-bound haemoglobin is more suitably oriented to Zn(II) binding, or Zn(II) binding orients haemoglobin to bind with oxygen more favourably.

   It has been shown that two Zn(II) are bound to the tetramer of haemoglobin. In deoxyhaemoglobin, the Zn(II) binding site is located at the interface between the tetramers in the region, where histidines 116 and 117 and glutamate, on the β_1 chain of one tetramer, are close to lysine 16 and glutamate 116 on the α_2 chain of a neighbouring tetramer. Thus, Zn (II) binding between tetramers keeps them separated, and thus, the uptake of oxygen by each tetramer is accelerated.

   This may also be one of the factors for the inhibition of sickling of blood cells by Zn (II). Zn (II) weakens the interaction between the haemoglobin tetramers, and thus, polymerisation of the haemoglobin tetramers is retarded. [The readers should first refer to the structure of haemoglobin, discussed in Chapter 5, for better appreciation of the above (2) and (3) roles of zinc.]

4. **Insulin:** This is a hormone present in β-cells of the pancreas in a crystalline form. Scolt first observed that insulin contains significant amount of zinc, indicating that synthesis and potency of insulin depends upon the amount of zinc, available to the body. The most important role of insulin is lowering of blood glucose. It has been observed that the Zn(II)-deficient animal has lower glucose tolerance.

Zn(II) has an important role in production of insulin in pancreatic β-cells. Zn(II) ions assist in the formation of insulin leading to hexameric structure. Insulin remains stored in the β-cells in the inactive hexameric form and is released for glucose metabolism in active monomeric form. Insulin is a protein hormone formed in the β-cells of the pancreas. Such cells amount to 60%–80% of the cells of pancreas.

Insulin is found not only in the body of animals, but also in plants like fungi.

The insulin found in human beings is a protein made of 110 amino acids and has a molecular weight of 5808 daltons. Insulin monomer is made of two chains, A and B, linked through two S atoms, that is S–S bridge.

Soon after the protein proinsulin is formed in the pancreas, two monomers with A and B chains dimerise. Three dimers of proinsulin come together in the presence of Zn²⁺ ions resulting in the formation of the hexameric insulin. This remains in the β-cells of the pancreas in the inactive form.

The structure of the hexamer was revealed by Dorothy Hodgkins on the basis of the X-ray crystal study of the hexamer. Six insulin molecules get arranged in threefold symmetry, with Zn ions, bound to the histidine units of the insulin. Zn(II) ions are located at the centre of the structures. Zn²⁺ ions hold the hexamer together, as shown in Figure 4.7, which is only an approximate picture. The number of Zn(II) in the hexamer is not stoichiometric. It is nearly three zinc per hexamer of insulin.

The hexameric structure of insulin is maintained throughout the subsequent processes. Zn(II) makes insulin insoluble and also plays an important role in the formation of micro-crystalline crystals which results in reducing contact of insulin with surrounding membrane containing other enzymes, and thus, insulin is not affected by them. Zn plays an important role in the synthesis, storage and secretion of insulin. As it maintains the hexameric structure of insulin, and deficient supply of Zn(II) affects the capacity of the pancreatic islet cells to produce and secrete insulin. It has also been shown that Zn(II) deficiency reduces response of cells to reducing glucose metabolism by insulin. Thus, Zn supplementation has beneficial effect on glucose homeostasis. Zn(II) also has an antioxidant activity on the insulin receptor and the glucose transporters. Many of the complications of diabetes are related to the oxidant and free radicals in the cells. Zn(II) and Zn(II)-dependant proteins control these adverse effects and control diabetes. However, the exact role of Zn(II) supplements in control of diabetes is not known.

Though tests have shown that the serum insulin content depends on the supply of Zn, it has been questioned whether Zn(II) has direct role in increasing the content of insulin or it indirectly controls the release of insulin. It has been suggested by Reinhold that Zn(II) stabilises the structure of insulin. In the case of Zn(II) deficiency, insulin degradation is increased, leading to lower decomposition of glucose. Alternatively, it can be argued that Zn(II) only affects the

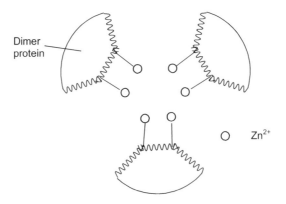

FIGURE 4.7   Insulin zinc hexamer formed in pancreas.

capacity of the cell membrane to absorb insulin. In the case of Zn(II) deficiency, the tissues are less efficient in absorbing insulin.

However, it has been observed that the content of Zn(II) and insulin in pancreatic β-cells is identical. During the decomposition of glucose, the insulin content in β-cells is lowered down, accompanied by the lowering of zinc content. This shows that Zn(II) is an essential component of insulin, which is necessary for the breaking down of glucose.

However, the equivalents of zinc required per molecule of insulin are not known. Though the primary structure of insulin is known, the role of Zn(II) in determining the ternary and quaternary structure of insulin is not known. The actual function of the hormone in glucose metabolisation is also not clear.

Phosphate signal cascade is a process of transfer of information from one cell to another, as a result of transfer of a phosphate group from one protein to another. If this signal cascade mechanism breaks down because of interference due to the presence of an external agent, it can cause diseases like diabetes and cancer.

Diabetes is caused, either due to the failure of pancreas to produce the hormone insulin, leading to type 1 diabetes, or the inability of the cells to absorb insulin, causing type 2 diabetes. Insulin regulates the phosphate signal cascade which leads to breakdown of glucose to glycogen. Hence, if there is inadequate supply of insulin, there is an increase in the blood sugar level.

In the process of activating sugar metabolism, monomeric insulin gets bound to the insulin receptor kinase (IRK) of the insulin receptor (IR) cell. The kinases are the enzymes which assist the phosphorylation of the biomolecules. IRK spans over the whole membrane of the insulin receptor cell. Binding with insulin changes the conformation of IRK, and thus, its kinase activity is enhanced, and thus, the phosphorylation process is activated. In the IR cell, insulin initiates a signal transduction which regulates the protein molecules that transport glucose molecules inside the cell, through the glucose channel (see Figure 4.8). Thus, the sugar concentration inside the IR cell increases. On binding with insulin, several tyrosine residues of the protein molecules of the membrane get phosphorylated by the formation of phosphate mono ester. The hydrolysis of the ester bonds (dephosphorylation) results in transfer of PO_4^{3-} from one protein to another in cascade form. Formation of phosphate monoester (phosphorylation) on the serine threonine and tyrosine residues of the proteins and the hydrolysis of these ester bonds (dephosphorylation) result in transfer of PO_4^{3-} from one protein to another in cascade form, called phosphate signal cascade. This process is helpful in conveying signal from one cell to another and is known as signal transduction.

With the phosphorylation of several regulatory tyrosinase, the phosphorylation of other biomolecules is activated many folds. In this process, glucose in the cells is phosphorylated at carbon atom 6 to form glucose 6-monophosphate. This leads to the metabolic process called glycogenesis and formation of glycogen and fats.

There is another enzyme protein tyrosine phosphatase (PTP), which causes dephosphorylation of the tyrosine phosphate of the IRK. Consequently, kinase activity, that is, phosphorylation of other molecules by the kinase, is reduced. Thus, PTP opposes the phosphorylation activity of kinase, which is activated by binding of IRK with insulin.

This antagonism of IRK and PTP activities maintains a balance of the extent of phosphorylation. Thus, insulin regulates the phosphate signal cascade and controls the sugar metabolism, sugar storage, gene expression and lipid storage. So, if there is lowering in the supply of insulin by the pancreas or if insulin is not effectively absorbed by the cells, the phosphate signal cascade is disrupted and the breakdown of the sugar is reduced causing diabetes. A simplified qualitative picture of the process in IR cell and the outline of the mechanism of phosphate transfer cascade are shown in Figures 4.8 and 4.9.

On intake of food, glucose level goes up in the blood. Insulin present in the body removes the excess sugar from the blood in the form of glycogen and finally converts it to fat, and it is stored. Whenever blood sugar level falls, the sugar is released by the breakdown of the stored glycogen by the process of glycogenolysis.

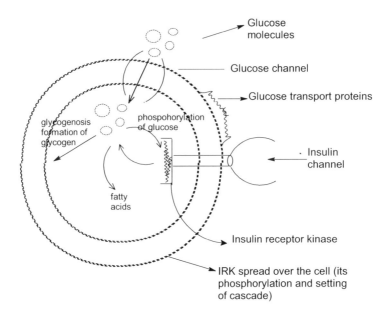

FIGURE 4.8   Reactions in insulin receptor (IR) Cell.

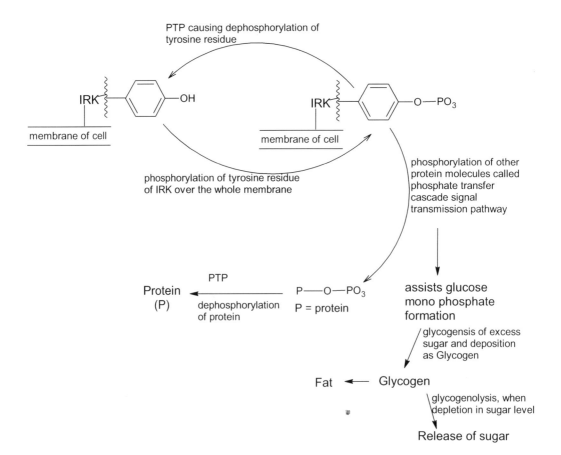

FIGURE 4.9   Steps in phosphorylation reaction in cell.

5. **Zinc fingers:** It has been observed that in the formation of nucleic acid and protein assemblies, important in biological reactions, metal ions bind with the protein, to help it in getting into an active conformation, to interact with nucleic acid efficiently. The metal ion selected by nature is Zn(II), because of its abundance and also because there is no redox chemistry, so that the damage of the DNA, due to oxidative degradation, is avoided.

A class of zinc proteins, involved in genetic factors, was discovered in 1983. It was observed by Klug and coworkers that such proteins have finger-like structural domain and hence were called zinc fingers.

A total of 427 types of zinc fingers with known functions have been identified, whereas 456 zinc fingers of other types have also been discovered, but their functions are not yet known.

Each protein contains two to three zinc ions. In the zinc protein, called transcription factor III, there are 7–11 Zn(II) ions per protein molecule. On removal of the metal ions by dialysis, the nucleic acid binding capacity of the protein is lost. It is regenerated on addition of Zn(II), or similarly behaving metal ion Co(II). This confirms the role of the metal ion in inducing the required conformation on the protein to combine with the nucleic acid.

The transcription protein III consists of 30 amino acids, in repeating sequence of a set of amino acids. In each repeating unit, Zn(II) gets bound to two cysteine S and two histidine N. Thus, a folded structure of the protein is stabilised. Each fold of the peptide unit, bound to the Zn(II) ion, forms a domain to bind efficiently with the nucleic acid and has a finger-like structure and is known as zinc finger (Figure 4.10). Thus, the metal ion orients the proteins to specific individual structures, such that they can recognise the complementary sites on the nucleic acid. Thus, the Zn(II) does not get bound to the nucleic acid, but orients the protein to interact with it strongly.

Unlike the other Zn proteins, which bind with DNA in a three-dimensional way, utilising the double-helix structure of the DNA molecule, zinc finger proteins recognise the nucleic acid sequence of different length by binding together in a two-dimensional way. The zinc fingers can combine in different ways to recognise DNA and RNA.

Zinc fingers are present in significant amounts in the genes of human genomes, up to 3%. It has been shown that the number of zinc fingers in the human genes may be much higher, up to 10% of the genome.

Zinc finger proteins organise protein subdomain for interaction with DNA or other proteins. They are structurally diverse and perform function in various cellular processes, such as replication and repair, transcription and translation. Zinc finger motifs bind to a wide variety of compounds, such as nucleic acids, proteins and small molecules. Zinc finger motifs are classified into eight groups, based on the different structural properties at the site of Zn binding with the protein. Majority of the zinc fingers belong to the three classes called C_2H_2 finger, treble chef finger and zinc ribbon.

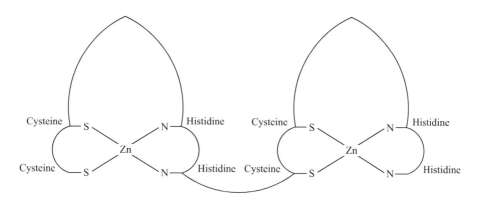

FIGURE 4.10 Transcription protein III zinc finger.

The zinc fingers have different orientation to target specific genes. Zinc fingers can also help the process of switching on and off the genes by getting bound to the activation area or repression area of the gene. Thus, zinc fingers can be used for controlling or modifying the genome. This in turn leads to the possible use of zinc fingers for the treatment of various genetic diseases. Several synthetic zinc fingers of the type named TFIIA have been synthesised, and their structures have been determined. Attempts are made to see the possibility of such synthetic zinc fingers being used as drugs.

Other Roles of Zinc Proteins

As discussed in the earlier part of this chapter, Zn(II) ions are known to act as a cofactor in a limited number of enzyme proteins. However, now it is known that the number of such Zn protein enzymes is much more and they have various catalytic activities, such as oxidation and reduction (47 Zn protein enzymes), transferase (167), hydrolysis (397), lysases and isomerase (24).

Besides the role of Zn(II) as cofactor of enzymes and as constituent of zinc fingers, which control the activity of the genes, Zn(II) ions are present endogenously in the tissues and exhibit regulatory roles, like control of cellular calcium.

Several proteins get bound to Zn(II) and assist the transport of zinc in biochemical systems. The total number of zinc proteins in biochemical systems is large, and they can be classified into three types:

  i. Zn proteins with known structure and known domains (areas of activities).
  ii. Zn protein with known structure but with unknown domain.
  iii. Zn protein with unknown structure but known domain.

With the knowledge of the presence of large number of Zn proteins in biochemical systems and their role, it is realised that Zn(II) ion is essential for life processes and hence the necessity of intake of zinc, as a micronutrient is necessary.

Some of the other important biochemical processes in which Zn(II) and zinc proteins are involved, besides as metalloenzyme and zinc finger, are as follows:

  1. **Control of cellular zinc**: In the biochemical system, there are many proteins which bind with the excess zinc in the body, so as to rule out the adverse role of excess free Zn^{2+}. These proteins transport the zinc in a cell through cell membrane and in between the cells. The zinc is stored at the proper site and is released, whenever needed. The different roles of these zinc bound proteins can be summarised as follows.

     i. **Sensing process of excess zinc and transporting it**: A protein with two bidentate coordinating sites is present in the cytoplasm. This has the potency to sense the amount of zinc in the cytoplasm. It undergoes a change in orientation, suitable for binding with two Zn^{2+}, and the binuclear complex is transported through the cell membrane. There is also a zinc sensing zinc finger protein in eukaryotes which is known as transcription factor MTP-1. MTP-1 has six zinc fingers, which have low affinity for zinc and hence can sense and bind with zinc only when it is present in larger concentration. Thus, only excess zinc is eliminated.

     ii. **Storage and release of intracellular zinc**: The concentration of free zinc in the intracellular region is controlled from two storage processes:

        a. As discussed (see Chapter 6, (ii) in section "Copper Transport"), thioneins are the proteins which bind with four or eight metal ions in the form of a cluster. Thioneins bind to the metal through the sulphur atoms of the cysteine residues. Wherever there is an excess of particular metal ion, it gets bound to the thionein forming a complex. In the body of the human beings, there are thirteen different thioneins. These get bound to the excess zinc forming a multinuclear complex, with seven zinc ions bound to the thionein. Each of these seven zinc centres is coordinated to the four cysteine S atoms. Thus, all the zinc centres seem to be equivalent in nature with respect to binding with thionein. However, the strength of binding of these seven zinc ions with the protein differs, due to which the Zn ions are released in steps. This is because of the redox activity of the Zn protein complex. Zn(II) is known to be insensitive to redox processes. However, the coordinating S atoms of thionein are redox active and undergo different extent of oxidation. Depending on this, the S atoms become non-equivalent as coordinating site. The S sites, which undergo more oxidation, become weaker coordinating site. Thus, the extent of binding of the different zinc atoms with the protein S atoms is different. The excess Zn(II) in the biological system gets bound to thionein and is stored. Zn(II) is released as per requirement in the body in the order of the strength of Zn–S bond. Thus, the oxidation reduction process on thionein controls the concentration of stored and free Zn(II) in the cell. The stored Zn in the form of Zn thionein complex is released under the influence of oxidative stress (discussed above) or due to electrical stimulation or replacement of Zn^{2+} by another more strongly binding metal ion like Hg^{2+}. Cellular zinc release occurs in several endocrine glands like pituitary glands, pancreatic β-cells of the islets of Langerhan's and prostate glands.

b.   Zinc is also stored in the vesicles of the cells, called zincosomes, in the form of a complex of Zn bound to various biochemical ligands. The exact ligand and the form of the zinc complex in the zincosome are not known. Zinc, from the zincosomes, is released to the interior of the cell in the free Zn^{2+} form, depending on the biochemical requirement of the cell. Thus, in the cells, zinc is not present fully in the zinc complex form, but there is sufficient concentration of free Zn(II) also.

In this intracellular release of Zn^{2+}, there is release of zinc from the endosome reductase (ER) to the cytoplasm of the cell. This release occurs because of a stimulus within the cell which initiates the zinc transport protein Zip7, and leads to the release of Zn^{2+} within the cell.

The release of Zn from the vesicle may also be of extracellular type. In this, there is release of zinc(II) from the zincosomes to the extracellular space outside the cell.

2.   **Role of released Zn^{2+}**: Following are the roles of free Zn(II) in biochemical system.

i.   **As signal**: on release of Zn(II) from zinc thionein in cells or Zn(II) complexes from the zincosomes (vesicles) of the cell, there is increases in the concentration of Zn^{2+} in the cell, compared to the concentration of Zn(II) at stationary state.

The most important function of the Zn(II) released is to act as a signal. The excess released Zn(II) in the cell attacks the specific proteins with Zn(II) binding sites, called the target proteins. Zn(II) binds with the different target proteins, depending on their binding site, and orientation. Thus, the Zn(II) signals on binding with the target proteins bring change in their activity, depending on the strength and mode of binding of the Zn^{2+}. This indicates that Zn(II) signal has different roles for different target proteins.

The main function of Zn(II) signal is in intracellular and intercellular information transfer.

ii.   **Treatment of Wilson's disease**: Zn(II) is also used in the treatment of Wilson's disease. As discussed in Chapter 6, Wilson's disease is caused by accumulation of excess copper in the liver, brain and kidney, due to a disorder in the metabolism of Cu, so that the liver fails to excrete extra Cu(II). The administration of Zn(II) orally showed lowering in the accumulated copper, especially in the liver. This suggests that Zn(II) assists the liver and other organs so that they can excrete Cu(II) more efficiently, and thus, the accumulation of Cu(II) is reduced.

iii.   **Treatment of Parkinson's disease**: The amount of Zn(II) present in the brain has significant influence in mental balance, and inadequacy of Zn(II) causes mental dementia. It is seen that the zinc concentration in the brain of a patients of Parkinson's disease is low. As Zn(II) maintains interelement relations in the brain, through Zn(II) signals, lowering of Zn content in brain affects the interelement relations and leads to imbalance in brain activity and causes Parkinson's disease.

iv.   **Treatment of Alzheimer's disease:** Another condition of mental dementia, that is Alzheimer's disease, is also affected by the Zn(II) content in brain. In case of Alzheimer's disease, however, there is an increase in the Zn(II) concentration in the brain. This causes element homeostasis[1], and there is multiple sclerosis. This causes mental dementia and Alzheimer's disease.

v.   Zn(II) secreted in the mammary gland of lactating mothers provides Zn(II) to the child as a nutrient in the mother's milk.

vi. In normal concentration in biochemical systems, Zn behaves as an antioxidant, though it is redox inactive. It acts as an anti-inflammatory agent. However, in high concentration Zn(II) activates oxidation and is proinflammatory. The concentration dependent role is not normal. Hence, while administering Zn(II) as a health supplement, there should be caution about the quantity of the Zn(II) intake.

     When not required, the free Zn(II), that is, the signal is deactivated by binding with the protein thionein and is stored back in the cell for later release.

vii. **Role of Zn on growth:** Zn(II) has an important role in the development of tissues and animal growth. Zinc participates in proliferation of cells. The mechanism of action of zinc is dependent on the regulation of the hormones involved in cell growth by Zn(II). Zinc also regulates DNA synthesis directly. The role of zinc in biochemical systems is dependent on several Zn(II)-containing enzymes involved in different metabolic processes. Zn(II) also interacts with hormones such as testosterone, thyroid hormones, insulin and vitamin D_3 and fortify them to form perfect bones. Zinc is essential for the induction of cell proliferation by insulin like growth factors (IGF-I). Inadequate Zn supply deactivates the signalling system of the membranes and the intracellular messengers, which increases the IGF-I content in cells. Lowering IGF-I causes lower cell proliferation and growth. It can, therefore, be said that inadequacy of Zn(II) disturbs the homeostasis[1] for normal growth and may result in impaired weight and height of human beings.

QUESTIONS

1. What is the importance of zinc in living systems?
2. Discuss the factors that affect absorption of zinc from diet.
3. What is phytate?
4. Describe various roles of zinc in metalloenzymes in biological systems.
5. What is the difference between the mammalian zinc amino peptidase and corresponding microbial enzyme?
6. What are alkaline phosphatases and why are they called so?
7. Zinc alkaline phosphates are active at what pH? What is the evidence for the same?
8. Which enzyme catalyses reversible hydration of carbon dioxide?
9. Explain the importance of hydration of carbon dioxide in human body.
10. Name the enzyme that catalyses the reversible oxidation of primary or secondary alcohols to aldehydes or ketones, respectively.
11. Which metal ion can substitute several Zn(II)-containing enzymes, reviving their activity?
12. Give reasons for exchangeability of Zn(II) and Co(II) in enzymes containing Zn(II).
13. Which chemical species is a precursor for the synthesis of pyrimidines in the biochemical systems?
14. Does zinc play catalytic role in the activities of DNA and RNA polymerases?
15. How does zinc influence oxygenation of haemoglobin?
16. What are zinc fingers?
17. Give an account of storage and release of intracellular zinc in biological systems.

---

[1] The word "homeostasis" is made of two Greek words "homeo" – similar, "stasis" – stable. It means that the systems have a typical integral regulation, so that a stable, relatively constant condition of the features is maintained.

5

Iron in Biochemical Systems

Among trace metals, iron is found in great abundance in animals. In the body of an adult human, there are about five grams of iron. Almost all cells have iron complexes in them acting as cofactors assisting the biochemical reactions occurring inside them. This happens because Fe(III) can change its coordination structure (square planar or octahedral). Hence, iron can easily bind with proteins and can mediate in catalytic processes as cofactors. It can also transfer electrons and bind with O_2, thus assisting oxidation reactions and O_2 transport.

More than 70% of the iron is in the form of an iron protein with haem as the prosthetic group. Haem is an Fe(II) complex of porphyrin. All the derivatives of porphine (Figure 5.1a) are collectively called porphyrins. Haems, with different derivatives of porphyrin, have been further classified.

Haem b is the Fe(II) complex of the most common porphyrin in nature called protoporphyrin IX with $-CH=CH_2$ at X and Y positions, and $-CH_3$ at position Z as shown in Figure 5.1b. This forms the prosthetic group in a large number of iron proteins, like haemoglobin, to be discussed later.

Haem a has a formyl group at position Z, a long phytol chain at position X (Figure 5.1b) and $CH_2=CH$ at position Y, in the porphyrin ring.

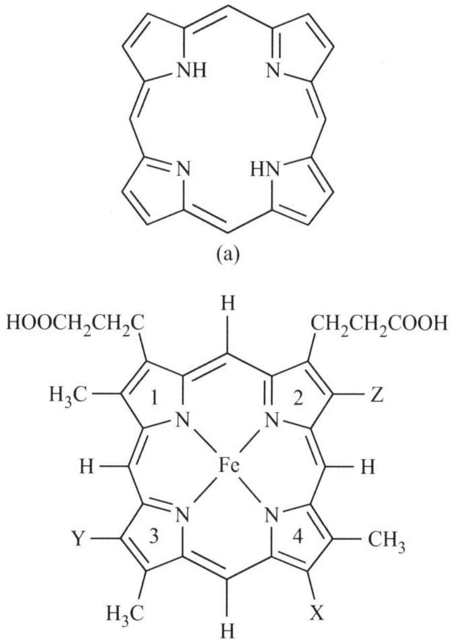

(a)

	X	Y	Z
Haem a	$C_{17}H_{29}O$	$CH_2 = CH$	CHO
Haem b	$CH = CH_2$	$CH = CH_2$	CH_3
Haem c	CH_2–CH–protein	CH_2–CH–protein	CH_3

(b)

FIGURE 5.1   (a) Porphine (b) Haems.

In haem c, the porphyrin part is covalently bound to a protein through S sites of the cysteine residues at positions X and Y (Figure 5.1b). There is a CH_3 group at position Z. Haem b and haem c prosthetic groups are present in cytochromes, to be discussed later.

The porphyrin rings in haem a, b and c consist of four pyrrole rings, linked through –CH groups.

In haem d, the double bond of the pyrrole ring I is dihydro-reduced.

In the formation of haem, the four nitrogens of protoporphyrin IX get coordinated to Fe(II) through σ bonds. There is deprotonation of two NH groups. Thus, the 16-membered macrocyclic ring gets two negative charges and forms a neutral haem complex with Fe(II). The resultant chelate has four six-membered rings with double bonds, leading to a stable structure. Furthermore, there is a macrocyclic effect (favourable enthalpy, entropy and multiple juxtapositional fixedness (MJF) effects, (see Chapter 2), resulting in further stabilisation of the macrocyclic chelate ring.

Besides the formation of N → Fe(II) σ bonds, there is Fe → N π bond formation, due to the interaction between metal dπ orbitals and the delocalised π orbitals of the ligand ring. Thus, a very stable haem complex is formed.

Unlike that expected in strong field ligand complexes of Fe(II), the metal ion in haem exists in high spin state. This is because Fe(II) is above the plane of the ring.

The oxidised form of haem, that is Fe(III) protoporphyrin complex, is called hemin. Several complex proteins, with haem or hemin as the prosthetic group, perform various functions in biochemical systems, which are discussed next.

Oxygenase Activity

Cytochrome P-450

This is a protein with Fe(III) protoporphyrin IX (haem b type) as the prosthetic group. It is found in the liver, intestine and lung tissues of mammals. It is bound to the membrane of the tissue and assists in transferring one oxygen atom of the molecular oxygen to a substrate, resulting in its monooxygenation. The role of the enzyme involves conversion of water-insoluble, aliphatic or aromatic hydrocarbons into soluble alcohols or phenols and of olefins into epoxides.

$$RH \; + \; O_2 \; \xrightarrow{\; 2e \; + \; 2H^+ \;} \; ROH \; + \; H_2O$$

(5.1)

It also causes oxidation of amines $\left(\overset{>}{\underset{/}{}}N:\right)$ to amine oxide $\left(\overset{>}{\underset{/}{}}N \to O\right)$ and of sulphide $\left(\overset{>}{\underset{/}{}}\ddot{S}:\right)$ to sulphoxide $\left(\overset{>}{\underset{/}{}}S \to O\right)$, and assists oxidative dealkylation ($Ph - OCH_3 \to PhOH + HCOH$).

Cytochrome P-450 has a molecular weight of ~50,000 daltons. The different forms of microsomal cytochrome P-450, in human beings, have the same prosthetic group, but differ in their protein backbone and assist the hydroxylation of various substrates such as drugs, steroid precursors, pesticides and halogenated hydrocarbons (like DDT), making them water soluble, so that they are excreted through urine. Thus, the enzyme is a part of the body's detoxification system. Cytochrome P-450 is also helpful in lipid metabolism and synthesis of carticosteroids by the monooxygenation process. However, it also has an adverse role of converting the pre-carcinogens into carcinogenic form. Liver cytochrome P-450 converts vinyl chloride into epoxide, which can cause tumours in liver. The lung cyt P-450 aggravates the formation of tumours in the lungs, due to benzpyrene in cigarette fume.

The structure determination of cyt P-450 has been difficult, as it is membrane-bound and insoluble. However, bacterial cyt P-450, which is a component of camphor 5 – monooxygenase, has been obtained in soluble form and has been characterised by X-ray study. The structure of the enzyme consists of a single α-helical protein chain with one Fe(III) protoporphyrin IX. There is no covalent link between the protein and the porphyrin ring. One axial site of Fe(III) is bound to the sulphur of the cysteinyl residue of the protein. The nature of the ligand occupying the sixth position is not definitely known (Figure 5.2). It may probably be water. Thus, Fe(III) has a strong ligand field around it and hence exists in low spin state in the native form of the enzyme.

In the enzymatic activity of cyt P-450, the substrate (RH) gets bound to the enzyme in a protein pocket, close to the haem, and the water or any other group, present at the sixth position, is replaced. As RH binding is noncovalent and weak, effectively Fe^{3+} enzyme–RH complex is penta-coordinated, and Fe(III) is in high spin state. Due to the binding with RH, the reduction potential of Fe^{3+} enzyme–RH becomes more positive, and it is more easily reduced to Fe^{2+} enzyme–RH, by receiving electron from the biochemical reducing agent NADH or NADPH. The electron is transferred from NADH to Fe^{3+} enzyme–RH, through electron transfer iron proteins. In the mammal microsomal cyt 450, only flavoprotein is involved in electron transfer process, whereas in bacterial cyt 450, one flavoprotein and one iron sulphur protein are involved.

The Fe^{2+} enzyme–RH species binds with O_2 at the sixth position. If CO is supplied, it can occupy the sixth position. In the electronic spectrum of this carboxy form, there is a characteristic absorption band at 450 nm and that is the reason for the nomenclature of the enzyme.

The oxygenated form of the enzyme is not capable of oxidising the substrate and hence leads to the generation of the active oxidant. For this, the oxygenated form takes another electron from NADH or NADPH through the electron transfer system. The iron centre accepts and transfers two electrons to the molecular oxygen and gets back to Fe(III) form. O_2 is reduced to O_2^{2-} form. The uptake and transfer of electron by the iron centre, leading to the formation of peroxide, can be considered to be taking place in two steps, with the formation of superoxide in the first step. The peroxide takes up one H^+, and a hydroperoxy Fe(III) form of the enzyme substrate complex is generated.

The oxygenated form of the Fe(II) enzyme shows optical spectrum typical of oxygenated Fe(II) haem and is like oxyhaemoglobin. It is, therefore, questionable how oxyhaemoglobin carries stable O_2 molecule, whereas Fe(II) form of cytochrome P-450 activates the O_2 molecule. This may be due to the difference in axial coordination, of cysteinyl S in cyt P-450, and of histidine N in haemoglobin.

The Fe(III) hydroperoxo complex undergoes a heterolytic cleavage of $^-$O–OH, resulting in O and O^{2-}. Fe(III) O gets converted to Fe(V) O, and O^{2-} combines with two H^+ forming H_2O. The oxo species is green in colour and is the activated oxygen form of cyt P-450.

There is, however, no satisfactory explanation of why there is heterolytic cleavage of the hydroperoxo group. The vicinity of amino acid chains, which develops charge separation in $^-$O–O$^-$ and causes heterolytic cleavage in catalase and peroxidase (to be discussed next), does not exist in cytochrome P-450. Probably, the cysteinyl S$^-$ at the axial position in cyt P-450 helps in heterolytic fission, and hence, charge separation may not be necessary. The sequence of events leading to the oxygenation of the substrate is shown in Figure 5.3.

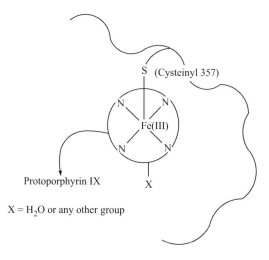

X = H$_2$O or any other group

FIGURE 5.2   Cytochrome P-450.

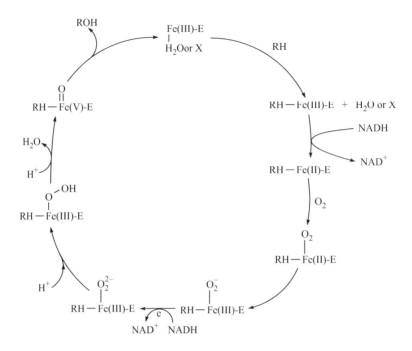

FIGURE 5.3   Oxygenation of substrate by cyt P-450.

The problem in the elucidation of the structure of the intermediates in cyt P-450 activity is that they cannot be physically observed. The formation of the monooxo cation Fe(V) O is supported by the fact that in the monooxygenation by cytochrome P-450 of substrates, using molecular oxygen and a reductant (as in the natural system) or using a monooxygen oxidant iodosyl benzene PhIO, the product obtained is same. This shows that the reactive intermediate in both the reactions must be same. As PhIO has only one oxygen atom, the intermediate must be a monooxygenated species (Figure 5.4a). Furthermore, in the case of oxygenation reaction by PhIO, the active intermediate has been experimentally shown to be Fe(V)O. Thus, it can be inferred that in the case of cyt P-450, using molecular oxygen and a reductant also, the active intermediate must be Fe(V)O.

The spectroscopic studies of the monooxo intermediate, in model system Fe(III)TPP (TPP – tetrapheyl porphyrin) complex, indicated that the species is not Fe(V)O porphyrin^{2-}, but is Fe(IV)O porphyrin^{1-}. One electron is lost from Fe(III) and one from porphyrin dianion, forming porphyrin monoanion radical (As porphyrin is in dianionic form, after deprotonation of two NH groups, the loss of one electron forms

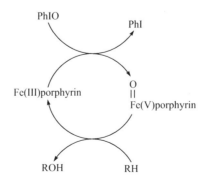

FIGURE 5.4a   Catalytic cycle showing Fe(III)oxo intermediate in oxidation of substrate using PhIO.

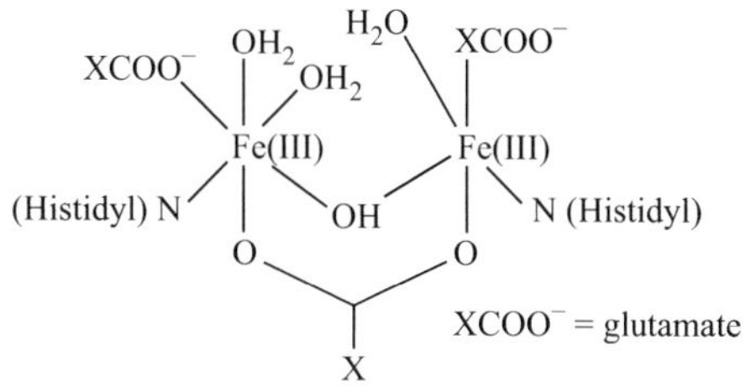

FIGURE 5.4b Structure of active part of MMO.

the monoanion). This is supported by the short Fe–O distance, as revealed by EXAFS studies. This is likely as Fe(IV)O bond is expected to have double-bond character. Besides O^{2-} to Fe(IV) σ bond, there is π bond formation due to back donation of pπ electrons of O^{2-} to the vacant dπ orbitals of the metal ion. Furthermore, Mössbauer (M. B.) spectrum of the oxo species is similar to Fe(IV) porphyrin complex. The green colour of the oxo species is also due to $\pi \to \pi^*$ transition in the porphyrin monoanion. The dianion exhibits red colour, as in haemoglobin.

The above conclusion, about the formation of the intermediate Fe(IV)OP⁻ species (P = porphyrin), is based on the oxygenation studies in model systems. However, the models differ from the natural enzyme, in the fact that there is no axially coordinated S⁻ group in them. Hence, the formation of Fe(IV)OP⁻ species as intermediate in the oxygenation of cytochrome P-450, is not conclusively established.

In the cycle (Figure 5.3), finally the oxo cation species transfers the oxygen to the substrate, regenerating Fe(III) enzyme, and causes two electron oxidation of the substrate:

$$\begin{matrix} Fe(V)OP^{2-} \\ \\ or \qquad \xrightarrow{\;RH_2\;} Fe(III)P^{2-} + RHOH \\ \\ Fe(IV)OP^- \end{matrix} \qquad (5.2)$$

The mechanism of the transfer of oxygen atom, from the oxo cation to the substrate, has been proposed. In the first step, there is abstraction of a hydrogen atom from the substrate, forming a free radical and a Fe(IV) hydroxide complex. In the second step, the –OH fragment recombines with the substrate radical, forming the hydroxylated product. This is shown in the following equation:

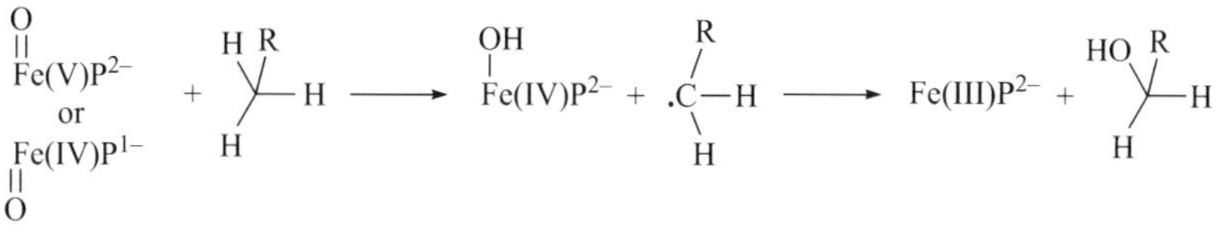

$$(5.3)$$

It is interesting to observe that the enzyme binds with the molecular oxygen only after the substrate comes in its vicinity. This prevents the formation of the active oxo cation in the absence of the substrate, and thus, the possibility of oxidative degradation of the enzyme protein structure, by molecular oxygen, is avoided.

Methane Monooxygenase

This is a soluble iron protein, affecting monooxygenation of methane. Another membrane-bound insoluble particulate copper protein also acts as methane monooxygenase (MMO). However, the iron protein is better studied, and the name "methane monooxygenase" refers to the iron protein, only.

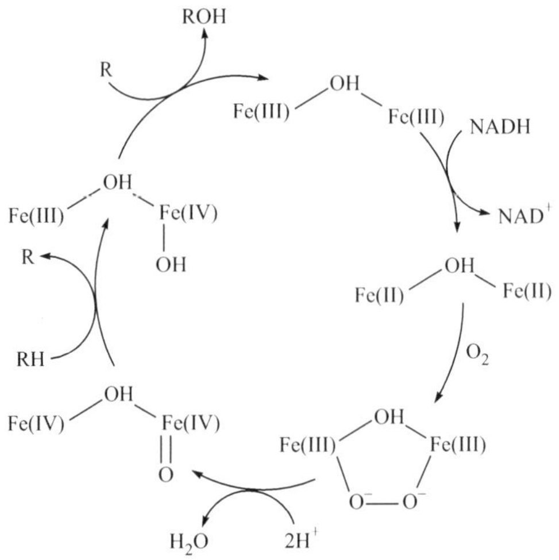

FIGURE 5.5   Catalytic cycle showing mechanism of MMO.

MMO is found in methanotrophic bacteria, which is responsible for converting methane into soluble form, and hence, its release in atmosphere is restricted. MMO is now known to cause oxygenation of a wide range of hydrocarbons and thus has a large number of practical applications, such as functionalisation of the hydrocarbons to more useful compounds, and the removal of oil spills by converting the hydrocarbon to soluble form.

The active part of MMO consists of a binuclear non-haem iron (III) protein (molecular weight ~251,000 daltons). Each Fe(III) site is bound to a nitrogen atom of histidyl residue of the protein. The other coordination positions are bound to one glutamate and two water molecules in the case of one Fe(III), and to two glutamate and one water molecule in the second case. Two Fe(III) sites are also bridged through an –OH group and a glutamate. There is an anti-ferromagnetic interaction between the two centres, by the super exchange through the bridging –OH (Figure 5.4b).

The active protein is associated with two other proteins: the protein reductase (molecular weight 38,600 daltons) which helps in the transfer of electrons to the Fe(III) centres from its NADH site, and coupling protein (molecular weight 15,500 daltons) which regulates the overall activity of MMO.

The mechanism of the activity of the enzyme involves transfer of two electrons from NADH to the two Fe(III) centres, resulting in two Fe(II) centres. These two Fe(II) get bound to one O_2 molecule. Each Fe(II) transfers one electron to the O_2 molecule, resulting in two Fe(III) centres, bridged through a peroxide. Thus, a hydroxide- and peroxide-bridged, Fe(III)-paired centre is formed. There is heterolytic fission of O_2^{2-}, forming O and O^{2-}. The latter combines with $2H^+$ and is removed as H_2O. The neutral O oxidises both iron centres to Fe(IV) and remains attached to one as O^{2-}. The substrate RH attaches at the oxo cation site. The design of the hydrophobic cavity, in which the active binuclear site of the enzyme is located, is such that the hydrocarbon is attached very close to the oxo cation. One H atom is abstracted from the substrate RH forming radical R. The abstracted H atom transfers one electron to the other Fe(IV) without oxygen and reduces it to Fe(III). The resulting H^+ combines with the oxo cation centre, to form $-OH^-$, which transfers the electron to the Fe(IV) centre to reduce it to Fe(III). The substrate radical R combines with the neutral OH, to form the product alcohol, and the binuclear Fe(III) form of the enzyme is regenerated (Figure 5.5).

Catalase and Peroxidase Activity

These are related iron proteins, containing Fe(III) protoporphyrin (haem b type) prosthetic group. They catalyse the reactions of hydrogen peroxide, leading to its decomposition. H_2O_2 is generated in several biochemical reactions. It is a strong oxidising agent and is likely to be toxic to the cells, causing their oxidation and degradation. Hence, it is necessary that the generated H_2O_2 has to be decomposed, so that its concentration does not increase.

Catalase is found in the peroxisome and the liquid inside the cytoplasm of the cells. It catalyses the decomposition of H_2O_2 into water and oxygen, thus causing its disproportionation. Thus, this enzyme mainly prevents the build-up of the concentration of H_2O_2.

$$2H_2O_2 \xrightarrow{\text{Catalase}} 2H_2O + O_2 \tag{5.4}$$

Peroxidase catalyses the oxidation of a variety of organic and inorganic substrate, using H_2O_2 as an oxidant [5.5]. Alternatively, the oxidised form of the enzyme may assist the oxidation of a redox partner like Mn^{2+} or Cl^-, which in turn can oxidise the substrate Eqs (5.6 and 5.7). Such enzymes are called Mn peroxidase and chloroperoxidase, respectively. Cytochrome c peroxidase is another enzyme which causes oxidation of cytochrome c to oxocytochrome c, which causes oxidation of the substrate [5.8].

$$\text{Peroxidase} + H_2O_2 + AH_2 \longrightarrow 2H_2O + A \tag{5.5}$$

$$\text{Mn-peroxidase} + H_2O_2 + 2Mn^{2+} \longrightarrow 2Mn^{3+} \xrightarrow{+AH_2} \text{Oxidized product of } AH_2 \tag{5.6}$$

$$\text{Chloroperoxidase} + H_2O_2 + Cl^- \longrightarrow ClO^- \xrightarrow{+AH_2} \text{Oxidized product of } AH_2 \tag{5.7}$$

$$\text{Cytochrome c peroxidase} + H_2O_2 + \text{Cyt c} \longrightarrow \text{OXO Cyt c} \xrightarrow{+AH_2} \text{Oxidized product of } AH_2 \tag{5.8}$$

In both catalase and peroxidase, Fe(III) is in high spin state, and located deep inside the protein molecule. The crystal structure of beef liver catalase shows that one of the axial sites is occupied by phenolate O^- of the tyrosyl residue of the protein and the sixth position has water molecule.

The crystal structure of horse radish peroxidase shows that Fe(III) is bound to histidinyl N of the protein at the fifth position. Mn peroxidase and cytochrome c peroxidase also have histidinyl N at the fifth position. However, chloroperoxidase is bound to the S, of the cysteinyl residue of the protein, at the fifth position.

In the catalytic reactions, H_2O_2 occupies the sixth position, displacing water molecule and coordinating through one of the O. The H atom bound to the coordinated O gets hydrogen-bonded to the N atom of the imidazole N of the histidinyl residue of the protein chain, thus developing δ^+ on the coordinated oxygen. The other oxygen atom develops δ^-, which is further accentuated by electrostatic interaction with positive charge of the arginyl residue of the protein. Thus, there is a charge separation between two peroxo oxygens. This leads to heterolytic fission into neutral oxygen and O^{2-}. The latter combines with $2H^+$ forming water molecule. The neutral O takes two electrons forming O^{2-}: one electron comes from Fe(III) forming Fe(IV) and the other from the porphyrinate (-2) ring forming porphyrinate (-1) ring, which is a π radical. This intermediate radical may be stabilised by the transfer of negative charge from the protein side chain, due to π contact or H bonding. Thus, the oxygen of the oxocation is closely surrounded by the protein. This is called compound I and is the active oxidant (Eq. 5.9).

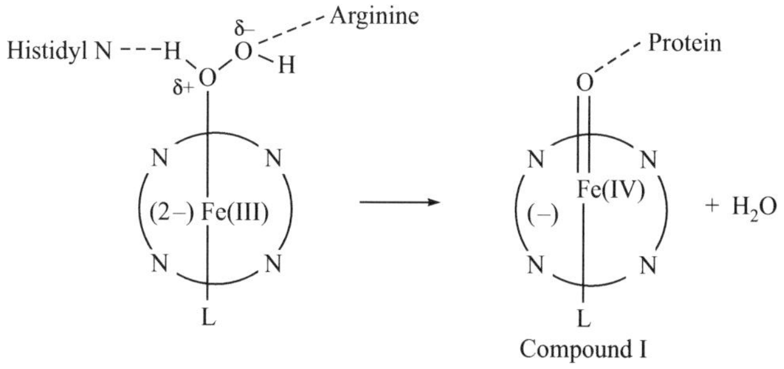

$$\tag{5.9}$$

Compound I

In the peroxidase reaction, two electrons are transferred in steps from the substrate molecule, AH_2, to compound I, forming A and $2H^+$. These two H^+ combine with O^{2-} of the oxocation forming H_2O, and Fe(III) haem enzyme is regenerated (Eq. 5.10).

In the catalase activity of decomposition of H_2O_2, it can be considered that one molecule of H_2O_2 takes part in the formation of the compound I. The O_2^{2-} of the second H_2O_2 molecule loses two electrons in steps to compound I, forming O_2, and the two H^+ combine with the O^{2-} of the compound I to form H_2O (Eq. 5.11).

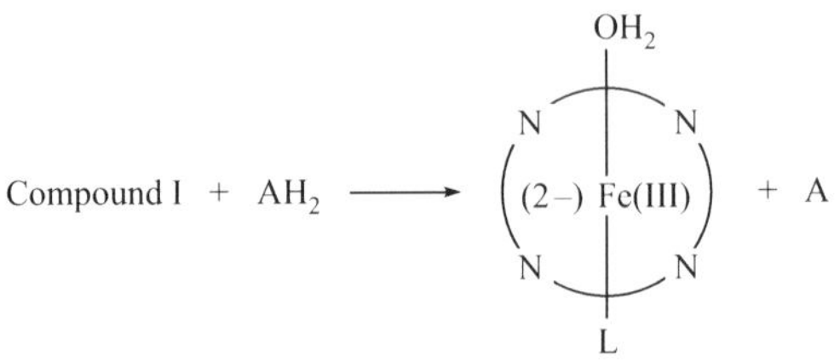

$$\text{Compound I} \; + \; AH_2 \; \longrightarrow \; \left(\begin{array}{c} OH_2 \\ N \quad | \quad N \\ (2-)\; Fe(III) \\ N \quad | \quad N \\ L \end{array}\right) \; + \; A \qquad (5.10)$$

$$\left(\begin{array}{c} OH_2 \\ (2-)\; Fe\,(III) \\ L \end{array}\right) \xrightarrow{H_2O_2} \left(\begin{array}{c} O \\ (-)\; Fe(IV) \\ L \end{array}\right) + H_2O \xrightarrow{H_2O_2} \left(\begin{array}{c} (2-)\; Fe\,(III) \\ L \end{array}\right) + O_2 + 2H_2O \qquad (5.11)$$

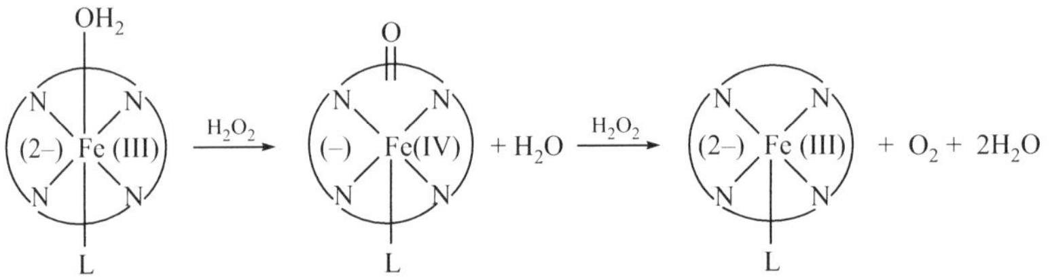

$$\text{Net reaction:} \quad 2H_2O_2 \; \longrightarrow \; 2H_2O \; + \; O_2$$

In the catalase reaction, the second H_2O_2 molecule plays the same role as the substrate AH_2 molecule in peroxidase activity. Both the oxygen atoms of the O_2 molecule come from the second molecule of H_2O_2. This has been confirmed by using labelled $H_2^{18}O_2$. The O_2 molecule obtained has molecular weight 32 or 36 dalton, showing that both the oxygen atoms of O_2 are either from the labelled or unlabelled molecule. If two O atoms of O_2 would have been from two different H_2O_2 molecules, there would be a possibility of formation of O_2 with a molecular weight of 34 dalton. This is not formed.

It is interesting to observe that in the enzymes cyt P-450, catalase or peroxidase, the intermediate is the same oxo cation, but in the case of the first, oxygen is transferred to the substrate and the enzyme acts as a monooxygenase (Eq. 5.2), whereas in the latter two, two electrons are transferred from the substrate to the oxo cation, resulting in the oxidation of the substrate. Thus, catalase and peroxidase exhibit oxidase activity.

The intermediate oxo cation performs two different types of activities, because the substrate can come very close to the intermediate oxo cation in cyt P-450, and hence, the oxygen can be easily transferred to the substrate, bringing about its oxygenation. In the case of catalase and peroxidases, the site near oxygen of the oxo cation is blocked by the side groups of the protein; hence, the substrate cannot come close to the oxygen. The substrate can interact only with the haem edge of the oxo intermediate of the enzyme and hence can be oxidised only by the transfer of electrons to the metal centre through the porphyrin. Hence, there is oxidation of the substrate by peroxidase, instead of oxygenation.

Electron Transfer Process

Cytochromes

Cytochromes, meaning cell pigments, were discovered by Keilin. These are Fe(III) haem proteins, occurring in cells, and act as electron carriers in mitochondrial oxidation process.

They are involved in the terminal stage of the respiratory chain, that is carrying electron from glucose to molecular oxygen, leading to its decomposition and also transporting electrons in the photosynthetic process. The electron transfer process involves reversible change of haem iron from oxidation state III to II. The cytochromes differ in their reduction potentials and are arranged in sequence in the electron transport process, in the increasing order of the reduction potentials. Therefore, the cytochromes affect sequential stepwise transfer of electron from the substrate to molecular oxygen, leading to its reduction to water. This avoids highly exothermic one step oxidation of the substrate, by transfer of electron directly to O_2, which can cause harm to the tissues.

The difference in the reduction potentials of cytochromes arises, due to the type of porphyrin ring with different substitutions, that is the type of the haem (discussed earlier) and also the nature of the axial ligand coordinated to the iron centre.

The cytochromes are of the types a, b, c and d. The nomenclature is based on the historical order of the discovery or the decreasing order of the λ_{max} in the absorption spectrum of Fe(II) form of the cytochrome.

580–590 nm (a), 555–560 nm (b)
548–552 nm (c), 600–620 nm (d).

In cytochromes b and c, the axial positions of Fe^{3+} are occupied by two strong field ligands, and so, they are low spin in nature. As they have no coordination position vacant, they are not capable of binding with O_2 molecule. As cytochromes a_3 and d have the sixth coordinated site vacant, they can bind with O_2 and transfer electron to it. Thus, they act as terminal oxidase in the chain.

The structure of cytochrome c has been studied most extensively. In the horse heart cytochrome c, the protein chain has 104 amino acid residues and has a molecular weight of 12.4 K daltons. The two axial positions of Fe(III) in the prosthetic group of cyt c are bound to the N atoms of histidyl 18 and the S atom of the methionine 80 residue of the apoprotein. The cytochrome c with differing molecular weights due to change in protein is named cytochrome c_X, where X can have values 0, 1, 3, 4, 5, 550 to 555. Cyt 552 is also called cyt c6f. There is also a cytochrome c'.

Cytochromes c_3 and cyt c552 (or cyt c6f) have four and two haem prosthetic groups. These cyt c have different reduction potentials and are involved in various processes such as respiration, photosynthesis, reduction of sulphates to sulphides and denitrification.

Among cytochromes of type c, only cyt c and c_1 are involved in the respiratory chain. The haem part of cyt c is buried deep inside the hydrophobic pocket of the apoprotein. Only an edge part of the haem is near the surface. The fact that there is a ring of lysine residues of the protein surrounding the exposed part of the haem, shows that the electron transfer of cyt c occurs through the small solvent exposed edge of the haem. The model studies have shown that cyt c interacts with the inorganic redox partners through the exposed haem edge. Consequently, the reduction potential of cyt c is also dependent on the stability and the solvent accessibility of the haem crevice and the hydrophobicities of the amino acid residues that form the line around the haem crevice. Another important feature is that the aromatic residues of the protein part of cyt c such as 10 phe, 48 Tyr, 59 Trp, 67 Tyr, 74 Tyr, 82 Phe and 97 Tyr are conserved in cyt c from different sources and Tyr 67 and Tyr 74 are parallel to each other.

Path of Electron Transfer in Cytochromes

In the respiration process, glucose is oxidised to CO_2 and H_2O with the electron being transferred to NAD^+ or fumarate, resulting in the formation of the reductants NADH or succinate, respectively. These reductants transfer electrons to iron sulphur clusters, called complexes I and II.

These in turn transfer electrons to coenzyme quinone (CoQ), reducing it to phenolic form. Then, the electron moves to cyt b, cyt c and cyt c, in the order of their increasing reduction potentials, as shown in Figure 5.6.

The rate of electron transfer of cyt c, that is K(M/s) for cyt c_1 Fe(II) to cyt c Fe(III) electron transfer reaction, is thousand times slower than similar iron complexes. The slow electron transfer indicates that the transfer takes place through the small exposed edge. As the haem part is closely wrapped with the hydrophobic protein, the electron donating cytochrome cannot have direct contact with the electron accepting cytochrome, resulting in slow electron transfer.

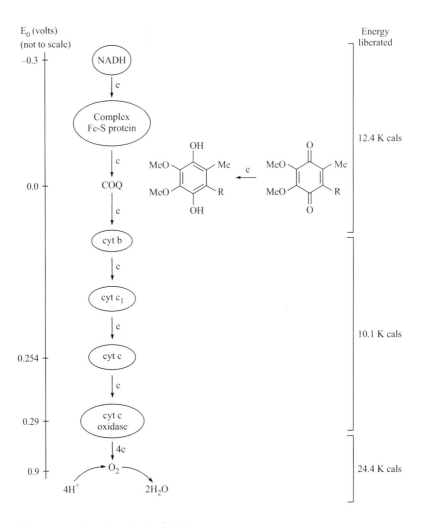

FIGURE 5.6   Electron transfer pathway in cytochromes.

The haem prosthetic groups of the two proteins are at a distance of 17 Å. The question arises, how does the electron transfer take place between the two distant haems. Most likely it must be through outer sphere electron transfer.

Crystal structure study shows that both the oxidised and reduced forms of cyt c have a very small difference in the structure, only in the protein part. Iron in both the forms is in the low spin configuration, which is suitable for outer sphere electron transfer.

A probable mechanism of outer sphere electron transfer, as proposed by Dickerson and Winfield, is as follows.

The Fe(III) centre of cyt c accepts an electron from the OH group of tyrosine 67 residue of the apoprotein through S of the axial methione 80 residue. The tyrosine 67 loses the proton and forms a π radical cation. There is a stacking interaction between Tyr 67 π radical and the parallel π ring of the tyrosine 74 residue, and an electron is transferred from tyrosine 74 to Tyr 67 π radical, through cloud overlap. Tyr 67 binds with the protons and comes back to original form. Tyrosine 74 in turn forms p radical and accepts electron from the donor Fe(II) of reduced cytochrome c_1, and oxidises it to Fe(III). Tyrosine 74 gets back to original form. In effect, electron is transferred from Fe(II) of reduced cyt c_1 to Fe(III) of cyt c, reducing it to Fe(II). Thus, the electron transfer takes place between the two distant haem centres through the protein molecules of cyt c. A simplified picture of outer sphere electron transfer process is shown in Figure 5.7.

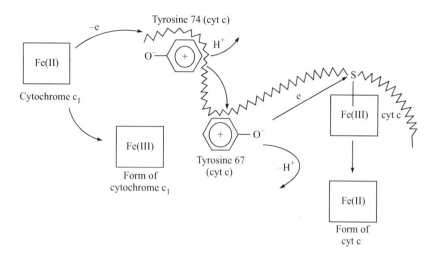

FIGURE 5.7   Outer sphere mechanism of electron transfer mediated by cyt c.

Cytochrome c transfers electron to cyt c oxidase. A similar outer sphere process may be operative for transfer of electron from cyt c to cyt c oxidase. Thus, both cyt c_1 and cyt c get into Fe(III) form to accept the electron and continue the chain.

Cytochrome c oxidase

This enzyme is the last in the respiratory chain and is located in the mitochondrial membrane of animals, plants, fungi and bacteria. In the electron transfer process, cyt c in the reduced form transfers electron to the cytochrome c oxidase. Ultimately, the electron is transferred to O_2 by reduced cyt c oxidase, and there is four-electron reduction of O_2 to water. In the transfer of electron from NADH (formed by oxidation of glucose) to O_2, large amount of energy is released. However, the energy is liberated in a controlled manner in steps. This establishes a transmembrane proton gradient through pumping of proton across the membrane, and it ultimately proves to be the driving force for the synthesis of ATP.

Cytochrome c oxidase has 13 subunits and has a complex protein structure. There are four redox active metal centres. There are two different sites. (i) cyt a and Cu A, and (ii) cytochrome a_3 and Cu B.

In the first site cyt a, the Fe(III) haem prosthetic group is six coordinated, with histidine residues bound at the axial sites, and it is in a low spin state. CuA is separated from it by nearly 13–26 Å (Figure 5.8W). The Cu(II) is bound to histidine ligands. Initially, it was thought that CuA is a mononuclear complex Cu(Hist)$_2$(Cyst)$_2$.

It was considered similar to type I copper with a charge transfer (C T) band in the electronic spectrum in the visible region, leading to deep blue colour, and electron paramagnetic resonance (EPR) spectrum corresponding to S = ½ with small hyperfine splitting, similar to type I copper. However, a closer look at EPR spectrum shows that the g value is less than 2, though most mononuclear Cu(II) complexes have g > 2. This indicates that there is a weakly coupled binuclear Cu(II) complex with two S bridges of cysteine residues: one Cu(II) being bound to one histidine and one methionine and the other bound to one histidine and one carboxylate (Figure 5.8X).

In the second site, the Fe(III) haem prosthetic group of cytochrome a_3 is penta-coordinated and is in high spin state. The fifth coordinated site is bound to histidine N. The sixth coordination site is bound to the ligand forming bridge with Cu B (Figure 5.8Y). Cu B is bound to two histidine N and one cysteine S, or to three histidine, as per a recent study, with the fourth position bound to the bridging ligand (Figure 5.8Z).

The magnetic study shows that Fe(III) of cyt a_3 (S = 5/2) and Cu(II) of Cu B (S = ½) are coupled. The resultant S = 2 indicates that Fe(III) and Cu(II) are strongly anti-ferromagnetically coupled, suggesting that there is a single atom bridge, though there is no agreement about the nature of the bridging atom. The proposed naturally occurring bridging groups are Cl⁻, S²⁻, imidazole, phenoxide, oxide or hydroxide. Since this site has S = 2, no EPR signal is seen.

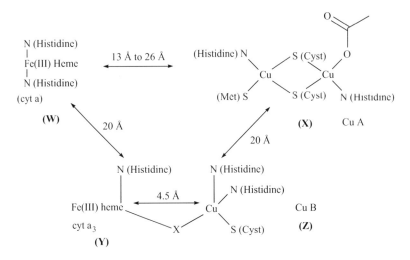

FIGURE 5.8 Electron transfer mediated by cyt c oxidase.

In recent refined structural study, it has been observed that there is delocalisation of electron density over the Fe(III) and Cu(II) metal ions and delocalisation is only possible by propagation through the bridging atom. This accounts for strong anti-ferromagnetic character, due to super exchange through the bridge.

On one electron reduction of the Cu B site, the resultant Cu(I) has S = 0. Hence, anti-ferromagnetic coupling with Fe(III) is broken. Fe(III) high spin site has S = 5/2, and its EPR signal can be seen. The magnetic moment corresponds to S = 5/2. Thus, unlike normal expectation, an increase in magnetic moment is observed on reduction of cyt a_3–Cu B.

The cyt a_3 and Cu B sites are separated by 4.5 Å, and both are separated from cyt a and Cu A by 20 Å, as shown in Figure 5.8.

Cytochrome a

Cu A is the low potential site and hence is the site receiving the electron. Cyt a and Cu A exhibit anti-cooperative redox behaviour. The reduction of haem a site makes the reduction of Cu A difficult, either due to electrostatic effect or structural effect. Moreover, the reduction of cyt a–Cu A brings in structural changes in the site, so that an electron transfer path is opened for the electrons to be transferred to cyt a_3–cyt B site. The structural change also opens up the cyt a_3– cyt b site to bind with O_2 and transfer electrons to it. Thus, cyt a_3–cu B site can lose electron more easily compared to cyt a–Cu A site.

Cyt a–Cu A site transfers the two electrons to cyt a_3–Cu B site. On reduction, the bridging ligand is removed from Fe(II) a_3–Cu(I) B pair. However, iron(II) and copper(I) remain close to each other in a geometry, suitable for O_2 binding. This results in the binding of O_2 with the reduced form of the enzyme forming an oxygenated complex. O_2 is mainly bound to Fe(II) centre, forming a oxyhaemoglobin HbO_2 type of structures. Two electrons are released from Fe(II)a_3–Cu(I) B to bound O_2 forming μ-peroxo complex Fe(III)–O^-–O^-–Cu (II) B.

The Fe(III) of the μ-peroxo complex receives one more electron from cyt a–Cu A to form Fe(II) a_3–Cu(II) B, resulting in the cleavage of O^-–O^- bond into O and O^{2-}. The oxygen takes two electron from Fe(II) forming Fe(IV)O^{2-} ion, and the other O^{2-} gets bound to two protons to form water, which binds to Cu(II) B. Transfer of one more electron from cyt a–Cu A to Fe(IV)O^{2-} results in the formation of

$$\underset{Fe(III)\, a_3}{\overset{\displaystyle O}{\overset{\|}{\rule{0pt}{0pt}}}}\!\!\!\!\text{—— Cu B – OH}_2 \text{ complex. Transfer of one proton from } H_2O \text{ of Cu B to } O^{2-} \text{ of Fe(III) } a_3 \text{ results}$$

in the formation of $\underset{Fe(III)\, a_3}{\overset{\displaystyle OH}{\overset{\|}{\rule{0pt}{0pt}}}}$ – Cu B – OH. Further protonation of the two –OH on the Fe(III) and Cu B

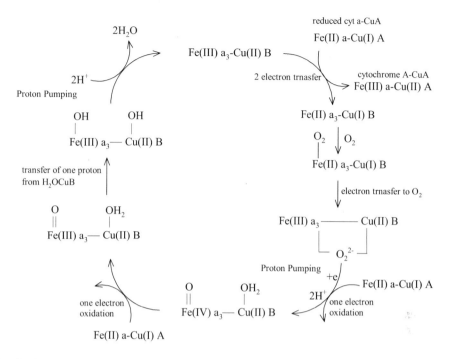

FIGURE 5.9 Catalytic cycle showing electron transfer in cyt c.

leads to the liberation of two water molecules and regeneration of the cyt a_3–Cu B. The electron transfer process is shown in Figure 5.9.

As mentioned earlier, the function of cytochrome c oxidase is not only to reduce $O_2 \rightarrow 2O^-$ but also to pump protons across the membrane. The protons binding with the two O^{2-}, formed by reduction of O_2, are the protons generated in the system, by the oxidation of the substrate. The energy released by the reduction of $O_2 \rightarrow 2O^-$ and the formation of H_2O is used to drive the protons across the membrane, against the concentration gradient. The energy released by the reduction of O_2 also assists the formation of ATP from ADP. However, the mechanism of proton pumping is not definitely known.

Out of the two components of cytochrome c oxidase, only cyt a_3–Cu B part combines with the ligands like CO or cyanide ion. This is because Fe(III) in cyt a is hexa-coordinated, and Fe(II) in cyt a_3 is a penta-coordinated site.

CN^- binds with the reduced Fe(II) of cyt a_3 and thus blocks the site from binding with molecular oxygen. Furthermore, cyanide stabilises Cu(I) oxidation state of Cu B. Transfer of electron from reduced cyt a_3–Cu B to O_2 is blocked and the terminal oxidation process is stopped. Thus, cyanide acts as a poison. The fatal effect of cyanide is not due to the blocking of the sixth coordination site in haemoglobin or myoglobin, as these sites are hydrophobic and have least affinity for binding with charged cyanide ion (see the "Haemoglobin and Myoglobin" section).

Iron Sulphur Proteins

These are non-haem iron proteins, which, like cytochromes carry electron in biological systems, from a molecule with lower redox potential to one with higher redox potential. They are found in anaerobic and aerobic bacteria, algae, fungi, higher plants and animals. In plants, they act as electron carrier for reduction of N_2, in nitrogenase activity, and in photosynthesis, for electron transfer from PS I to PS II. In the animals, they assist electron transfer process in the mitochondria. These proteins are similar to cytochromes, in the fact that they also involve one electron process in the Fe(III)/Fe(II) couple.

Iron sulphur proteins are classified into two types: rubredoxin and ferredoxin. In the latter, the prosthetic group is $(Fe–S)_n$, with no organic component, and is bound to a low-molecular-weight protein, which does not incorporate complex amino acids. This simplicity of the protein part shows that ferredoxins are older in the evolutionary process, than cytochromes, which have more complex prosthetic groups and apoproteins. However, iron sulphur proteins were discovered later than cytochromes.

In the $(Fe–S)_n$ part of ferredoxin, the iron cations are bound to the inorganic sulphide ion S^{2-}, which are acid labile and are liberated as H_2S on treatment with acid. $(Fe–S)_n$ prosthetic group is bound to the protein, through the S of the cysteinyl residue. This sulphur is called organic sulphur.

Rubredoxin differs from ferredoxin in the fact that it does not have labile inorganic sulphide.

Following is the detailed discussion of the two types of iron sulphur proteins:

1. **Rubredoxin:** It is found in aerobic and anaerobic bacteria and consists of one iron per protein of molecular weight ~ 60 K daltons. The protein is constituted of only 54 amino acid residues. X-Ray study shows that Fe(III) is bound to the S atoms of the four cysteine residues of the protein at Cyst – 6, Cyst – 9, Cyst – 39 and Cyst – 42. Fe–S distances have a small difference, the average Fe–S distance being 2.26 Å. The S–Fe–S angles show that the S of the cysteine residues is located at the corners of the tetrahedron (Figure 5.10). This is supported by the observation of magnetic moment 5.8 BM and two peaks in the ESR spectrum at g = 4.31 and 9.42, corresponding to the high spin e^2t^3 configuration. In the MB spectrum, there is very small quadrupole splitting, as high spin Fe(III) has spherical electronic charge distribution. The protein exhibits red colour due to $S^- \rightarrow Fe(III)$ charge transfer.

   The electron transfer by the protein involves the redox process $Fe(III) + e \rightarrow Fe(II)$. The reduction potential of rubredoxin (–57 mV) is higher (more positive) than the ferredoxin. Magnetic moment of the reduced form of rubredoxin is observed to be 5.05 BM, corresponding to four unpaired electrons, with the electronic configuration $e^3t_2^3$. The d–d absorption band $e^3t_2^3 \rightarrow e^2t_2^4$ is observed at 625 nm. The Fe(II) ion with S = 2 shows no electron spin resonance (ESR) signal, due to large zero field splitting. The Mössbauer (MB) spectrum shows quadrupole splitting, due to asymmetric electron distribution in e^4t^2 configuration.

   The role of rubredoxin in aerobic bacteria is to provide electron to the enzyme hydroxylase, which acts as a monooxygenase, affecting the hydroxylation of aliphatic acids. Rubredoxin receives the electron from NADH, gets reduced to Fe(II) form and passes on the electron to the hydroxylase, to get converted back to Fe(III).

2. **Ferredoxin:** The types of ferredoxin are as follows:

   i. **[2Fe–2S] Ferredoxin:** These proteins are found mainly in plants and are involved in the electron transfer process in photosynthesis. Hence, they are called plant ferredoxin. X-Ray crystal study of (Fe_2S_2) protein in *Spirulina platensis* (blue green algae) shows that the prosthetic group contains two Fe(III), bridged through two sulphides. Each Fe(III) of the Fe_2S_2 unit is bound to the protein of molecular weight 11,000 daltons, through two S atoms of the cysteine residues, located at 41, 46, 49 and 79 positions (Figure 5.11).

      Thus, each Fe(III) is bound to four sulphurs, located at the corners of a tetrahedron. These two tetrahedra share one edge with bridging sulphides. The six sulphurs (both sulphide and cysteinyl S) are H-bonded to the six peptide NH and one –OH of serine 40, leading to the stability of the $Fe_2(m\ s)_2$ unit and its protein complex.

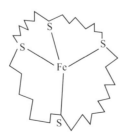

FIGURE 5.10   Tetrahedral arrangement of S atoms around Fe in rubredoxin.

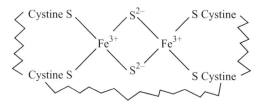

FIGURE 5.11 Fe_2S_2 unit of ferredoxin in Spirulina platensis.

Mössbauer spectrum shows the presence of Fe(III). There is no quadrupole splitting, supporting the presence of Fe(III) in high spin state. However, the protein is diamagnetic. This is due to strong super exchange between the two Fe(III) centres, through the two bridging sulphides. The anti-ferromagnetic coupling between the five unpaired electrons, on each of the two Fe(III) centres, is very strong, leading to S = 0. It is also ESR silent.

The Fe_2S_2 undergoes one electron reduction. The redox potential of the protein is −250 to −420 mV, indicating reducing nature of the protein. The redox states of the protein can be represented as follows:

$$Fe^{3+} - Fe^{3+} \xrightarrow{\ e\ } Fe^{3+} - Fe^{2+} \tag{5.12}$$

In the reduced form, there is a high spin Fe(III) centre (S = 5/2) and a high spin Fe(II) centre (S = 2). However, the four unpaired electrons of Fe(II) couple with the four unpaired electrons of Fe(III), due to anti-ferromagnetic coupling, through the sulphide bridges, leading to S = ½. This is supported by the observation of magnetic moment, corresponding to one unpaired electron and the ESR spectrum with three g values 1.88, 1.94 and 2.04, characteristic of a rhombic structure, with S = ½.

Though there are two Fe(III) centres present in this Fe–S protein, it transfers only one electron as in rubredoxin, with one Fe(III) centre. It mainly serves as the electron donor system to cyt P-450 monooxygenase, for the hydroxylation of steroids and also to cyt P-450 camphor monooxygenase.

ii. **[4Fe–4S] proteins:** The Fe_4S_4 proteins occur in bacteria, plant and mammals. They are known to act as electron transport system in photosynthesis and in the activity of nitrogenase, hydrogenase and other redox enzymes. They act as one electron transfer proteins and are classified into two types.

The first type are the Fe_4S_4 protein, which have redox potential ~ −600 mV, and are called normal ferredoxin. The second type of proteins have high redox potential +350 mV and is called high potential iron proteins (Hi PIP).

Both the types have a [4Fe–4S] prosthetic group, bound to a low-molecular-weight protein (6,000 daltons). On being treated with a proper solvent and a strongly binding ligand L, 4Fe–4S unit is extracted out.

$$Fe_4S_4 \text{ protein} + 4L^- \longrightarrow [Fe_4S_4L_4]^{2-} + \text{apoprotein} \tag{5.13}$$

The apoprotein loses its activity, which is regenerated on addition of [4Fe–4S] unit.

The X-ray crystallographic studies have established that Fe_4S_4 unit has a cubane structure with four Fe and four S, located at the corners of a cube. Each iron is bound to one cysteine S of the protein. Thus, each iron is at the centre of a tetrahedron, bound with four sulphur, three acid labile inorganic sulphides and one organic sulphur (Figure 5.12).

The thiocubane structure has three possible oxidation states $[Fe_4S_4]^{3+}$, with three Fe(III) and one Fe(II), $[Fe_4S_4]^{2+}$, with two Fe(III) and two Fe(II), and $[Fe_4S_4]^{1+}$, with one Fe(III) and three Fe(II).

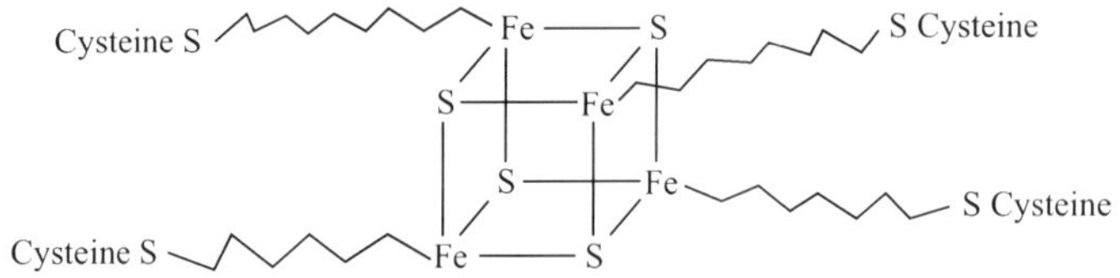

FIGURE 5.12 The cubane 4Fe-4S unit of ferredoxin protein.

The redox interconversion of the three states can be shown as follows:

$$[Fe_4S_4]^{3+} \overset{+e}{\underset{}{\rightleftharpoons}} [Fe_4S_4]^{2+} \overset{+e}{\underset{}{\rightleftharpoons}} [Fe_4S_4]^{1+}$$

$$3Fe(III)\ 1Fe(II) + 350\ m\ V\ 2Fe(III)\ 2Fe(II) - 600\ m\ V\ 1Fe(III)\ 3Fe(II)$$

(5.14)

Hi PIP exists in the $[Fe_4S_4]^{3+}$ oxidation state, with 3Fe(III) and 1Fe(II), and is reduced at high potential to $[Fe_4S_4]^{2+}$, with 2Fe(III) and 2Fe(II). Normal ferredoxin exists in $[Fe_4S_4]^{2+}$ state 2Fe(III) 2Fe(II) and is reduced to $[Fe_4S_4]^+$ at a lower potential.

The native forms of Hi PIP $[Fe_4S_4]^{3+}$ with 3Fe(III) 1Fe(II) or the reduced form of ferredoxin $[Fe_4S_4]^+$ 1Fe(III) 3Fe(II) are paramagnetic corresponding to one unpaired electron. This is because there is anti-ferromagnetic interaction between two Fe(III) and one Fe(III) and one Fe(II) in the first case, $[Fe_4S_4]^{3+}$, whereas in $[Fe_4S_4]^+$, there is an anti-ferromagnetic interaction between two Fe(II) and one Fe(II) and one Fe(III).The coupling between two Fe(III) or two Fe(II) results in complete quenching of paramagnetism, whereas coupling between one Fe(III) and one Fe(II) leads to only one unpaired electron. Both $[Fe_4S_4]^{3+}$ and $[Fe_4S_4]^+$ show EPR signals, with three g values, in the case of $[Fe_4S_4]^{2+}$, corresponding to rhombic structure, and two g values in $[Fe_4S_4]^{3+}$, corresponding to tetragonal distortion.

The reduced form of Hi PIP and native ferredoxin have the same oxidation states of iron in the Fe_4S_4 prosthetic group, that is, 2 Fe(III) and 2 Fe(II). Both have even number of electrons and are diamagnetic due to anti-ferromagnetic coupling between pairs of two Fe(III) and two Fe(II), involving super exchange through bridging sulphides. However, reduced Hi PIP and native ferredoxin differ in their redox behaviour. Ferredoxin can be reduced to $[Fe_4S_4]^+$ at −600 mV, but further reduction of reduced form of Hi PIP $[Fe_4S_4]^{2+}$ to super reduced state $[Fe_4S_4]^+$ (that is, the same as one electron reduced product of ferredoxin) causes its partial breakdown and denaturing.

This difference in the redox behaviour is due to the difference in protein structure. In both the cases, the cysteine S of proteins are hydrogen-bonded to the backbone of the protein. The redox behaviour of the iron centre is controlled by the extent of H bonding. Hi PIP cluster is situated at a deeper position in the protein than ferredoxin. In normal ferredoxin the H bonding between S^{2-} and −NH of protein is stronger, inhibiting oxidation and hence ferredoxin stays in the lower two oxidation states. In Hi PIP, hydrogen bonding is weaker and hence, higher redox couple is stabilised.

It is interesting to note that the oxidation states in $[Fe_4S_4]^{2+}$ are not localised on specific iron ions as in the case of Fe_2S_2 type of ferredoxin. There is delocalisation over the whole Fe_4S_4 cluster. Thus, all four iron ions are equivalent and exhibit the same average oxidation state of +2.5. In the oxidised form $[Fe_4S_4]^{3+}$, that is Hi PIP, there is small non-equivalence of the four centres, but on the whole, they have an average oxidation state of +2.75. The MB spectrum of Fe_4S_4 also corresponds to the average value of the combination of different oxidation states of iron present and thus supports the equivalence of the iron centres.

Model studies

Various model studies have been carried out to understand the structure and the redox properties of Fe_4S_4 proteins. Various reactions of iron and sulphur compounds have been carried out. They lead to the formation of the most stable core Fe_4S_4. Complexes of the type $[Fe_4S_4]\ X_4$, where X = chloride, $^-OC_6H_5$ or $^-S–CH_2–C_6H_5$ have been synthesised. The reduced compounds $[Fe_4S_4]^{n+}$ with n = 2, and 1 have also been synthesised.

It is seen that Fe_4S_4 units in the above complexes have thiocubane structure with deformation. In most $[Fe_4S_4]^{2+}$ compounds, there is tetragonal compression along z axis, so that there are eight longer Fe–S bonds and four shorter Fe–S bonds along z axis.

However, $\{[Fe_4S_4](S\ C_6H_4)_4\}^{3-}$, with reduced $[Fe_4S_4]^+$ core, reveals elongation distortion, with four long Fe–S and eight short Fe–S bonds.

It is observed that in the synthetic models also, the iron centres are equivalent, showing that there is delocalisation of electron between the iron centres, as in the biological systems. However, $[Fe_4S_4]^{n+}$ with n = 4, 3, 2, 1 and 0 could be electrochemically synthesised, in model systems. Thus, Fe_4S_4 with 4Fe(III) and also with 4Fe(II) could be prepared synthetically, whereas, in biological systems, this is not attained. This shows that biochemical systems have specificity of oxidation states, due to the different proteins, to which Fe_4S_4 is bound. There can be non-bonded interaction of the metal centres with the side groups of the proteins, and this affects the chemical properties of the iron centres in the natural systems.

Three Iron-Centred Proteins

These proteins have been discovered recently. The protein of this type was first found in Ferredoxin I of the nitrogen fixing bacteria *Azotobacter vinclandii*, called AvFdI in short. This was followed by the discovery of Fe_3S_4 sites in Fd II protein from beef heart. Presence of Fe_3S_4 is now known in dozens of proteins.

Initial X-ray study of AvFdI suggested that Fe_3S_4 contained a Fe_4S_4 cluster and Fe_2S_2 centre. However, later refinement of X-ray structure showed that it is an almost flat, open six-membered ring, Fe_3S_3, with alternate iron, and S^{2-} and has a twisted boat conformation (Figure 5.13).

However, study of Fe II by ESR and MB techniques and its subsequent X-ray study showed that Fe_3S_4 has a closed thiocubane structure with one iron missing in the cube. Repeated X-ray study of AvFd I also proved the planar structure to be wrong and suggested the closed thiocubane structure. Each iron is bound to the protein through a single cysteinyl thiolate. Thus, each iron is in a tetrahedral environment of sulphur atoms (Figure 5.14).

All three iron centres in Fe_3S_4 are in +3 oxidation state and are equivalent and EPR active. Fe_3S_4 undergoes one electron reduction, and the reduced form has 2Fe(III) and 1Fe(II).

The role of three iron atom clusters is not definitely known. The fact that Fe_3S_4 is easily converted into Fe_4S_4 indicates that the three Fe clusters may be intermediate in the biosynthetic route of the formation of Fe_4S_4. It has been suggested that in the activity of aconitase, the Fe_3S_4 thiocubane combines with Fe(II) to complete the cube, and thus, the active form of the enzyme, that is Fe_4S_4, is formed.

8Fe–8S ferredoxin

These are small proteins with a molecular weight of nearly 6 kDa and have two 4Fe–4S clusters situated at a distance of 12 Å. Each Fe_4S_4 has the bonding same as in 4Fe–4S proteins, and these two units are linked through the protein. The cysteinyl S at 8, 11, 14 and 45 positions of the protein bind with four iron of one Fe_4S_4 and the 18, 35, 38 and 42 cysteinyl residues bind at the other Fe_4S_4.

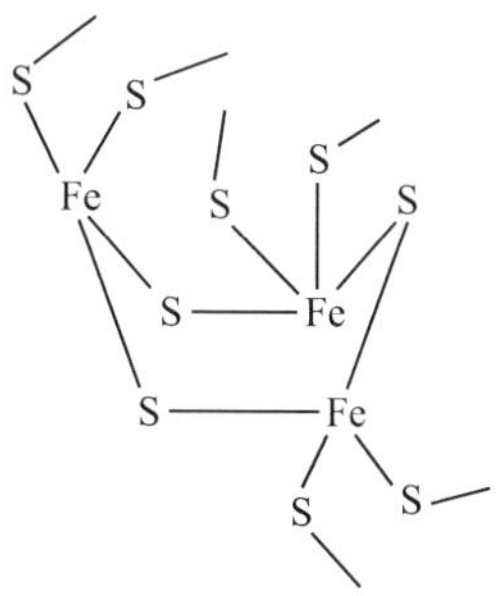

FIGURE 5.13 Twisted boat conformation of Fe_3-S_3 unit in AvFdI.

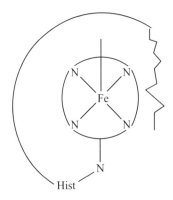

FIGURE 5.14   Closed thio cubane structure with Fe$_3$-S$_4$ unit in AvFdI.

Each Fe$_4$S$_4$ has 2Fe(III) and 2Fe(II) and can undergo one electron reduction to 1Fe(III) and 3Fe(II) as in [4Fe–4S] ferredoxin. Thus, the whole protein acts as two electron acceptor. When one Fe$_4$S$_4$ accepts electron, there is a change in the tertiary and quaternary chains of the protein. This prompts the other Fe$_4$S$_4$ to accept the second electron.

The native protein exhibits low magnetic moment (1.2 BM) per Fe$_4$S$_4$ centre due to incomplete anti-ferromagnetic spin coupling, between Fe(III)–Fe(III) and Fe(II)–Fe(II). On reduction, in the resultant structure, containing 1Fe(III) and 3Fe(II), there is coupling between Fe(II)–Fe(II) and Fe(III)–Fe(II). This leaves one uncoupled electron and hence, there is an increase in magnetic moment, on reduction.

Transport of Oxygen

Haemoglobin and Myoglobin

In human beings and bigger animals, O$_2$ is transported from the lungs to the different parts of the body through the arteries. The transport process is carried out by haemoglobin, present in the RBC. In the muscles, the same function is carried out by myoglobin (from the Greek word 'mus' meaning muscle).

The structure of oxygen carrier molecule should be such that the oxygen molecule is held with sufficient strength, but not too strong to irreversibly oxidise Fe(II) to Fe(III), so that oxygen can be released and delivered at a rapid rate, at the proper site.

Both haemoglobin and myoglobin have a structural unit haem, that is Fe(II) bound with protoporphyrin IX. There is an extensive delocalisation of dp orbital electron of Fe(II) into the π orbital network of the ring, thus making the haem iron site different in electronic properties and reactivity from non-haem environments.

Haem is held within the cleft of the associated protein molecule globin, with 150–160 amino acid residues. There is an extensive hydrophobic interaction between the porphyrin part and globin molecule, besides the covalent interaction between Fe(II) and an imidazole nitrogen of the histidine residue of the protein (Figure 5.15).

FIGURE 5.15   Haem unit in the cleft of globulin part in haemoglobin and myoglobin.

Myoglobin has molecular weight 17,500 daltons and has only one haem centre. Haemoglobin has a molecular weight of 64,450 daltons and is a tetramer, containing four Fe(II) protoporphyrin units, located within hydrophobic pockets in the globin, consisting of four chains, two α (140 amino acid residues) and two β (146 amino acid residues). Thus, the tetramer has two different types of haemoglobin, differing in amino acid composition of the protein part (Figure 5.16).

There is no covalent bonding between four subunits. There is a long-range electrostatic $(COO^- \ldots NH_4^+)$ or H-bonded interaction $(O – H \ldots O^- OC)$ between one a and one b chains. This results in two identical dimers $a_1 b_1$ and $a_2 b_2$. There is also a network of electrostatic and H bonding interaction between these two dimers. Due to these bondings, there is constraint in the structure of haemoglobin. Hence, the overall rate of binding of O_2 with haemoglobin is less than with myoglobin. Consequently, haemoglobin can take up O_2 at higher pressure, a condition available in the lungs (higher animals), gill (fishes) or skin (smaller organisms). Myoglobin can take up O_2 at lower pressure in the muscles.

Deoxyhaemoglobin is penta-coordinated to four protoporphyrin N and one N of the histidyl residue of globin. As the Fe(II) haem is enclosed in a hydrophobic pocket of globin, water molecule or any polar solvent cannot occupy the sixth coordination position. It is free to be occupied by non-polar molecule O_2 or CO. Thus, the hydrophobic globin chain inhibits the displacement of O_2 by water molecule, present in the blood.

Globin chain has another role to play. In its absence, Fe(II) haem is irreversibly oxidised to Fe(III) haem hydroxide, which cannot transport O_2. The reaction proceeds through intermediate formation of m peroxo and subsequently μ^- oxo Fe(III) haem.

$$Fe(II) \text{ haem} + O_2 \longrightarrow \text{haem } Fe(III) – O – O – Fe(III) \text{ haem}$$

$$\longrightarrow \text{haem } Fe(III) – O – Fe(III) \text{ haem} \qquad (5.15)$$

$$\longrightarrow 2Fe(III) \text{ haem} – OH$$

Globin chain in Hb and Mb helps to keep the haem groups apart, haem μ^- oxo Fe(III) haem dimmer cannot be formed and the oxidation of Fe(II) is inhibited. Hence, oxyhaemoglobin remains stable.

$$\text{Globin } Fe(II) \text{ haem} + O_2 \longrightarrow \text{Globin } Fe(II) \text{ haem} – O_2 \qquad (5.16)$$

There is a formation of stable $Fe(II)–O_2$ bond due to s bonding between sp^2 lone pair of O_2 with the sixth d^2sp^3 hybrid orbital of Fe(II). Furthermore, there is an interaction between the filled dp orbitals of Fe(II) (dxy and dyz) with π^* $(p_z–p_z)$ orbital of O_2 (back bonding). Thus, there is synergistic stabilisation of $Fe(II)–O_2$ bond.

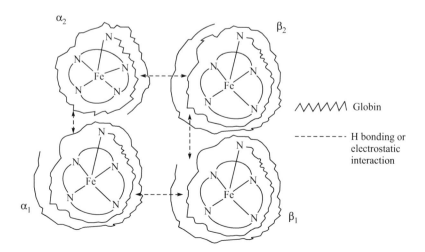

FIGURE 5.16 Haemoglobin tetramer with four Fe(II)protoporphyrin units.

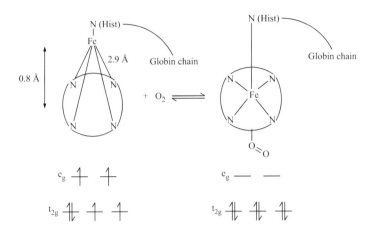

FIGURE 5.17 Change in spin states and position of iron centre in oxy- and deoxy- haemoglobin.

In deoxy Hb or myoglobin, Fe(II) is in high spin state, though bound to strongly coordinating ligands. Perutz and Kendrew carried out the X-ray study of deoxyhaemoglobin and showed that Fe(II) is 0.8 Å above the plane of porphyrin ring nitrogens. Thus, the distance of ligand nitrogens from Fe(II) is 2.9 Å, and hence, the four N atoms create a weak field and Fe(II) is in high spin state (Figure 5.17).

When O_2 gets bound to the Fe(II) at the sixth position, the metal ion is pulled down to the centre of the plane of the macrocyclic ring. As a result, Fe(II)–N distance is reduced to 2 Å, and hence, a strong field is created, and Fe(II) goes to a low spin state.

In high spin state, Fe(II) is bigger in size and cannot be accommodated in the porphyrin ring. In low spin state, Fe(II) is smaller in size and can be accommodated in the ring. Thus, the 16-membered porphyrin ring has very specific discrimination of high and low spin Fe(II), and is, therefore, preferred in haemoglobin, for its efficient functioning.

On binding of O_2 with Fe(II) haem, there is downward movement of Fe(II). In turn, the coordinated histidine also moves towards the porphyrin ring. This brings a conformational change throughout the protein chain, resulting in the rupture of some $COO^-...NH_4^+$ interactions. Thus, the constrained haemoglobin (tensed or T form) becomes relaxed on binding of O_2 with one haem centre, and in the relaxed form (R form), the remaining haem groups become exposed to O_2 and bind with it easily. Thus, there is a cooperative effect between the four units. They take up and liberate O_2 together.

$$Hb_4 + O_2 \rightleftharpoons Hb_4O_2 \quad K_1 = \frac{[Hb_4O_2]}{[Hb_4][O_2]}$$

$$Hb_4O_2 + O_2 \rightleftharpoons Hb_4(O_2)_2 \quad K_2 = \frac{[Hb_4(O_2)_2]}{[Hb_4O_2][O_2]} \quad (5.17)$$

The order of stepwise take-up of oxygen by the four haem units of tetrameric haemoglobin is as follows: $K_4 > K_3 > K_2 > K_1$. This is opposite to the statistical expectation. The opposite order is, because, when one O_2 gets bound to one haem centre, the other haem centres, in the relaxed structure, of the tetramer are more favourably exposed to O_2, and hence, binding of O_2 to the subsequent centres is facilitated, resulting in higher K_n values. Thus, though O_2 uptake by haemoglobin is less at lower pressure, due to its strained (Tense T) structure, the uptake increases quickly with increasing pressure. This phenomenon in which binding of one molecule with protein increases the binding of subsequent molecules with the protein is called homotropic allosteric interaction.

As stated earlier, myoglobin with no constraint in structure is converted to MbO_2 at low oxygen pressure. The nature of uptake of O_2 by haemoglobin and myoglobin at biological pH 6.5, with increasing pressure, is shown in Figure 5.18. This explains why in tissues, where pressure is low, (40 mm) oxyhaemoglobin transfers oxygen to myoglobin easily.

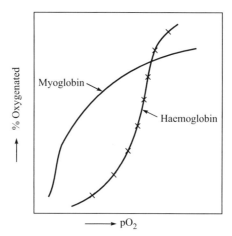

FIGURE 5.18 Oxygen uptake (% oxygenation) as function of O_2 pressure in case of haemo- and myo-globin.

As the oxygenation of haemoglobin involves the breaking of $COO^-\ldots NH_4^+$ bonds, oxygenation is pH dependent. With increasing pH, there is more breaking of bonds and O_2 uptake is favoured, whereas at low pH, the electrostatic salt bridges are strengthened, resulting in lowering of the O_2 uptake Thus, deoxygenation is favoured at low pH. This effect is termed Bohr effect, after the name of the discoverer, Christian Bohr, father of the physicist, Neil Bohr.

The pH at which the O_2 binding affinity is lowest is 6–6.5, and the working tissue provides this pH. The tissues consume O_2 producing CO_2 and H_2O, by oxidation of glucose.

CO_2 dissolves in water to form carbonic acid, assisted by the enzyme carbonic anhydrase. Furthermore, when the supply of oxygen goes down, glucose undergoes incomplete oxidation to lactic acid. This lowers the pH and this leads to further release of oxygen from oxyhaemoglobin.

Besides CO_2 and proteins, 2,3-diphophoglycerate (2,3-DPG), present in RBC, also controls the affinity of haemoglobin for O_2. Foetal haemoglobin binds less strongly to 2,3-DPG, compared to adult haemoglobin, and hence has higher affinity for O_2. This facilitates transfer of O_2 from mother's oxyhaemoglobin to the haemoglobin of the foetus, across the placenta.

As stated earlier, Fe(II) is in low spin state in oxyhaemoglobin and it is diamagnetic, though molecular oxygen is known to be paramagnetic. To explain this, two different modes of binding of O_2 molecule have been suggested.

Pauling's model: O_2 is bound to low spin Fe(II) in a bent manner (Figure 5.19a) from one oxygen atom. On coordination, the triplet O_2 is reduced to singlet state. Thus, O_2 is in excited state. The Fe–O_2 bond is considered to be stabilised due to H bonding between the O_2 molecule and the globin imidazole NH group, coordinated to Fe(II) in the axial direction.

Griffith's model: O_2 is bound to Fe(II) from both the oxygen atoms as ethylene (Figure 5.19b). Fe(II) is in low spin state, and O_2 molecule is in the excited singlet state. However, this "side on" geometry of O_2 has been recently ruled out for O_2, as this should lead to the formation of Fe(III) superoxo- or Fe(IV) peroxo structure.

Both the structures suggested by Pauling and Griffith consider $O_2 \rightarrow Fe$ σ bond and $Fe \rightarrow O_2$ π bond formation.

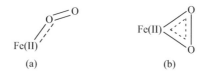

(a) (b)

FIGURE 5.19 Suggested oxygen binding modes in haemoglobin (a) Pauling's model (b) Griffith's model.

A third model was suggested by Weiss. It considers that O_2 binds with iron in a bent manner, as suggested by Pauling. However, it further considers that one electron from Fe(II) is transferred to O_2 forming super oxide O_2^-. Fe(III) formed is in low spin state. The unpaired spins over Fe(III) and O_2^- get paired up, resulting in diamagnetism.

Experimental observations support first two models. IR spectrum of oxygen in oxy haemoglobin corresponds to O_2 molecule and not to O_2^-.

However, MB spectrum of oxyhaemoglobin corresponds to Fe(III) and supports Weiss structure. Though, this can also be explained by Pauling's model. As there is Fe $\rightarrow$ O_2 dπ–pπ interaction, the electron density around Fe(II) is reduced, and MB spectrum looks similar to that of Fe(III).

The bent coordination of O_2 molecule to Fe(II) haem was finally established by the low temperature crystallographic study of oxy myoglobin. The bent coordination is preferred by O_2 because of sp^2 hybridisation with the lone pair, and the double bond being at an angle of 120°. Nature has provided the structure in the hydrophobic pocket of haemoglobin, such that O_2, coordinating in a bent manner is preferred. Carbonmonooxide, coordinating in straight fashion, due to sp hybridisation is not selected by the cavity.

However, CO still binds more strongly (250 folds) with haemoglobin than O_2, because of its strong π acidic character. The formation of carboxy haemoglobin (HbCO) prevents O_2 transport and hence has toxic effects. The lethal concentration of CO is 750–1,000 ppm, when 60%–65% of carboxy haemoglobin formation takes place. Lower concentration of CO above 10 ppm causes headache, dizziness and loss of consciousness.

Cyanide, a π acidic ligand, is toxic to all forms of life. The lethal dose is very small, 0.1–1 ppm. As very small amount causes death, it cannot be due to cyanohaemoglobin formation. It has been suggested that the toxic effect of cyanide is because of its binding with cytochrome or cytochrome c oxidase, resulting in the inhibition of the electron transfer process in terminal respiration, as discussed earlier under cytochromes.

It is possible to treat cyanide poisoning by administration of $NaNO_2$ or amyl nitrite, which oxidises Fe(II) of haem to Fe(III) haem. As Fe(III) of oxidised haem has a stronger affinity for cyanide, it pulls out the CN$^-$ over cytochrome or cytochrome c oxidase, thus making them free of cyanide.

Alternatively, the CN$^-$ over cytochrome or cytochrome c oxidase can be converted to SCN$^-$ by treatment with $S_2O_3^{2-}$. SCN$^-$ being weaker ligand is less toxic. In biochemical systems same function, of converting CN$^-$ to SCN$^-$, is carried out by the enzyme rhodonase.

Another natural way, of detoxification of cyanide, is the formation of cyanocobalamin, (vitamin B_{12} coenzyme) using cyanide bound to cytochromes.

Synthesis of Haem

In the synthesis of haem in biological systems, first there is condensation of glycine with succinyl CoA in the presence of the enzyme 5-amino levulinic acid synthetase, in the mitochondria. The condensation reaction results in the formation of δ amino levulinic acid (ALA). The acid then moves from mitochondria to the cytosol, where the condensation of two ALA molecules takes place, assisted by the enzyme called ALA dehydratase (also known as porphobilinogen synthetase), leading to the formation of porphobilinogen. This molecule has a pyrrole ring. Four porphobilinogen molecules undergo head-to-tail condensation to form a molecule with four pyrrole rings, called hydroxymethylbilane. The condensation reaction is catalysed by an enzyme porphobilinogen deaminase (PBG deaminase). The linear molecule so formed undergoes cyclisation to form a cyclic compound called uroporphyrinogen. This reaction can take place without enzyme but is catalysed in the presence of an active catalyst, which is made up of uroporphyrinogen synthetase and a coenzyme. These compounds undergo decarboxylation, losing CO_2 molecule to form coporphyrinogen III. The reaction is catalysed by the enzyme uroporphyrinogen decarboxylase.

Coprotoporphyrinogen III is transported back to the mitochondria, and here, it loses two CO_2 molecules out of the propionate residues, catalysed by the enzyme coprotoporphyrinase, and results in the formation of protoporphyrinogen IX. This is further oxidised, catalysed by the enzyme protoporphyrinogen oxidase to form the macrocyclic ring protoporphyrin IX. This is a ring with alternate double and single bond; hence, there is complete delocalisation of electrons in the π bonds. The molecular orbital diagram shows that due to hyperconjugation, the bonding and anti-bonding π orbitals are close in energy. The $\pi \rightarrow \pi^*$ is a low-energy transition and results in deep red colour.

Fe(II) entering in the biochemical system through intestinal absorption, or formed by the breakdown of transferrins or haemoglobin, is oxidised to Fe(III) and is taken up by the apotransferrin to reform transferrin. The Fe(III), in transferrin, gets reduced to Fe(II) and is liberated (see Figure 5.36 and the related text). This gets bound to the protoporphyrin IX, present in the reticulates, catalysed by the enzyme ferrochelatase, and forms haem b, after liberation of $2H^+$. The haems are of three types – a, b and c, with small difference in the structure of the protoporphyrin in the case of haem a and haem c (see Figure 5.1).

The haem b gets bound to the protein globin to form monomeric myoglobin and tetrameric haemoglobin. Other forms of haem form other enzymes, cytochrome P-450, catalase and peroxidase and cytochrome c, discussed earlier in this chapter.

In the case of poisoning due to intake of large excess of heavy metal, this metal can get bound to the enzymes and inhibit their activity in the different stages of synthesis of protoporphyrin and haem. Synthesis of haem is thereby retarded.

Another factor inhibiting the formation of haem is the presence of Zn(II) in the biochemical system. This being of a suitable size to bind with protoporphyrin combines with it and does not allow Fe(II) to form haem and hence haemoglobin. This leads to inadequacy of haemoglobin level and thus anaemic condition.

In both the above cases of toxic metal interference, excess free Fe^{3+} is left in the system, and this gets deposited in the mitochondria in the form of siderosis.

An outline of the synthesis of haemoglobin is shown in Figure 5.20a.

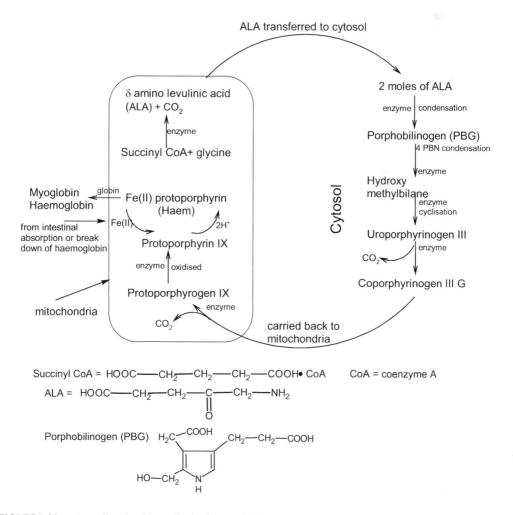

FIGURE 5.20a An outline showing synthesis of haemoglobin.

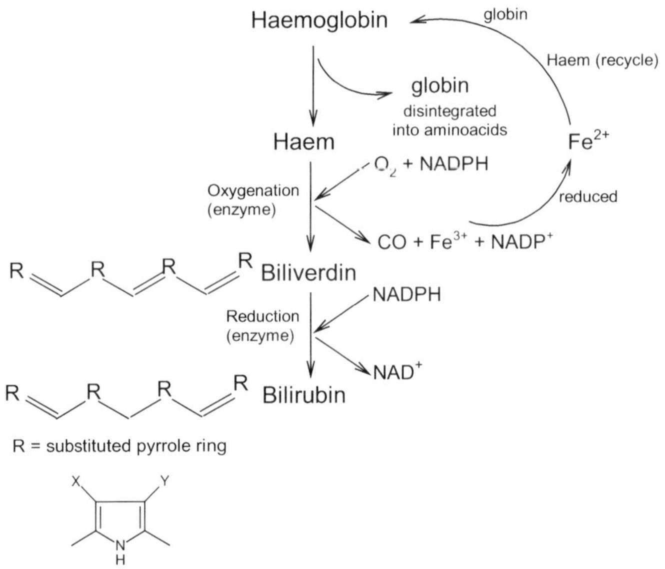

FIGURE 5.20b   Haemoglobin breakdown.

Breakdown of Haemoglobin

After carrying out the cycle of oxygen transport for nearly three months, haemoglobin undergoes disintegration. The protein globin first gets dissociated and breaks down into the component amino acids. In the separated haem, the porphyrin part is hydrophobic. Hence, it is converted to the water-soluble form by oxidation with O_2, associated with reaction with NADPH. This oxidative disintegration of haem leads to the formation of Fe^{3+}, CO, $NADP^+$ and H_2O. This oxidative reaction is catalysed by the enzyme haem oxygenase. The disintegration proceeds in steps. First, the ring compound porphyrin breaks down to the open-chain compound biliverdin, which is further reduced to bilirubin by NADPH $\rightarrow$ $NADP^+$ reaction, assisted by the enzyme biliverdin reductase. The Fe^{2+} of haem is oxidised to Fe^{3+}, and because of its weak binding with the open-chain compound, bilirubin Fe^{3+} is liberated. It is further recycled for the synthesis of haemoglobin (Figure 5.36). It can be seen that this reaction is exceptional in the fact that in this, CO is liberated, which is known to be poisonous in biochemistry. The liberated CO is expelled out of the lungs.

In the first stage of disintegration, the liberated compound biliverdin is unsaturated with long-range hyper conjugation. Hence, there is a possibility of low-energy $\pi \rightarrow \pi^*$ transition, and this results in blue green colour in the compound. In the next step, there is reduction of the middle bridging –CH of biliverdin, resulting in the less unsaturated compound bilirubin with consequent less π electron delocalisation. Hence, $\pi \rightarrow \pi^*$ transition occurs at higher energy, and the colour of bilirubin is yellow red. Bilirubin is carried to the liver along with albumin and gets disintegrated there.

In the case of liver infection or other factors lowering the efficient activity of the liver, the bilirubin disintegration is retarded or completely inhibited. Hence, bilirubin appears in the blood and urine, showing up a yellow colour on the skin and in the urine. This medical condition is called hepatitis or jaundice.

An outline of the breakdown of haemoglobin is shown in Figure 5.20b.

Synthetic O_2 Carriers as Models

Synthetic Co(III) Complexes

Salen complex of Co(II) (Figure 5.21) is known to take up O_2 at higher pressure, forming μ-peroxo-Co(III) binuclear complex. On releasing the pressure, O_2 is liberated, and Co(II)salen is reformed. The complex was used in the Second World War to separate O_2 from air, for use in oxygen cylinders in hospitals.

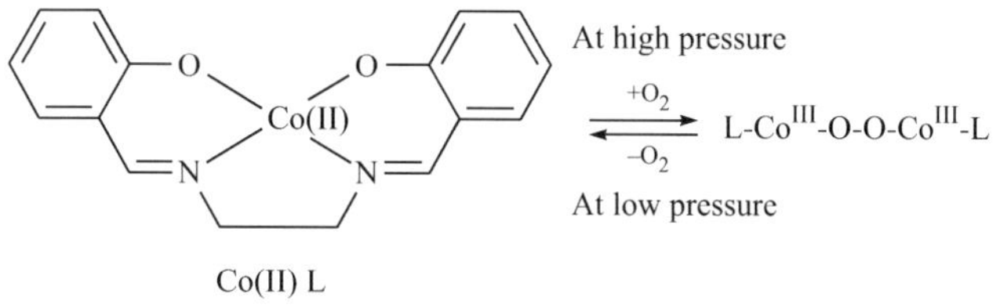

FIGURE 5.21 Co(II)salen.

Similar synthetic Fe(II) complexes, of salen, do not normally function as oxygen carriers, due to the irreversible oxidation to Fe(III)-μ-oxo-dimer.

$$L\ Fe(III) - O - Fe(III)\ L$$

As discussed earlier, the efficiency of haemoglobin to transport O_2 is due to the hydrophobic environment around Fe(II). Hence, the model Fe(II) haem complexes, synthesised, are such that the protoporphyrin ring has side groups, which provide hydrophobic shielding around Fe(II). They inhibit the approach of hydrophilic groups at the sixth coordination position of Fe(II) and also mitigate the close vicinity of the complex molecules and thus inhibit the formation of the dimeric Fe(III)-μ-oxo- complex.

Some of the complexes synthesised are as follows:

i. Chiang and Taylor synthesised an Fe(II) porphyrin complex with a side amide chain, on the ring, which coordinates with Fe(II) through imidazole nitrogen at the fifth position (Figure 5.22)

   It binds O_2 reversibly in dichloromethane at −15°C. The efficiency decreases with time, due to irreversible oxidation to Fe(III)-μ-oxo, as number of cycles increase.

ii. Another Fe(II) N_4 complex, of a macro cyclic ligand, with four phenyl groups attached on the macrocyclic ring, at a distance from Fe(II) centre, has been synthesised (Figure 5.23). It is found to bind with O_2 at low temperature.

iii. **Picket fence model:** Colman synthesised a Fe(II) tetraphenylporphyrin-type complex with trimethyl acetamide chain on each phenyl ring, protruding upwards from the plane of the porphyrin ring (Fig. 5.24).

   This was called picket fence model. The fifth coordination position of Fe(II) is occupied by the added base *N*-methyl imidazole, mimicking the globin coordination. The presence of the bulky trimethyl groups creates a hydrophobic environment around Fe(II) and does not allow the formation of μ-oxo-dimer. This complex is the best model and undergoes reversible oxygenation at room temperature.

iv. Another complex of Fe(II) in which the porphyrin has side groups has been synthesised, such that it looks capped (Figure 5.25). The hydrophobic cavity formed is large enough to incorporate O_2, but excludes the polar solvent molecules.

v. The complex Fe(II) haem, with two ethyl propionate side groups in the ring and one N⁻, (2-phenyl ethyl) imidazole at the fifth coordination position, has been synthesised (Figure 5.26). It transports O_2 at low temperature.

   Another technique used to avoid dimerisation of haem into μ-oxo-Fe(III) complex was to immobilise the Fe(II) haem complex on a solid surface. Using this method, Fe(II) tetraphenyl porphyrin complex, with imidazole, supported on silica, occupying the fifth position has been synthesised (Figure 5.27). As the complex molecules are at a distance and thus separated on the support, the possibility of μ-oxo dimer formation is lowered, and the complex can hold molecular oxygen.

   However, none of the model complexes could be tried as a substitute for haemoglobin. Instead, it has been shown that perfluorodecaline, perflurotripropyl amine emulsion in polyoxy ethylene polyoxy propylene polymer surfactant, with no features of haemoglobin, transports oxygen efficiently. However, it lacks properties of blood, like clotting, immunodefence mechanism, enzymatic activity and maintenance of pH.

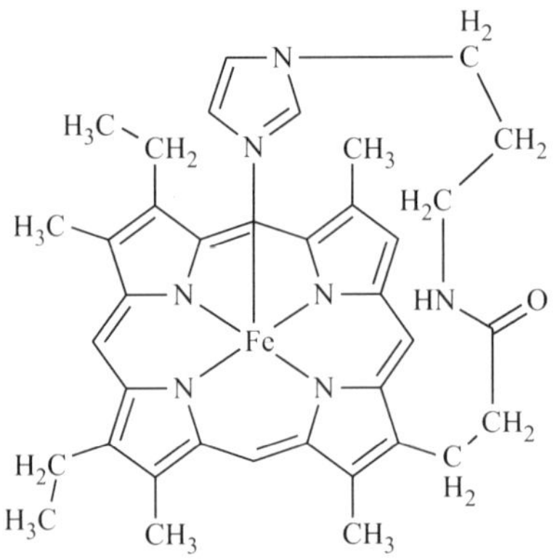

FIGURE 5.22   Synthetic Fe(II)porphyrin with amide side chain having imidazole coordinating at 5th position.

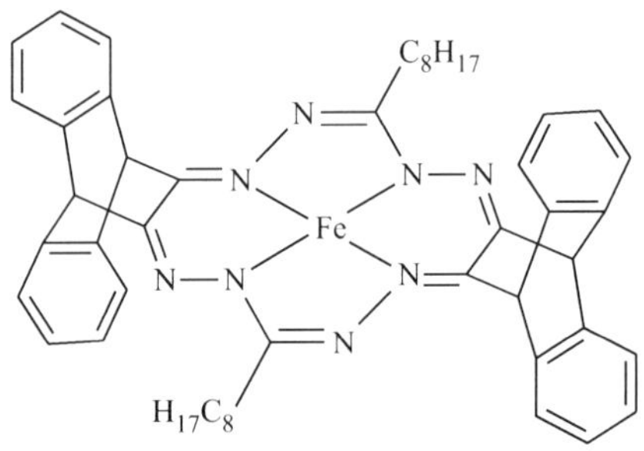

FIGURE 5.23   Synthetic macrocyclic Fe(II)N$_4$ complex.

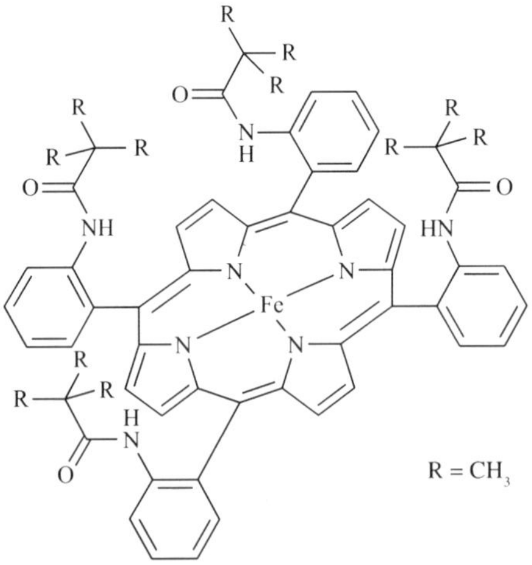

FIGURE 5.24   Picket Fence model of Fe(II)porphyrin type synthetic complex.

Hemerythrin

Though the name of this oxygen carrier involves the term "haem", it is a non-haem iron(II) protein, which transports O$_2$ in lower organisms. It does not occur in vertebrates, but is found in worms such as annelids, molluscs and arthropods.

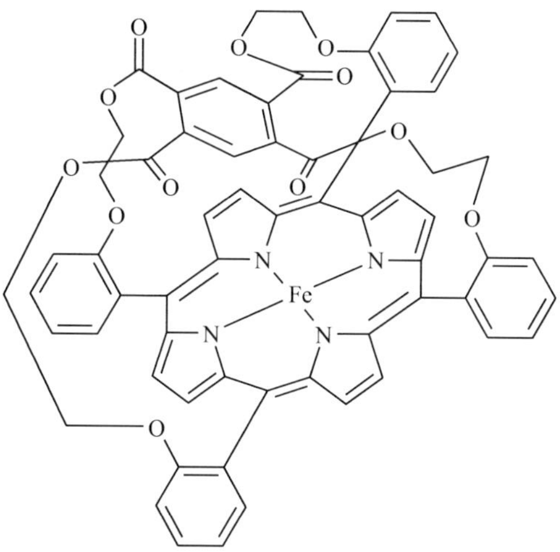

FIGURE 5.25   Capped model of synthetic Fe(II)porphyrin.

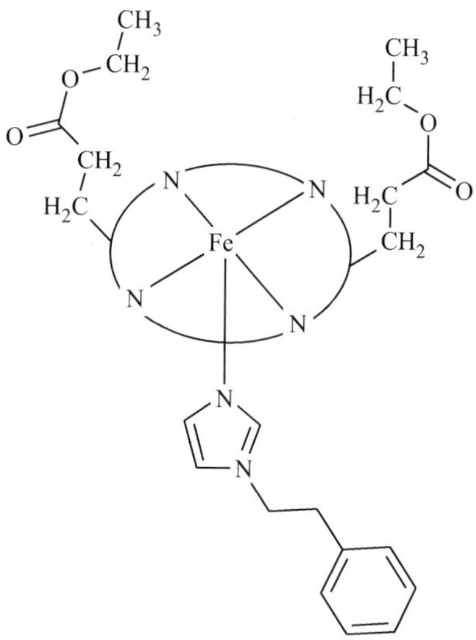

FIGURE 5.26   Synthetic complex Fe(II)haem with two ethyl propionate side groups in the ring.

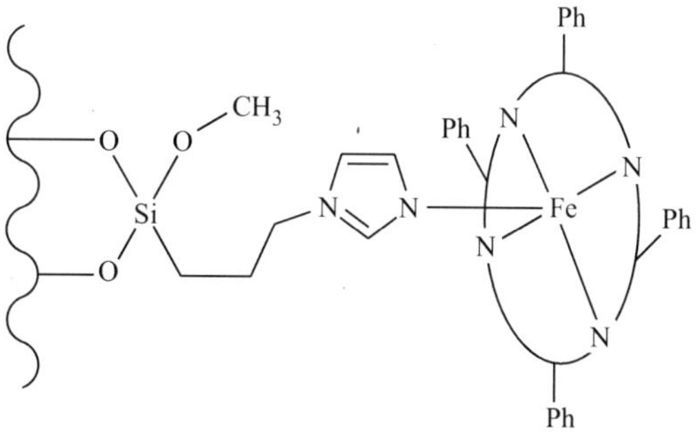

FIGURE 5.27   Fe(II)haem complex immobilised on solid surface.

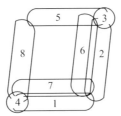

FIGURE 5.28 Eight subunits of hemerythrin Fe(II)protein isolated from isolated from the blood of Golfingia Gouldii.

Hemerythrin exists in monomeric (myohemerythrin), trimeric and octomeric forms. Hemerythrin, isolated from the blood of Golfingia Gouldii (a marine worm), has eight subunits, arranged in a symmetrical way, with four fold symmetry (Figure 5.28).

In deoxyhemerythrin, each subunit has two Fe(II) ions, bound to protein of molecular weight 135,000 daltons, containing 113 amino acid residues. One Fe(II) is bound to three histidine N of the protein, and the other Fe(II) is bound to the N of two histidine residues. Two carboxylate groups of the glutamate and aspartate residues of the protein form bridges between the two iron(II). The two Fe(II) are also bridged through a –OH group. Thus, one Fe(II) centre is hexa-coordinated, and another is penta-N-coordinated, as shown in Fig. 5.29a.

The Fe(II) centres are in high spin state, as revealed by doublet in MB spectrum. There is a weak anti-ferromagnetic interaction between the two high spin Fe(II) centres, by super exchange through the bridging –OH group, leading to lowering in magnetic moment, than expected from the presence of four unpaired electrons, on each Fe(II) centre.

The presence of –OH bridge is supported by the observation in the IR Spectrum of $v_{Fe-O-Fe}$ band at 486 cm^{-1} and its lowering by 10 cm^{-1}, when bridged –OH is exchanged by labelled $^{18}OH_2$.

Hemerythrin differs from haemoglobin in following ways:

i. It has no haem prosthetic group.

ii. Two Fe(II) are simultaneously involved in oxygen transport.

iii. Two Fe(II) combine with one O_2 molecule getting oxidised to Fe(III), and O_2 is reduced to peroxide form. Thus, unlike haemoglobin or myoglobin which binds with neutral O_2 molecule, retaining iron in Fe(II) form in hemerythrin binding of O_2 is associated with redox process.

iv. Though there are eight subunits in hemerythrin, in the protein extracted from the organisms, cooperative effect between subunits is minimum. However, higher cooperative effect (n~2.5, that is, up to 2.5 molecule of O_2 is taken up by the eight Fe(II) centres at a time) is seen in the hemerythrin of the coelomic cells of the worms, between the inner membrane lining of the digestive tract and the outer membrane.

v. There is no biologically controlled pH dependence of O_2 uptake and release (Bohr effect).

However, the similarity between hemerythrin and haemoglobin is that the peroxide ion O_2^{2-} formed in the oxyhemerythrin is not irreversibly bound to Fe(III), but is transferred at the required site as O_2 and Fe(III) comes back to Fe(II).

Four possible models of binding of O_2 with two Fe(III) have been suggested, as shown in Figure 5.29b.

Resonance Raman study shows v_{O-O} at 844 cm^{-1}, corresponding to O_2^{2-}, and $O_2^{2-} \rightarrow$ Fe(III) charge transfer transition is observed at 500 nm. These observations are in agreement with all the four structures.

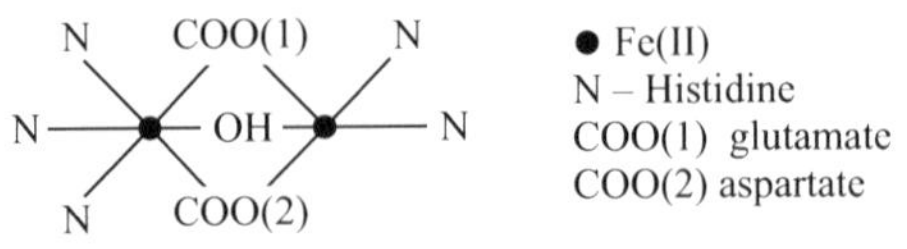

$$
\begin{array}{ccc}
N & COO(1) & N \\
 & \diagdown | \diagup & \\
N\!\!-\!\!\!\!\!\times\!\!\!-OH\!\!-\!\!\!\!\!\times\!\!\!-N \\
 & \diagup | \diagdown & \\
N & COO(2) & N
\end{array}
$$

● Fe(II)
N – Histidine
COO(1) glutamate
COO(2) aspartate

FIGURE 5.29a Dimeric Fe(II) centre in deoxy-haemerythrin.

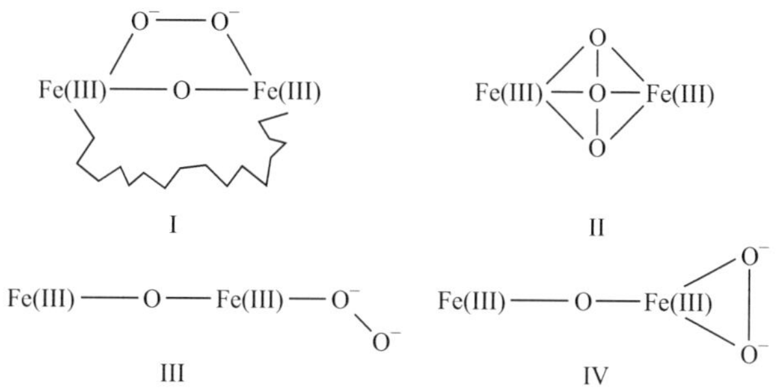

FIGURE 5.29b Possible models of Fe(III)-O_2 binding in hemerythrin.

However, ν_{O-O} in oxyhemerythrin is 100 cm^{-1} higher than that in oxyhemocyanin, where O_2^{2-} is typically bound to two Cu(II). This shows that in oxyhemerythrin, O_2^{2-} is bound to only one Fe(III) centre. Furthermore, MB spectrum shows two singlets, indicating that the two iron centres are high spin Fe(III), but are not equivalent.

Labelled O_2 ($^{16}O-^{18}O$) was used to confirm the structure. Two symmetric O–O stretches were observed in the resonance Raman spectrum. This is in agreement with structure III with monodentate O_2^{2-} coordination, because ν_{O-O} should have different values, for coordination of the monodentate O_2^{2-} with the Fe(III) from ^{16}O or ^{18}O ends. If O_2^{2-} is bidentate, as in structure IV, both ^{16}O and ^{18}O would be coordinated simultaneously to two Fe(III) and only one ν_{O-O} should be observed. The structure III has finally been confirmed by high-resolution X-ray crystallography.

In the formation of oxyhemerythrin, the O_2 molecule binds with the penta-coordinate Fe(II) centre of hemerythrin. Two electrons are transferred from the two Fe(II) centres. Thus, the binuclear Fe(III) site is generated with the peroxide formed being bound with one Fe(III) centre. The proton of the OH bridge of hemerythrin is transferred to the coordinated peroxo forming O–OH, and thus, –OH bridge is converted to μ-oxo bridge.

Though the MB study shows that the two iron (III) centres are high spin in oxyhemerythrin, it exhibits paramagnetism, much less than expected. At 1.4–4.2 K, it becomes diamagnetic. This shows that there is a significant amount of anti-ferromagnetic spin coupling, through super exchange between the two metal ion centres. This supports the existence of O^{2-} bridge and strong super exchange through it.

Hemerythrin is colourless, but oxyhemerythrin exhibits purple violet colour. It shows a spectral band at 700 nm, assignable to d–d transition. However, the intensity of the band is much higher than expected for the spin and orbitally forbidden d–d transition in Fe(III) complex. Hence, this can be attributed to CT transition, between the coordinated O_2^{2-} and the metal ion. Alternatively, it can be suggested that spin coupling between two Fe(III) centres in oxyhaemerythrin results in different spin states (S = 5, 4, 3, 2, 1, 0). In spin states S = 4, 3, 2, 1, spin allowed d–d transition is possible. Such spin allowed transitions should be of higher intensity. For example, the spin allowed transition in the case of state with S = 4 is shown in Figure 5.30.

The electron transition can be better explained in terms of molecular orbital theory. The ground state electronic configuration, with one electron paired in bonding molecular orbital, is shown in Figure 5.31a. The one electron transition from bonding to anti-bonding molecular orbital (abmo) $b^6a^4 \rightarrow b^5a^5$ (Fig. 5.31b) is spin allowed and is of higher intensity.

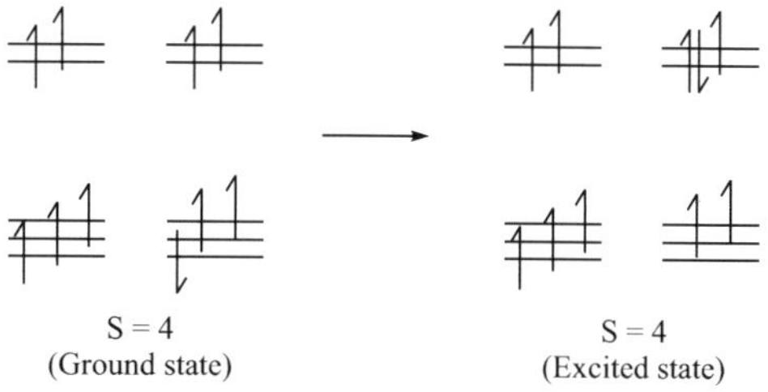

FIGURE 5.30 Spin allowed transitions in the case of state with S = 4, imparting colour to oxy-haemerythrin.

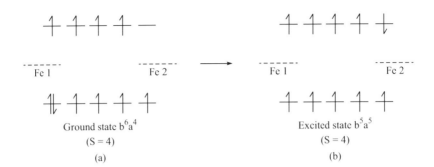

FIGURE 5.31   Electronic transitions in terms of molecular orbital theory for oxy-haemerythrin. (a) ground state b^6a^4 and (b) excited state b^5a^5.

Iron Storage

Iron changes easily from reduced Fe^{2+} to oxidised Fe^{3+} form. This leads to oxidative stress and can cause damage to cell tissues, causing diseases. Hence, excess iron is stored mainly as ferritin. However, Fe^{3+} is also stored in some cells as chelates of ligand with small molecular weight, like citrates, and smaller peptides, ATP and ADP. This forms a pool of iron chelates. This Fe(III) can also cause oxidative damage to cells. However, this is avoided by the conversion of the labile iron pool into ferritin. Deposition of excess ferritin can have harmful peroxidation effect, and so, a balance has to be maintained between labile iron pool and ferritin deposit.

Ferritin

Ferritin is an iron storage protein in mammals, plants, fungi and bacteria. In mammals, iron is stored in bone marrow, liver or spleen in the form of ferritin. It is a precursor of other forms of iron storage in living systems, such as hemosiderin, an iron protein in animals, magnetite (Fe_3O_4), and an inorganic oxide form of iron storage, found in magnetic bacteria, bees and humming pigeons. Ferritin serves as a temporary store of iron to prevent the toxic build-up of iron concentration in human body and also functions as a long-term mobilised iron reserve and can be drawn when needed.

The structure of ferritin consists of two parts, an inorganic iron core, inside a protein coat.

1. **Structure of the protein coat:** The protein consists of 24 subunits. Each subunit is made up of about 175 amino acids, and the subunits vary in molecular weight from 18 to 24 kDa. The exact sequence of amino acid in the protein is not known. However, very similar amino acid sequence is found in ferritin, obtained from animals or plants. Each protein subunit is folded in the form of an ellipsoid. The ellipsoidal 24 subunits are packed in cubic symmetry with fourfold, threefold and twofold axes. Inside the symmetrical protein coat, there is a hollow

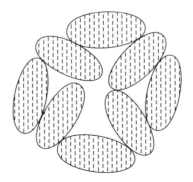

FIGURE 5.32   Top view of ferritin protein coat.

sphere of 100 Å diameter (Figure 5.32). The channels which have fourfold axis are lined with hydrophobic leucine residues, whereas channels with threefold axis are lined with hydrophilic aspartate, glutamate and serine residues.

The channels within the protein coat allow movement of iron in and out of the protein. The channels may be filled with varying amounts of solvent and iron. Thus, the iron content in ferritin can be very low or as high as 4,300 atoms per molecule.

2. **Structure of the iron core:** The gross composition of the iron core, present inside the channels of the ellipsoidals protein, is oxyhydroxy phosphate $FeO(PO_3H_2) \cdot 8Fe(OH)_3$. X-Ray study has shown that iron is present as the mineral ferrihydrate $5Fe_2O_3 \cdot 6H_2O$. The role of phosphate is not definitely known. It is, probably, involved in attaching the ferric oxide hydroxides units to each other and to the protein coat. The iron cores are poorly crystalline and are irregularly arranged within the protein. Only a small fraction of Fe(III) binds directly to the protein.

Ferritin, obtained from different sources, exhibits differences in the structure of the iron core, due to different anions, like phosphate or sulphate, being associated with the iron core. Anions affect the solvent composition and the properties of the protein coat.

The iron atoms in the core may be in the tetrahedral $[Fe(III)O_4]$ or octahedral $[Fe(III)O_6]$ form. The first support for octahedral structure was the observation of the low-energy band in the electronic spectrum, corresponding to the $6A_{1(s)} \rightarrow 4T_{1(G)}$ transition. This spin and orbitally forbidden transition is of lower energy in Oh complex, compared to the Td complex, as shown in Figure 5.33. Normally, the band in Oh complex is expected to be of higher energy, compared to analogous tetrahedral complex. This anomaly is because the transition is between the split up states of two different free ion terms.

The octahedral coordination has now been confirmed by MB spectral and X-ray structural studies. It has been suggested that there are hexagonal closed packed layers of oxygen with octahedrally coordinated Fe(III) (Figure 5.34). The average magnetic moment is lower than expected for the Fe(III) centres with S = 5/2. This is due to anti-ferromagnetic coupling between the Fe(III) centres by super exchange, through the bridging oxides. The amount of Fe(III) in the core varies, because all the layers may be irregularly incomplete. The ends of the layer are linked through phosphate.

FIGURE 5.33 Tanabe Sugano diagram for d^5 high spin case.

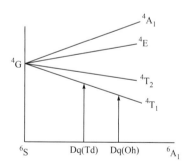

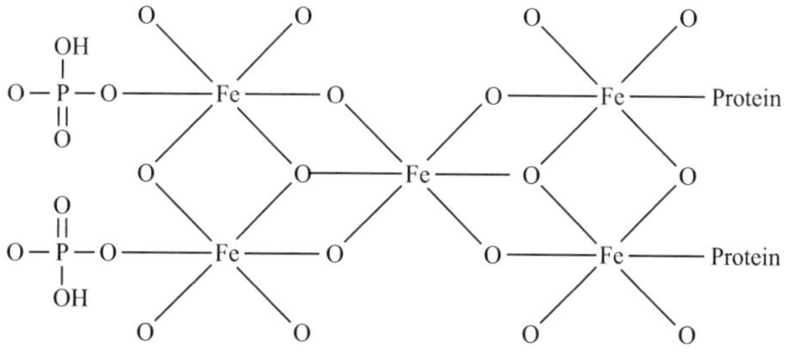

FIGURE 5.34 Octahedral coordination geometry of iron atoms in ferritin core.

Formation of Iron Core

Iron obtained from food or breakdown of haemoglobin enters the mucosal cell, in the form of Fe(II), probably, because Fe(III) is very stable and also Fe(II) compounds are kinetically labile. Fe(II) enters the shell of the apoprotein through the channels that have three-fold symmetry with hydrophilic lining. The nucleation process of the iron core occurs at the interior surface of the protein, where it gets bound to the carboxylate group of the glu or Asp residue of the protein. There is oxidation of Fe(II) to Fe(III), with the formation of ferric oxide. Subsequent hydrolysis of iron oxide by water and uptake of orthophosphate leads to the formation of big ferrihydrite phosphate, trapped within the protein. Further oxidation and hydrolysis at the ferrihydrite surface leads to iron core expansion. The formation of the core is prevented on the esterification of the carboxylate residues of the proteins, emphasising their essentiality for the oxidation of $Fe^{2+} \rightarrow Fe^{3+}$ and the nucleation of the ferrihydrite core.

Availability of iron regulates the synthesis of ferritin. Large amount of ferritin is synthesised, if there is excess of iron, and small undetectable amount is formed, if there is deficiency of the iron. The apoferritin of the mucous cells of the intestine binds with excess iron, and it is excreted into the intestine along with the exfoliated mucous cells. Thus, the body maintains a balance of iron and does not accept excess iron. Extra dose of iron is required, only if there is a loss of iron due to cut or haemmorage or during pregnancy. The occurrence of ferritin in human placenta suggests that there is release of iron from placental ferritin to regulate iron transport from the maternal plasma to the foetal plasma, for supplying iron for the biosynthesis of haemoglobin in the foetus.

Ferritin breaks down to release Fe(III), which is carried by transferrin and ultimately used, for the synthesis of haemoglobin and other haem proteins. For the release of iron from ferritin, Fe(III) has to be reduced to Fe(II). In model systems, reduction is achieved by ascorbic acid or cysteine and the release of Fe(II) is observed to be slower. It is much faster in the biological systems, where the reduction is brought about by the enzyme NADH or $FADH_2$. Probable mechanism of the reaction is that NADH reduces FMN to $FMNH_2$. This enters the protein coat through the hydrophobic channels of fourfold symmetry and reduces Fe(III) to Fe(II). The exact mechanism of the reduction reaction is not known.

Recently, it has been observed that ferritin may also have role other than iron storage. Ferritin from bacterial sources has type B haem bound on the surface of the ellipsoidal subunits and may mediate in electron transfer reaction.

Hemosiderin

This is another form of iron protein, which stores iron in human beings and animals.

Hemosiderin has structure similar to ferritin and can be considered as its other physical form. The iron is in Fe(III) form, and its content varies from 26% to 34%, depending on the availability of iron. The rest of the hemosiderin has the apoprotein (35%), octa-substituted porphyrin, mucopolysaccharide and fatty acid esters. The difference with ferritin is that the Fe(III) complex of the apoprotein in hemosiderin is insoluble.

Hemosiderin occurs in the liver and spleen in the form of granules. Whenever iron is required for the synthesis of haem proteins, it is released from hemosiderin, after reduction to Fe(II), as in the case of ferritin.

In the case of pernicious anaemia, where there is quick destruction of haemoglobin, releasing free iron, or if there is increased absorption of iron in the body, there is a formation of excess hemosiderin. The excess deposit of hemosiderin granules in the liver leads to the diseases like hemosiderosis or hemochromatosis.

Iron Transport

Under conditions of neutral or slightly alkaline pH, as in hemin system, iron exists in Fe^{3+} state. However, Fe^{3+} forms large complexes with anions, water and peroxide. They are less soluble, and their aggregation is harmful for health. Besides this, iron can combine with various macromolecules and change their structure and function. Hence, Fe^{3+} does not exist in free form, when transported, but remains bound to protein.

1. **Transferrin:** It is an iron (III) protein in the blood serum of animals of phylum chordata. It transports iron for haemoglobin synthesis. Only a small fraction of iron in the body is in transit at a time. Iron transfer protein is also present in the egg, called ovatransferrin and, in milk, called lactotransferrin. Human transferrin has molecular weight 80 K daltons, ovatransferrin 77 K and lactotransferrin 80 K daltons. Though not definitely known, this protein may have a role in sequestering the excess iron in the body.

Transferrins are iron glycoproteins. Human transferrin has 6% carbohydrate, bound to the protein. The carbohydrate affects the conformation of the protein and the recognition by it. The crystal structure shows that apotransferrin has ellipsoid structure, with two lobes. Each lobe gets bound to one Fe(III). The two Fe(III) centres in the protein have very similar coordination sites, but are distinguishable by kinetic studies. In both the lobes, the metal ion is very strongly bound to the protein, the stability constant K being 10^{31}. Titration of the apotransferrin shows that at low pH, three protons are liberated per molecule, corresponding to tyrosine OH, and one proton is liberated at pH 7, indicating the presence of histidine imidazole proton. This supports that Fe(III) is bound to three tyrosines and one histidine of the apoprotein.

Each Fe(III) is also associated with two carbonate or bicarbonate anions. There is a synergic uptake of the anion. X-Ray study has further revealed that each Fe(III) is also coordinated to the aspartate group of the protein.

A typical model for the iron site of transferrin is the complex of Fe(III) with the ligand *N,N'*-ethylene-*bis*-(2-hydroxyphenyl)glycine, shown in Figure 5.35, which binds with two Fe(III), and both get six coordinated.

The two sites in transferrin release iron at a different rate in a pH-dependent manner, C-terminal iron being acid stable and N-terminal iron is lost at less acidity. Thus in an acidic solution of transferrin, following different species are obtained: (a) transferrin with two Fe(III), (b) monodemetallated, with one Fe(III) bound to C-terminal, (c) monodemetallated with one Fe(III) bound to N-terminal, and (d) demetallated apotransferrin.

The apotransferrin takes up iron that is released as Fe(II), in blood from intestinal absorption, from ferritin and also breakdown of haemoglobin. The uptake of iron by apotransferrin involves the oxidation of Fe(II) to Fe(III). This oxidation is catalysed by plasma ceruloplasmin (copper protein). This is probably one of the reasons why copper deficiency in animals may affect iron metabolism, leading to anaemic condition.

In the process of iron transfer, transferrin gets bound to the cell forming a vesicle (endosome). The vesicle contains a piece of membrane with transferrin bound to it. At low pH, there is reduction of Fe(III) followed by demetallation of transferrin, and the liberated iron(II) is taken up by the immature blood cells (reticulates), which are very active in iron uptake and form haemoglobin. After the uptake of iron by the reticulates, the apotransferrin is released outside the cell, to get bound to another two Fe(III) ions and continue the process of iron transfer. Thus, transferrin can carry many cycles of iron transport in the tissues.

In vitro conditions, it has been observed that iron is released from transferrin on reaction with dithionate. This shows that iron is released due to reduction of Fe(III) to Fe(II). In the tissue, the breakdown of transferrin may be due to the release of reducing agent at low pH, and the reduction of Fe(III) to Fe(II) may lead to demetallation of transferrin. The mechanism of the release of iron at low pH is not definitely known. However, it has been indicated that the cell binding site of the transferrin, in the membrane, is responsible for the breakdown of transferrin and release of iron. The mechanism of storage and transfer of iron is shown in Figure 5.36.

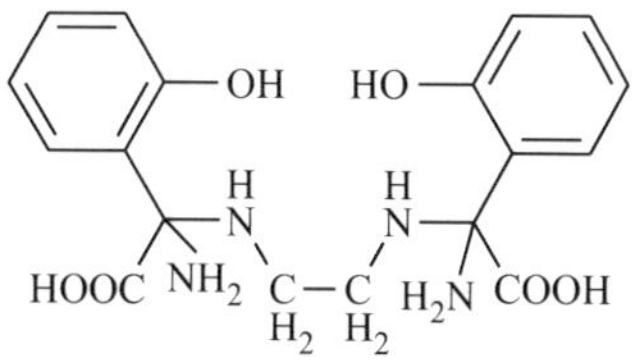

FIGURE 5.35 Synthetic model complex for iron site of transferrin.

2. **Siderophores:** Most of the bacteria fungi and other microorganisms need iron as an essential element. However, they do not have the capacity to biosynthesise high molecular complex proteins, like apotransferrin (found in higher animals), which can bind strongly to iron.

Hence, the nature has provided an alternative mechanism for iron uptake. When the microorganism faces iron deficiency, it secretes an organic compound which binds strongly with iron. The general name for such compounds is siderophore, which is a Greek word, meaning iron carrier. More than two hundred types of siderophores have been isolated from the microorganisms, and their structures have been determined. Most of the studies of iron uptake by siderophore have been carried out on the cells of the organisms, under iron deficient aerobic growth conditions. The iron uptake and transport in the bacteria *Escherichia coli* by the siderophore enterobactin has been studied, in detail, as a model for siderophore in other microorganisms.

The enterobactin has three domains in its structure: (i) the backbone, (ii) the domain with amide linkage and (iii) metal binding catecholate site (Figure 5.37).

The enterobactin binds the iron very strongly at the catecholate site, and then, it interacts with a specific site on the outer cell membrane of the bacteria (the structure of the receptor site not being known definitely), and the complex is taken inside the membrane by an active transport process. It has been suggested that in the process of passage through the cell membrane, there is a possibility of the reduction of

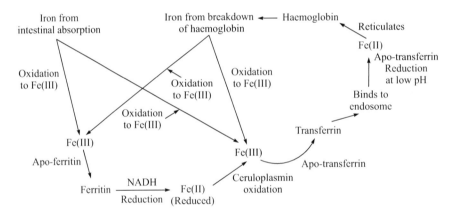

FIGURE 5.36   Mechanism of iron storage and transfer.

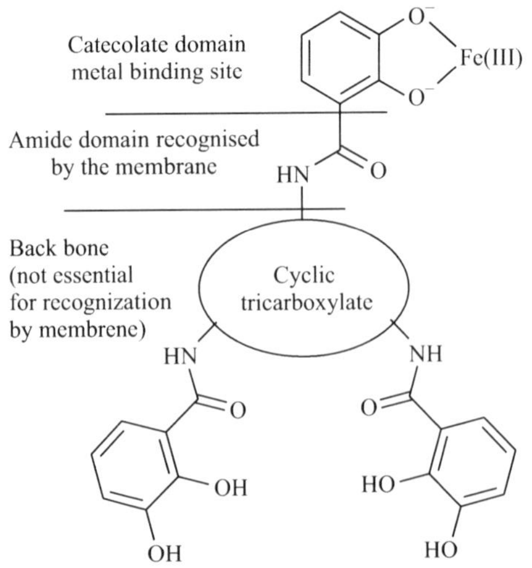

FIGURE 5.37   Iron uptake and transport in the bacteria E. Coli by the siderophore enterobactin.

Fe(III) to Fe(II) at low pH so that Fe(II) is released inside the cell. However, the exact mechanism of the release of iron from enterobactin is not known. Fe(III) complexes of compounds having structure similar to enterobactin have been administered to *E. coli*, and they are found to increase the iron supply to the bacteria, thus assisting its growth.

The study of the model complexes can help to know, which part of the ferric enterobactin is reorganised by the cell membrane.

Rh(III) complex of the synthetic compound MECAM having structure similar to enterobactin except the backbone (Figure 5.38) was found to inhibit the uptake of iron(III). The cell membrane recognises the Rh complex without the backbone, and it gets bound at the receptor site of the membrane. Thus, the binding of Fe(III) enterobactin to the membrane is blocked. This shows that the backbone of enterobactin is not essential for the recognition of Fe(III) by the cell membrane.

It had been suggested earlier that the iron catecholate part is recognised by the receptor site on the membrane and the remaining part is not recognised. However, later studies have shown that tris-catecholate Rh(III) complex (Figure 5.39) does not inhibit the uptake of iron. If the metal catecholate site of Fe(III) enterobactin is the only site recognised by the membrane, Rh(III) catecholate should have blocked the receptor site, inhibiting iron uptake. However, Rh(III) complex of dimethyl amide of catecholate (Figure 5.40) inhibits the enterobactin-mediated Fe(III) uptake. This shows that the amide site of the enterobactin is recognised by the membrane.

It has been concluded that the receptor site of the membrane has a pocket of specific size to accommodate the iron catecholate site of Fe(III) enterobactin. Around the pocket, there are proton donor groups which form H bond with the amide group of the second domain of Fe(III) enterobactin.

Thus, the catecholate site is for binding with Fe(III), and the amide site is recognised by the membrane for binding.

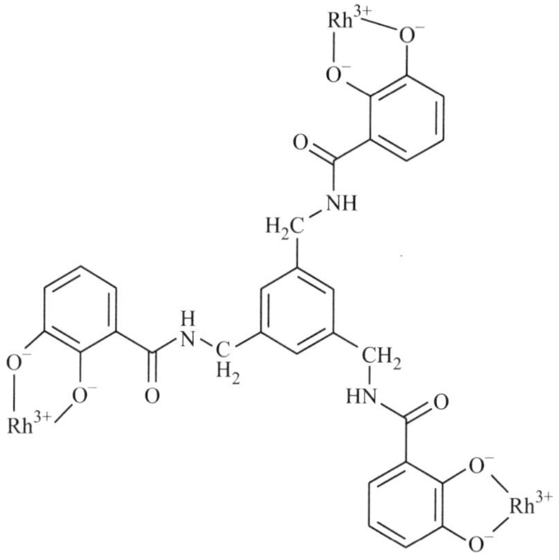

FIGURE 5.38 Rh(III) complex of the synthetic compound MECAM having structure similar to enterobactin except the backbone.

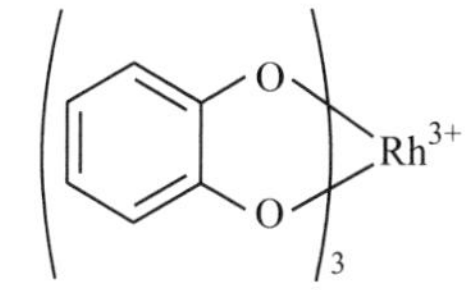

FIGURE 5.39 tris-catecholate Rh(III) complex.

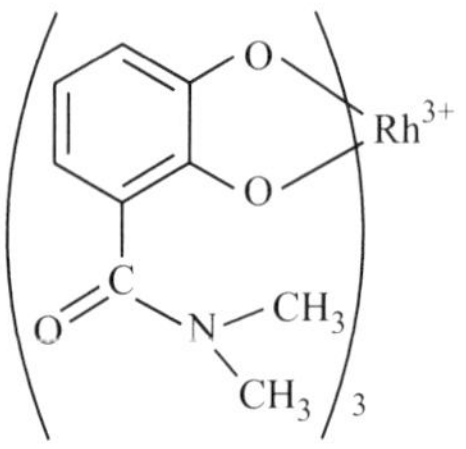

FIGURE 5.40   Rh(III) complex of dimethyl amide of catecholate.

QUESTIONS

1. Differentiate between haem a and haem b.
2. Why does the haem form very stable metal complex with Fe(II)?
3. Which enzyme in mammals is responsible for conversion of water-insoluble, aliphatic or aromatic hydrocarbons into soluble alcohols or phenols and of olefins into epoxides?
4. Cytochrome P-450 works in body as detoxification system. Justify.
5. Why is cyt P-450 named so?
6. Write the active iron species of cyt P-450.
7. How may metal centres are there in MMO, and what is their oxidation state?
8. Hydrogen peroxide is generated in some biochemical reactions. Due to its toxicity, it needs to be removed. Name the enzymes that are responsible for this.
9. Compare Cyt P-450, catalase and peroxidase for their catalytic activity.
10. What are cytochromes?
11. Why electron transfer process needs to occur in several sequential steps in respiration?
12. Apart from iron-containing cytochromes, which other proteins carry electrons in biochemical systems?
13. Name the two types of iron sulphur proteins.
14. How does rubredoxin differ from ferredoxin?
15. What is Hi PIP?
16. Which proteins are responsible for oxygen transport?
17. Give a flowchart showing path of electron transfer in cytochromes.
18. Which cytochrome has a copper centre along with an iron centre? Depict the mechanism of its action.
19. Describe experimental evidence for structure of rubredoxin.
20. What is difference between haemoglobin and myoglobin?
21. Whereas the iron in deoxyhaemoglobin is situated above the porphyrin ring, in oxyhaemoglobin it is in the plane of the ring. Explain why?
22. Which proteins are used for storage of iron?
23. Name the proteins responsible for uptake/transfer of iron in biological systems.

6

Copper in Biochemical Systems

Among the essential trace elements in the biochemical systems, copper is third in abundance, after iron and zinc. Though essential in trace amounts, excess copper, more than 80–120 mg, in human adult, has toxic effect.

The daily dietary consumption of copper is about 1.5 mg, and half of it is absorbed by the stomach and small intestine. This copper is transported to the liver and is processed there, and gets bound to the proteins apoceruloplasmin and albumin. These proteins sequester the Cu(II) so that it does not have toxic effect. Only 10% of the copper is retained in the liver, and the excess copper is thrown back in the intestine.

The inorganic copper present in drinking water is not absorbed in the liver, but directly enters the blood. This copper in free form can get transported to the brain and can have toxic effect. This unabsorbed Cu(II) and also the Cu(II) rejected by liver after homeostasis is thrown back in the intestine and is excreted.

The homeostasis of copper is regulated by the amount of copper absorbed and the amount of copper thrown by the excretions from the bile. If an excess of copper is absorbed, more of it is sequestered by the proteins and stored in the liver. However, if the availability of copper is less, more of copper is excreted by the liver.

Copper occurs in the form of copper proteins, which exhibit stereochemistry and properties, distinct from simpler copper complexes. These special features of copper proteins are due to the protein structure, which creates a specific stereochemistry and local dielectric constant, at the site of coordination of the metal ion.

Copper proteins participate in biochemical reactions, similar to that of iron proteins. The two metal ions function together in proteins, such as cytochrome C oxidase.

The protein coordinating site attributes specific characteristics to the copper, bound there.

On that basis, the copper, in the biological proteins, can be classified into the following types:

  i. Blue copper(II) type I.
 ii. Normal copper (II) type II.
iii. Coupled copper (II) type III.
 iv. Copper (I) type IV.

The copper proteins may have one or more types of copper centres and thus exhibit different kinds of roles. The main functions of copper proteins are

  a. Catalytic – electron transfer, oxidation and oxygenation of substrates and superoxide dismutation.
  b. Dioxygen transport.
  c. Copper transport and storage.

The different types of proteins, performing the above roles, are as follows.

Electron Transfer Proteins

These are smaller proteins, with low molecular weight, and occur widely in organisms, from bacteria to human beings. They contain a single type I copper centre, which is characterised by an extremely intense ($\varepsilon = 4,000$–5000 M^{-1} cm^{-1}) absorption band at 600 nm. In other words, these proteins exhibit intense blue colour {400 times that of $[Cu(H_2O)_6]^{2+}$} and hence are called blue copper proteins.

In the electron transfer process, the Cu(II) centre accepts electron from the donor biomolecule and gets reduced to Cu(I). In turn, it transfers the electron to the acceptor molecule and Cu(II) is regenerated. This way, it catalyses the electron transfer process, by providing a low-energy path.

Several different blue copper proteins have been derived from different sources. Three blue proteins shown in Table 6.1 have been studied more thoroughly.

The less characterised small blue proteins are rusticyanin (16.5 K daltons), umecyanin (14.6 K daltons) and plantocyanin (8 K daltons). It is presumed that their role is electron transfer. Rusticyanin is stable at low pH and participates in electron transfer in the respiratory chain of the bacteria, *Thiobacillus ferrooxidans*, at pH 2.

The apoprotein parts in the first three blue proteins show some similarity in structure, supporting the fact that they belong to the same class. For example, the following amino acid residues are present in all the three blue proteins: histidine, asparagine, valine, glycine, tyrosine, cysteine, proline and methionine.

It is observed that azurins have greater similarity in amino acid sequence with plastocyanin, indicating that these two types of proteins have same origin. Stellacyanin does not have much similarity in amino acid sequence with plastocyanin and azurin. Furthermore, 40% of this protein is carbohydrate, which are attached on the asparagines, at 28, 60 and 102 positions of the protein.

It is observed that all of the above proteins, obtained from different sources, retain some of the amino acid residues. For example, plastocyanin from different sources has conserved amino acid residues at positions 31–44 and 84–93. Azurins from different sources have three cysteine residues.

The protein part, in all the blue copper proteins, is coiled and copper (II) is located inside the cavity. In plastocyanin and azurin, the Cu(II) ion is bound to the nitrogen atoms of imidazoles of two histidine residues and to two sulphur atoms, of one cysteine and one methionine residue of the protein.

In plastocyanin, the protein is cylindrical in structure. Cu(II) ion is buried 6 Å deep inside the interior of the protein and is linked to hist 37, hist 87, cyst 84 and methionine 92. The second nitrogen of hist 87 is H bonded to water.

The structure of azurin from *Pseudomonas aeruginosa* has been determined by X-ray crystallography. In this, Cu(II) is bound to hist 46, his 117, cyst 112 and methionine 121.

In stellacyanin, the protein has no methionine residues. This shows that the ligand around Cu(II) in stellacyanin may be different from that in plastocyanin and azurin. There is evidence for a Cu(II) bond with two histidine N and one cysteine S. It has been suggested that instead of the methionine S link at the fourth position, there is binding of a S–S group.

Thus, in all the three proteins, Cu(II) is linked to two S ligands and two N ligands and the four ligand atoms are oriented around the Cu(II) ion in a distorted tetrahedral geometry. The skeletal structure of plastocyanin is shown in Figure 6.1.

TABLE 6.1

Blue copper proteins

Blue Copper Protein	Source	M.w.	No. of Amino Acids	Functions
1. Plastocyanin	Plant and blue and green algal chloroplast	10.5 K daltons	99	Membrane-bound electron carrier between photo systems I and II in photosynthesis
2. Azurin	Different species of bacteria pseudomonas	14 K daltons	130	Electron transfer in respiratory chain of bacteria
3. Stellacyanin	Japanese lacquer tree *Rhus vernicifera*	20 K daltons	107	Electron transfer in *Rhus vernicifera*

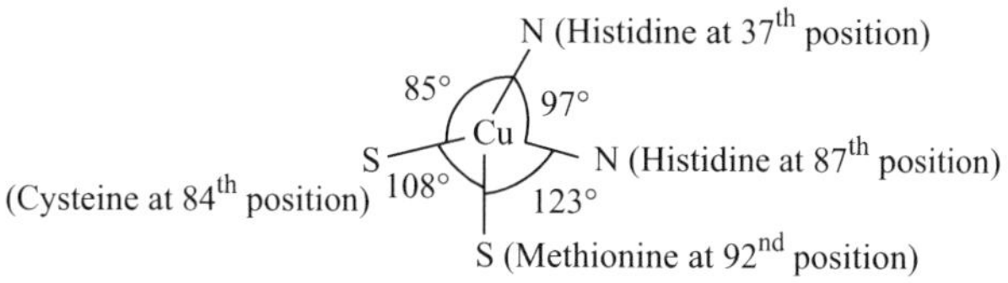

N (Histidine at 37th position)

85°   97°

Cu

(Cysteine at 84th position) S 108°   N (Histidine at 87th position)

123°

S (Methionine at 92nd position)

FIGURE 6.1   Skeletal structure of plastocyanin.

In both plastocyanin and azurin, there is a shorter Cu–cysteine sulphur bond (2.10 Å) and a longer Cu–methionine sulphur bond (2.25 Å). The electronic structure of all the three proteins exhibit three to four bands in the region 400–1,900 nm. The less intense bands, in higher wave length region, are due to d–d transitions. The occurrence of more than one d–d band is consistent with distorted tetrahedral geometry. The intense band, at 600 nm, is responsible for the intense blue colour. This was initially attributed to the tetrahedral structure, without centre of symmetry, leading to high-intensity d–d transitions. However, the intensity is 400 times that of Oh complex of Cu(II) and two to three times higher than tetrahedral Cu(II) complexes. Hence, the highly intense band has now been attributed to the Cu(II) → S charge transfer transition.

The ESR spectra of plastocyanin and azurin exhibit two signals ($g_\parallel$ and $g_\perp$), corresponding to a tetragonal structure. In stellacyanin, the ESR spectrum shows three g values, indicating that there is rhombic distortion. The characteristic of the ESR spectrum of these blue proteins is a low $A_\parallel$ value in hyperfine splitting, due to coupling with Cu(II) nucleus. This low value of $A_\parallel$ is attributed to Cu → S π bonding, which reduces the electron density on the Cu(II), leading to less delocalisation of electrons over Cu(II) nucleus. As a result, there is less electron spin nuclear spin coupling, and also low $A_\parallel$ values.

The ESR spectrum does not show any splitting due to electron spin coupling with the coordinated atom of the protein. The direct evidence of coordination of cysteine sulphur to Cu(II) is obtained by the observation that on removal of Cu(II) from plastocyanin, the apoprotein shows titratable – SH sites.

The structure of apoplastocyanin is very similar to Cu–plastocyanin, showing that the protein has a pocket with the coordinating atoms predisposed such that a distorted tetrahedral geometry is imposed on Cu(II). Therefore, on removal of Cu(II), the structure does not change significantly. In apoplastocyanin, only the orientation of hist 87 is changed and it gets bound to other water molecules more than in Cu–plastocyanin, and the N atom gets exposed. In the reversible binding of Cu(II), it combines first with the N atom of hist 87. This is followed by the rotation of the histidine ring, and the Cu(II) gets bound to the other three sites.

Besides the characteristic blue colour and small $A_\parallel$ value, the blue proteins exhibit high reduction potential of the Cu(II) centres, ~0.38 V in plastocyanin, 0.33 V in azurin and 0.30 V for stellacyanin. This has been attributed to two factors: (i) the coordination of Cu(II) to softer S ligand, resulting in lowering of electron density on Cu(II); (ii) Cu(II) is predisposed in a tetrahedral geometry in the protein, making its reduction to Cu(I) is easy. This is because Cu(I) prefers tetrahedral geometry. Thus, it is seen that in the blue proteins, the protein offers the metal ions a tetrahedral geometry and softer ligand atoms, so that the conversion of Cu(II) to Cu(I) and vice versa is facilitated.

However, the reducible Cu(II) centres in the blue proteins are situated deep inside the hydrophobic globular structure of the protein and are not accessible to solvents. As the distance between the redox centres of the substrate biomolecules and blue Cu(II) electron transfer site is more than 10 Å, the transfer of electrons to and from Cu(II) centre, to assist electron transfer process between two biomolecules, cannot take place by inner sphere electron transfer mechanism. Hence, the electron transfer in proteins must be by outer sphere process. The rate constant for the electron transfer in proteins cannot be expressed, only by the fundamental parameters, as given by the Marcus relationship (Chapter 2). The expression of the rate constant assumes the following form:

$$K_{et} = \text{constant} \cdot \exp\left[\beta(d - d_0)\right]\exp\left(-\Delta G^*/RT\right) \quad (6.1)$$

where $\Delta G^* = \left(\lambda + \Delta G^0\right)^2 / 4\lambda$

The relevant factors controlling electron transfer in biomolecules are d = distance between the donor and acceptor centres, d_0 = minimum internuclear distance, based on van der Waals radius, and β = conductivity of the protein matrix through which electron passes, which is a measure of distance dependence of electron coupling between the two centres and ΔG^*, ΔG^0 and λ, which have the same definition as in the Marcus equation (Chapter 2).

It is expected that long range electron transfer process should be less; hence, in long-chain proteins, electron transfer should be slow. Furthermore, as explained under Marcus equation (Chapter 2), λ contains both inner sphere (λ_i) and outer sphere (λ_0) components and λ_0, being dependent on changes in polarisation of the solvent molecules during electron transfer, decreases with decrease in solvent polarity. The static dielectric constant of the interior of the protein is ≈ 4, whereas that of water is ≈ 78. Hence, λ_0 of the redox component of the protein, which is buried deep inside the protein, is less. Therefore the electron transfer process in the proteins is slow.

For the outer sphere electron transfer, the protein should have regions at which electron transfer can take place with the redox partner. This has been studied in plastocyanin. The initial suggestion was that the Cu(II)-bound ligand histidine, which is ~6 Å from Cu(II) centre and is non-polar, may be involved in electron transfer. However, it was later suggested that Tyr – 83, though ~15 Å from Cu(II) centre, is exposed and may allow the electron transfer process. Specially, the presence of negatively charged carboxylates, at positions 42–45 and 50–61, makes the tyrosine site attractive for the positively charged redox reagents.

Oxidase Activity

These are multicopper proteins which catalyse the oxidation of the substrates by transfer of four electrons to dioxygen molecule, leading to its reduction to two O^{2-}, which form two water molecules. These proteins contain type I, II and III Cu(II) centres in different proportions in different oxidase enzymes. The characteristic features of the two major oxidase enzymes can be summarised as follows:

i. **Laccase:** It has a molecular weight 110 K daltons (from *Rhus vernicifera*) or 62 K daltons to 80 K daltons (from *Polyporus versicolour*) and contains one type I, one type II and one pair of type III Cu(II). It is obtained from the latex of Asian Lac tree. It catalyses the oxidation of *p*-diphenols and *p*-phenylene diamines to para-quinones.

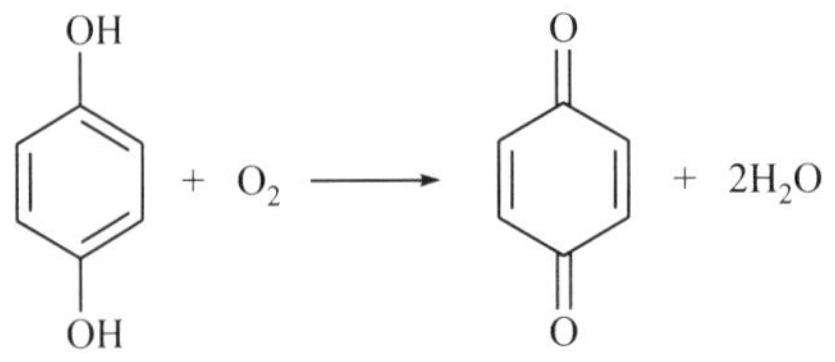

$$(6.2)$$

ii. **Ascorbate oxidase:** It is a protein obtained from plants and bacteria. It has molecular weight 140 K daltons with eight Cu(II) per unit, 4(types I and II) and 4(type III). Copper composition may vary in the protein. It catalyses oxidation of ascorbic acid to dehydroascorbic acid.

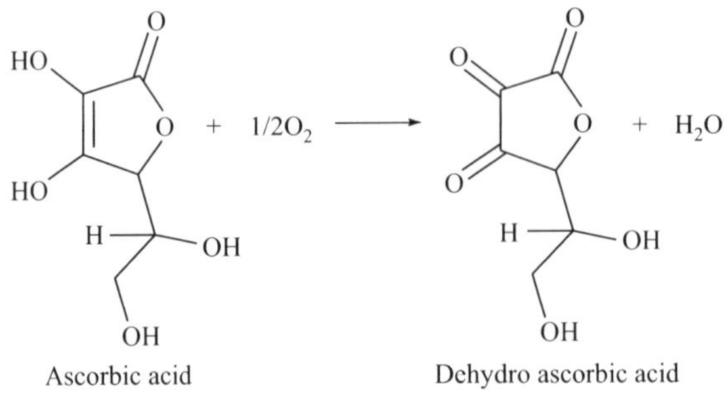

Ascorbic acid   Dehydro ascorbic acid (6.3)

In these oxidases, the type I Cu(II) has electronic spectral, electrochemical and ESR spectral characteristics, similar to that of electron transfer blue proteins. So, there too have intense blue colour and are called blue oxidases. The oxidase, which does not have type I Cu(II), is not intense blue and is termed non-blue oxidase.

In blue oxidase, the electronic spectral band of the type II Cu(II) is less intense and is obscured by the intense absorption of type I Cu(II) centres. However, in non-blue oxidases (dopamine-β-hydroxylase), the spectroscopic property of type II Cu(II) is found to be similar to the tetragonal Cu(II) complexes of lower molecular weight. This shows that type II Cu(II) has normal tetragonal structure, with no special mode of binding to the protein.

The presence of type II Cu(II) has been identified in blue Cu(II) oxidases by EPR studies. In the EPR spectrum of Laccase, at 9 GHz field, the signal due to type I Cu(II) is observed at 2,900 gauss, as four hyperfine lines with small $A_{\parallel}$ value, as in blue Cu(II) protein. However, there are two additional lines present at 2,650 and 3,000 gauss in the $g_{\parallel}$ region, indicating the presence of another type of Cu(II) centre. This is further confirmed by the addition of one equivalent of fluoride to the enzyme. No change is observed in the signal of type I Cu(II). However, in the case of the non-blue type II signal shifts and splits into two, due to the coupling of the Cu(II) electron spin with the fluoride nuclear spin. This shows that the fluoride binding is specific with type II Cu(II). On addition of more fluoride, each of the hyperfine lines of non-blue Cu(II) is further split into three components, with an intensity ratio of 1:2:1, as expected due to the coupling of type II Cu(II) electron spin with the nuclear spin of two equivalent fluoride, which get bound to the Cu(II) (Figure 6.2).

The remaining two type III, Cu(II), of laccase, exhibit an electronic band at 330 nm. They are diamagnetic and EPR silent. The possibility that the type III copper is in Cu(I) state, is ruled out by the fact that Laccase can accept four electrons, one per Cu(II) centre. This confirms that all the four coppers are in Cu(II) state. The ESR silence of the two Cu(II) centres can be explained by considering the existence of anti-ferromagnetism between the two type III Cu(II) centres. The spin coupling is through an endo-bridge atom X of the protein, resulting in diamagnetic character and ESR silence. The probable structure of the binuclear Cu(II) site is shown in Figure 6.3.

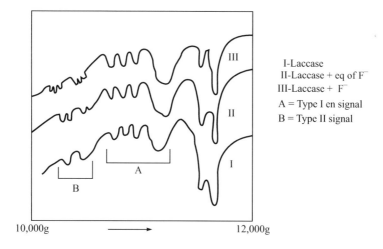

I-Laccase
II-Laccase + eq of F⁻
III-Laccase + F⁻

A = Type I en signal
B = Type II signal

10,000g ⟶ 12,000g

FIGURE 6.2 EPR spectra proving presence of copper(II) centres in laccase.

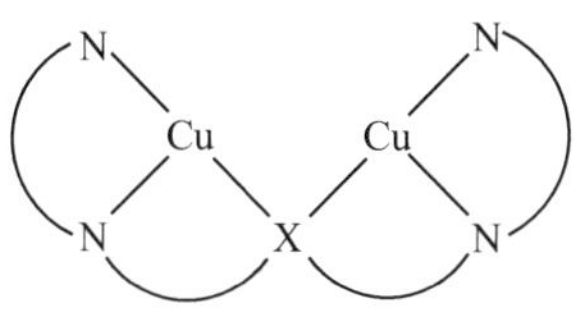

FIGURE 6.3 EPR silent endo bridged Cu(II) centres in laccase.

When type I Cu(II) is reduced to Cu(I), the low-energy, more intense bands, due to the tetrahedral Cu(II) centre, disappear. This supports that type II and type III Cu(II) do not have tetrahedral geometry.

It is interesting to trace the mechanism by which electron is transferred from the substrate to molecular O_2, mediated by the blue oxidase.

On reduction of laccase by three equivalents of ascorbate, it is observed that type I and type III Cu(II) centres are reduced, simultaneously, as revealed by the disappearance of the electronic spectral bands at 660 and 330 nm, characteristic of type I and type III Cu(II), respectively. On addition of further equivalent of ascorbate, type II Cu(II) is reduced. According to Malmstrom, there is transfer of electron from the substrate in steps. First, the electron is transferred to the type I Cu(II). This, in turn, is transferred to the type III Cu(II) through intramolecular process, with type II Cu(II) playing some role. The type III site is thus reduced to Cu(I)–Cu(I).

The oxygen molecule is bound at the binuclear type III site. Two electrons are transferred from the Cu(I)–Cu(I) pair to the O_2 molecule, reducing it to peroxide. Simultaneous transfer of two electrons does not allow the formation of the thermodynamically unstable superoxide O_2^-. The type III site is reoxidised to binuclear Cu(II), after transferring two electrons. Binuclear Cu(II) site further accepts two electron from type I site. These two electrons are transferred to the bridging peroxide, reducing it to two oxide (O^{2-}) ions. These combine with $4H^+$, liberated by the substrate, and two H_2O molecules are formed. Thus, at the type III Cu(II) site, the molecular oxygen is reduced to two water molecules, and the substrate parahydroxy phenol (in laccase activity) or ascorbic acid (in ascorbic acid activity) is oxidised by the loss of four electrons.

In model studies, it has been shown that on using mononuclear $Cu(NO_3)_2$, as a catalyst, in the oxidation of ascorbic acid by molecular oxygen, it is reduced to H_2O_2 only. Whereas on using binuclear Cu(II) acetate as catalyst, in ascorbic acid oxidation by molecular oxygen, it is first reduced to peroxide and then to $2O^{2-}$. This indicates that the binuclear type III Cu(II) site is essential for the activity of the oxidases. The oxidation reaction is shown in Eq. 6.4:

$$(6.4)$$

In the above mechanism, the role of type II Cu(II), in the transfer of electron to O_2, is not evident. The role of type II Cu(II) has been revealed by carrying out oxidation in the presence of one equivalent of fluoride. Under this condition, type I and type III are not reduced simultaneously, but are reduced in steps. As mentioned earlier, the fluoride ion binds with type II Cu(II) preferentially. This shows that the binding of F with type II Cu(II) affects the pattern of reduction of type I and type III Cu(II). This indicates that the type II Cu(II) has some role in the electron transfer process.

Furthermore, the above mechanism of transfer of two electrons, simultaneously, for the conversion of O_2^{2-} to two O^{2-}, involves high energy process of the breaking of O^-–O^- bond. Hence, Braden suggested an alternative mechanism of three electron transfer from type I and type III sites to O_2, forming O^{2-} and O^-, and the latter is further reduced to O^{2-}, by receiving one more electron from the type II Cu(II). However, the fact that O_2 binds at the binuclear Cu(II) site indicates that two electrons are transferred at a time and supports Malmstrom's model of formation of peroxide O_2^{2-}, in the first step, and then two O^{2-} ions, in the second step.

Recently, it has been suggested by Solomon, that the blue multi copper oxidase, Laccase, contains a tri nuclear site, with two type III and one type II copper. Similar trinuclear structure has been suggested in ascorbate oxidase also. Trinuclear structure has also been confirmed by the X-ray structural analysis of ascorbate oxidase by Messerschmidt. Trinuclear site is constituted of two type III and one type II Cu(II), which are at a distance of 3.9 Å from each other. The type I Cu(II) is situated at a distance of 12 Å from the trinuclear site. There is a strong anti-ferromagnetic interaction between the two type III Cu(II), through bridging OR group. There is also some electron spin exchange interaction between type III and type II Cu(II), through the bridging ligand, as shown in Figure 6.4.

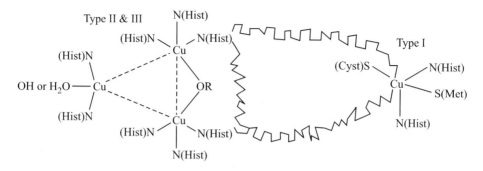

FIGURE 6.4   Trinuclear copper sites in the blue multi copper oxidase laccase.

It has been suggested that in the oxidation process, the substrate transfers the electron to the type I Cu(II), and from there, it is transferred to the trinuclear copper centre. At this centre, O_2 gets bound, and there is four-electron reduction of O_2 to $2O^{2-}$.

Monooxygenase Activity

The best known Cu(II) protein, with monooxygenase activity, is tyrosinase. It is widely distributed in nature. It catalyses the conversion of monophenols to diphenols and thus assists in the synthesis of the pigment melanin and other polyphenolic compounds. This monooxygenation is a monophenolase or cresolase activity. Besides this, tyrosinase also exhibits oxidase activity. It accelerates the oxidation of O-diphenols to O-diquinones, and thus exhibits diphenolase or catecholase activity.

This is a binuclear Cu(II) protein, with molecular weight 40–50 K daltons. In the resting form of the enzyme, also called met tyrosinase, the two Cu(II) centres are bridged through an endogenous bridging atom of the protein and also through an exogenous bridge. Each Cu(II) is bound to two N atoms of the histidinyl residue of the protein. The structure is shown in Figure 6.5.

The two Cu(II) centres are strongly anti-ferromagnetically coupled through the bridges; therefore, the enzyme shows very low magnetic moment and is EPR silent.

Electrochemical study shows that the two Cu(II) atoms are equivalent and are reduced simultaneously at $E_0 = 0.36$ V. The electronic spectrum of the enzyme shows a band at 700 nm, with normal extinction coefficient. This is attributed to the d–d transition in the Cu(II) centres, in the tetragonal field. This shows that the two Cu(II) centres in tyrosinase are different from the two magnetically coupled type III Cu(II) centres of the blue multicopper oxides.

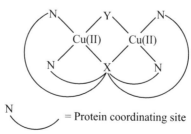

  = Protein coordinating site

X = endogenous bridging atom,
Y = exogenous bridge.

FIGURE 6.5   Resting form of the binuclear copper(II) protein tyrosinase.

The diphenolase (oxidase) and monophenolase (oxygenase) activities of the enzyme are related to each other and can be explained as follows:

i. **Diphenolase activity:** As stated earlier, in this activity tyrosinase molecules catalyse the oxidation of catechol to O-diquinone, reducing molecular oxygen to water.

In the process of oxidation, catechol first gets bound to the bicupric centre of met tyrosinase. It transfers two electrons to the two Cu(II) centres, forming dicuprous deoxytyrosinase, and itself gets oxidised to diquinone. There is no evidence of the formation of semi-quinone, indicating that two electrons are simultaneously lost by the diphenol to form diquinone. Deoxytyrosinase combines with one molecule of oxygen at the two Cu(I) centres. The metal centres transfer two electrons to O_2 molecule, forming a bicupric site, bridged through a peroxide. This is called oxytyrosinase. The electronic spectrum shows a prominent band at 345 nm, absent in met tyrosinase. This is assigned to $O_2^{2-} \rightarrow$ Cu(II) charge transfer. The low-temperature absorption spectrum of oxytyrosinase shows another shoulder at ~430 nm, which can be assigned to endogenous bridging atom X $\rightarrow$ Cu(II) charge transfer. This supports that the two Cu(II) centres are linked through an endogenous protein bridge atom.

Peroxy-bridged met tyrosinase is found to be ESR silent. This is attributed to anti-ferromagnetic coupling, through the endogenous bridging atom X. It can be assigned the structure shown in Figure 6.6.

Oxytyrosinase molecule further combines with another diphenol molecule. Two electrons are transferred by the diphenol to O_2^{2-}. Diphenol gets oxidised to diquinone, and O_2^{2-} is reduced to two O^{2-}. The reaction can be considered to take place in two steps. First, there is a formation of the hypothetical intermediate Cu(I)–O_2^{2-}–Cu(I), by the transfer of two electrons from the diphenol. In turn, two Cu(I) centres pass the two electrons to the bridged O_2^{2-}, reducing it to two O^{2-}, forming two H_2O molecules. Thus, met tyrosinase is regenerated for the second cycle. The cycle is shown in the upper half of Figure 6.7.

In the intermediate catechol-bound complex, the diphenol has to coordinate with the two Cu(II) centres. They are at a distance of 3.4 Å in tyrosinase, which is suitable for catechol binding.

A binuclear Cu(II) complex (Figure 6.8) has been synthesised, with two Cu(II) sites at suitable distance for catechol binding. The complex binds with catechol at the two Cu(II) centres (Figure 6.8) and thus serves as a model for the intermediate catechol binding, in the tyrosinase activity.

However, in the enzyme activity of natural tyrosinase, besides the binding of catechol with the two Cu(II) centres, its binding with the protein is also important, for the formation of the stable intermediate. This is supported by the observation that the presence of non-phenol benzoic acid inhibits catechol oxidation, assisted by tyrosinase. Though non-phenolic benzoic acid cannot block the Cu(II) binding sites of catechol, it blocks some positions in protein, essential for incorporating of the diphenol, and thus deactivates the enzyme.

ii. **Monophenolase activity:** Oxytyrosinase is the catalytic intermediate in the diphenolase activity. This is the active intermediate for monophenolase activity also. It combines with a monophenol substrate to form a ternary complex, with the enzyme bound to monophenol and peroxide ligands. One of the O^- of the peroxide loses electron and causes monooxygenation of the substrate monophenol forming diphenol. The second phenolic –OH gets directed at ortho positions, to the existing OH, due to the orientation of substrate monophenol and the peroxide, in the intermediate ternary complex.

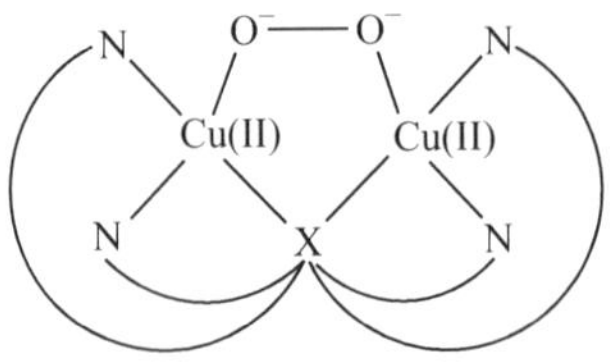

FIGURE 6.6 The EPR silent peroxy-bridged met tyrosinase.

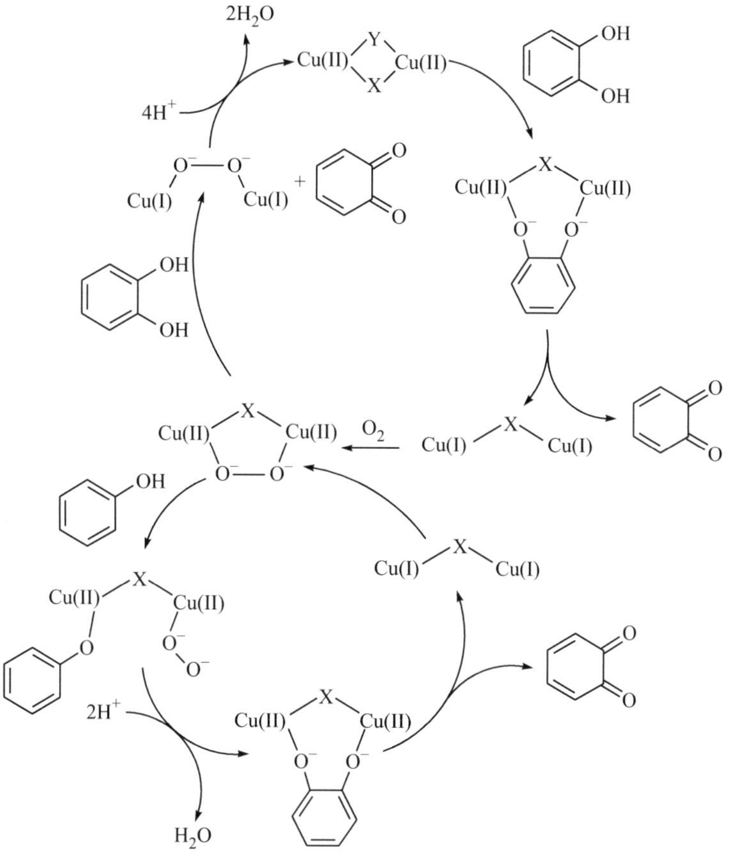

FIGURE 6.7 Catalytic cycle for diphenolase (upper part) and monophenolase (lower part) activity of tyrosinase.

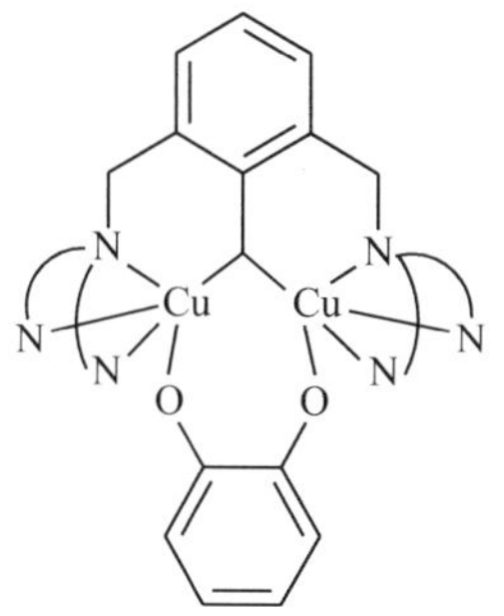

FIGURE 6.8 Binuclear Cu(II) complex bound to catechol – model for intermediate catechol binding in tyrosinase activity.

Hence, there is formation of O-diphenol. The other O$^-$ of the peroxide gets the electron from the first O$^-$ and is converted in O^{2-}. This combines with 2H$^+$ forming H$_2$O molecule. The study of the enzymatic activity, using labelled O$_2$, shows that the second oxygen introduced in the monophenol is obtained from the molecular oxygen. The resulting catechol gets bound to the two Cu(II) as shown in the first half of the lower cycle in Figure 6.7.

This intermediate complex is similar to catechol binding with met tyrosinase in the first half of the upper cycle (Figure 6.7) of the diphenolase activity. Hence, as in diphenolase activity, diphenol transfers two electrons to the two Cu(II) centres forming bicuprous deoxytyrosinase, and catechol is oxidised to di-quinone. Two Cu(I) of deoxytyrosinase get bound with one oxygen molecule, transfer one electron each and form bicupric oxytyrosinase with peroxo bridge, completing the second half of the cycle (lower cycle in Figure 6.7). Thus, the second half of the monophenolase activity corresponds to the first half of the diphenolase activity. This reveals the important link between the monophenolase and diphenolase activity, with oxytyrosinase being the connecting link, between the cycles.

It is seen that in monophenolase activity, diphenol is the product. However, the presence of diphenol is essential for the generation of the active compound oxytyrosinase. Therefore, it is observed that the monophenolase activity attains maximum velocity of the hydroxylation process, after an initial lag period, that is, till sufficient diphenol has been slowly generated. Thus, this is a case of autocatalysis. The lag period can be reduced if some O-diphenol, a reducing agent, is added externally. In the model system of ortho-hydroxylation of phenol, with oxygen, using copper amine complex as a catalyst, the reaction is initially slow. The time lag can also be eliminated, if H_2O_2 is used as the oxidant, so that oxytyrosinase type of active intermediate could be directly formed.

Because of the oxidase activity of copper, excess of copper has toxic effects on the liver and kidney. Cu(II) gives rise to active oxygen species, which causes oxidative damage to the tissues.

The reactions, which give rise to the free radical, causing oxidative damage are as follows:

$$Cu(II)L + H_2O_2 \longrightarrow LCu(I) + \bullet OOH + H^+$$

$$LCu(I) + H_2O_2 \longrightarrow LCu(II) + \bullet OH + OH^-$$

Non-blue Copper Protein – Dopamine-β-Hydroxylase

The term "non-blue" implies the absence of type I Cu(II) site in the protein. Dopamine-β-hydroxylase occurs in the medulla of animals and catalyses hydroxylation of dopamine, using molecular oxygen, as oxidant. This is a monooxygenase. The enzyme obtained from bovine medulla is a glycoprotein with molecular weight 290 K daltons. The enzyme consists of two types of protein, differing only by a NH_2 terminal tripeptide.

There are two atoms of Cu(II), per molecule of protein. Earlier studies had shown that the samples of dopamine-β-hydroxylate from certain sources, contained more number of Cu(II) centres. The Cu(II) in the enzyme is EPR detectable, showing that there is no anti-ferromagnetic coupling between them. Thus, the Cu(II) site is different from that in tyrosinase.

In the activity of dopamine-β-hydroxylase, the enzyme acts as a monooxygenase, and ascorbic acid acts as a reducing agent and hydrogen donor.

The reaction can be shown as the following equation:

$$\underset{\substack{\text{Dopamine-}\beta\text{-}\\\text{hydroxylase}\\\text{P = protein}}}{P - Cu(II)} + \underset{\text{Ascorbic acid}}{AH_2} \longrightarrow P - Cu(I) + A\dot{H} + H^+ \qquad (6.5)$$

Two mechanisms have been suggested for the hydroxylation reaction.

Goldstein considered involvement of one copper centre and formation of a superoxide intermediate.

Though, in this mechanism, super oxide has been considered as the active intermediate, the presence of superoxide dismutase (SOD), which decomposes O_2^-, does not have any effect on the rate of hydroxylation. This shows that super oxide may not be the intermediate (Eq. 6.6).

Alternative mechanism, considering active binuclear Cu(II) site, was suggested by Kaffman (Eq. 6.7). This involves formation of a peroxide intermediate, and the reaction pathway has been considered to proceed, as in monophenolase activity.

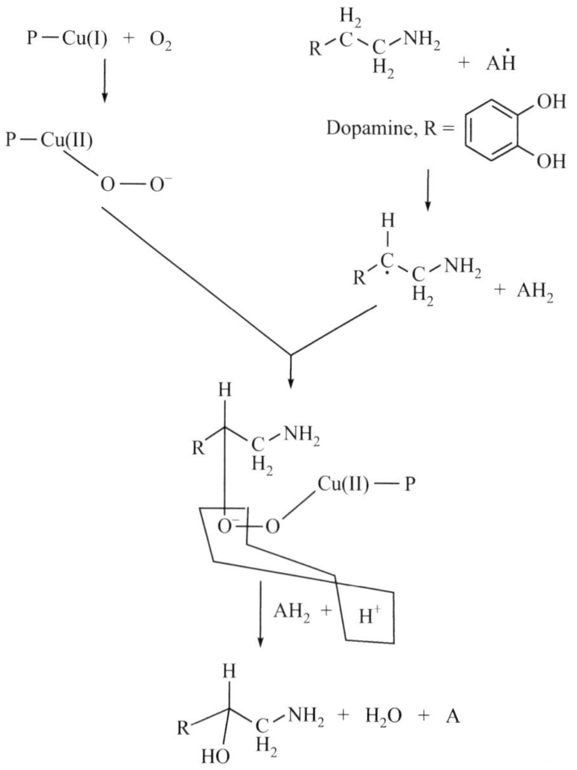

$$P-Cu(I) + O_2$$

$$P-Cu(II)$$
$$O-O^-$$

$$R-\overset{H_2}{\underset{H_2}{C}}-C-NH_2 + A\dot{H}$$

Dopamine, R =

(benzene ring with OH, OH)

$$R-\overset{H}{\underset{H_2}{\overset{|}{C}}}-C-NH_2 + AH_2$$

$$R-\overset{H}{\underset{\overset{|}{C}_{H_2}}{C}}-NH_2$$
$$Cu(II)-P$$
$$O^- \!\!\mid\!\! O$$

$$AH_2 + H^+$$

$$R-\overset{H}{\underset{\overset{|}{HO}}{C}}-\overset{}{\underset{H_2}{C}}-NH_2 + H_2O + A \tag{6.6}$$

$$Cu(II) \qquad Cu(II) + AH_2 \longrightarrow Cu(I) \qquad Cu(I) + A + 2H^+$$
$$P \qquad\qquad\qquad P$$

$$\downarrow O_2$$

$$R-\overset{H_2}{C}-C-NH_2$$

$$R-\overset{H}{\underset{HO}{C}}-\overset{}{\underset{H_2}{C}}-NH_2 + H_2O \longleftarrow Cu(II) \quad \overset{O-O}{} \quad Cu(II)$$
$$P \tag{6.7}$$

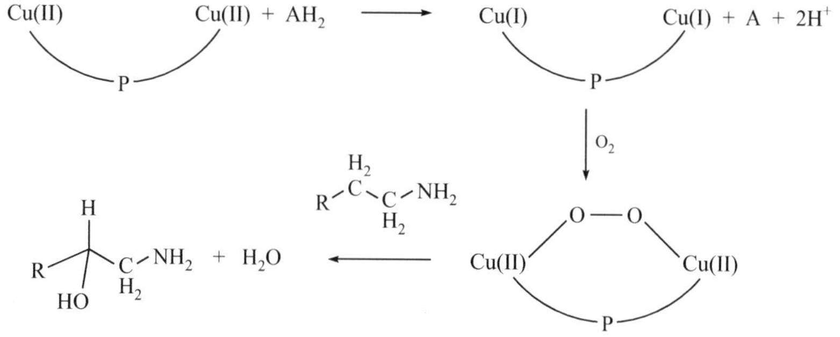

Other Type II Cu(II)-Containing Proteins

These are type II Cu(II)-containing proteins, which appear to act as oxidases. However, ESR spectral study shows that Cu(II) does not undergo change in oxidation state, during the oxidation of the substrate by molecular oxygen. Nevertheless, Cu(II) is essential for the oxidation process, catalysed by this enzyme. If a chelating agent is added, which sequesters Cu(II), the catalytic reaction is deactivated. This indicates that though Cu(II) does not participate in the redox process, it gets chelated with the substrate and acts as a Lewis acid catalytic centre and assists the oxidation process.

Following are some examples of this type of Cu(II) protein oxidases.

a. **Amine oxidase:** This is a Cu(II) protein which catalyses the oxidation of the amines to aldehydes, using molecular oxygen as oxidant. In the reaction, pyridoxal phosphate acts as a cofactor (Eq. 6.8).

$$R-CH_2NH_2 + O_2 \xrightarrow{\text{Amine Oxidase}} R-CHO + NH_3 + H_2O_2 \tag{6.8}$$

b. **Galactose oxidase:** This Cu(II) protein is obtained from polyporus circinatus and catalyses the oxidation of D galactose to D galactose hexo dialdose.

$$
\text{(structure)} + O_2 \longrightarrow \text{(structure)} + H_2O_2 \qquad (6.9)
$$

c. **Ureate oxidase:** This is a copper protein found in mammalian liver. It has one type II Cu(II), per protein, of molecular weight 12,000 daltons. This enzyme catalyses the oxidation of uric acid to allentrin.

Superoxide Dismutase (SOD) Activity

As the name indicates, superoxide dismutases are proteins, which cause the breakdown of superoxide, generated in the biological system, as a by-product of metabolic processes, involving molecular oxygen. Thus, they act as protective agent, against the aging and carcinogenesis.

Three types of SOD are known. They are characterised by having copper, manganese or iron as cofactors. The copper protein has Zn(II) as an essential component. Iron SOD and Manganese SOD subunits have only one metal ion. Thus, the three types have been named: CuZn SOD, Fe SOD and Mn SOD.

Cu Zn SOD

This protein was first isolated from bovine erythrocyte by Mann and Karlin in 1968. Its SOD activity was revealed in 1969 by Mccord and Fridovich. It deactivates enzyme xanthine oxidase, whose activity is known to be due to the superoxide, generated by it. The enzyme has also been obtained from other sources and has been given other names such as erythrocuprein, hepatocuprein and cerebrocuprein.

The enzyme obtained from bovine erythrocyte has a molecular weight of 31.5 K daltons. The molecule has a blue green colour, suggesting the presence of Cu(II). The X-ray study shows that the molecule has a cylindrical-shaped structure and consists of two equal subunits. The subunits are held by hydrophobic interactions. Each subunit has one Cu(II) and one Zn(II). The Cu(II) ion is coordinated to four nitrogen atoms of the imidazole nitrogens of hist 44, 46, 61 and 118 residues. In addition, water molecule occupies the fifth position, resulting in a square pyramidal structure.

Zn(II) is bound to the three nitrogens of the 61, 69 and 78 histidine residues of the protein. The fourth coordination position of Zn(II) is occupied by aspartate 81, giving an approximately tetrahedral geometry. Cu(II) and Zn(II), of a subunit, are bridged by the deprotonated imidazolate of hist 61 and are at a distance of 6.3 Å.

In the electronic spectrum, the enzyme shows a high-intensity transition ($\varepsilon = 155$ M^{-1} cm^{-1}) at 680 nm, in keeping with four nitrogen coordination of the Cu(II) centre. The high extinction coefficient supports the asymmetric penta-coordinated structure, with water as the fifth ligand (Figure 6.9). The EPR spectrum shows three g values, 2.265, 2.01 and 2.03, supporting a rhombic structure.

There is no change in the d–d transition band of Cu(II), on replacement of Zn(II) by other metal ions such as Co(II), Cd(II), Hg(II) and Ni(II).

The electronic spectrum of Cu–Zn SOD or mutated enzyme with Zn substituted by other metal ions, exhibit another band at 417 nm ($\varepsilon = 24,000$ M^{-1} cm^{-1}). This has been assigned to imidazolate $\rightarrow$ Cu(II) charge transfer (CT). The transition occurs at slightly lower energy than that in Cu(II) imidazolate complex, probably because of the second metal ion bound to the imidazolate bridge. The CT band disappears in complexes, where Zn(II) site is empty, because imidazolate bridge cannot exist in the absence of the second metal ion.

If Cu(II) and Zn(II) are removed from the enzyme, the apoprotein gets deactivated, but the activity is revived on addition of Cu(II) only. This shows that only Cu(II) is the active site of the enzyme. Zn(II) has predominantly structural role in the enzyme, and has all the four coordination positions occupied and does not have any accessible coordination site. Hence, it cannot participate in the enzymatic process.

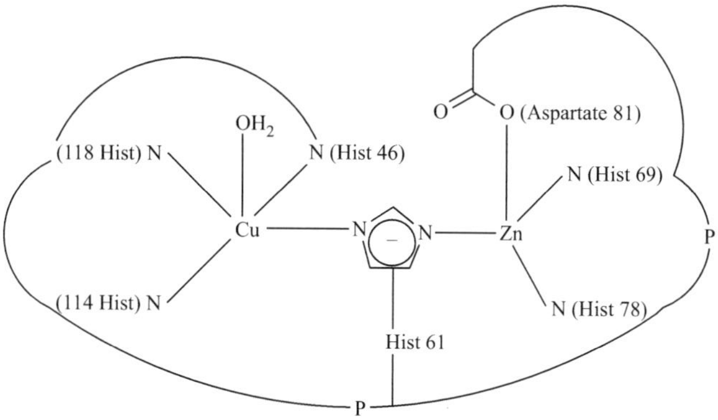

FIGURE 6.9 Cu Zn superoxide dismutase.

Cu(II) in the enzyme can bind to several anions such as CN^-, NO_3^- or halides. This is facilitated, because the positively charged side chain, arginine 141 of the protein, is 5 Å away from Cu^{2+}. The anions first get bound to arginine 141 by H bonding and thus are guided to bind with Cu(II) site. The same mechanism probably holds good in binding of superoxide with Cu(II). The anions or superoxide bind with Cu(II) by displacing coordinated histidine 46.

In the SOD activity, two superoxide ions are converted into one peroxide and one molecular oxygen, as shown in the following equation:

$$O_2^- + O_2^- + Cu, Zn - P \longrightarrow O_2^{2-} + O_2 + Cu, Zn - P \quad (6.10)$$

In the above process, the copper performs duel role, as an oxidising agent, converting O_2^- to O_2, and as a reducing agent, reducing O_2^- to O_2^{2-}.

This is possible because the reduction potential of O_2^- is −0.45 V and its oxidation potential is +0.98 V. These match with the potentials of the redox couple Cu(II)/Cu(I) in the synthetic Cu(II) complexes.

The reduction of Cu(II) in the protein to Cu(I), by one O_2^-, is followed by uptake of one proton at the bridge imidazolate. This means that the imidazole bridge breaks at Cu(I) centre. The fact that the protonation takes place in the enzymatic systems is supported by the observation that in vitro model systems, the reduction potential of the Cu(II) centre is pH dependent.

The second peroxide ion takes up the electron from Cu(I) and gets reduced to O_2^{2-}. Because of the reformation of Cu(II), proton is liberated from the histidine NH and the N^- gets bound to Cu(II) forming the bridge. The base O_2^{2-} combines with the liberated proton and goes away as O_2H^-. The mechanism of the reaction is shown in Figure 6.10.

The mechanism discussed above necessitates the presence of Zn(II) to hold the protein complex, when the imidazolate bridge with Cu(I) is broken. However, studies have shown that the enzyme activity is not affected, when Zn(II) is completely removed. This raises doubt about the above proton splitting mechanism.

Since O_2^- can combine directly with Cu(II) or Cu(I), the electron transfer in the redox process is by inner sphere mechanism.

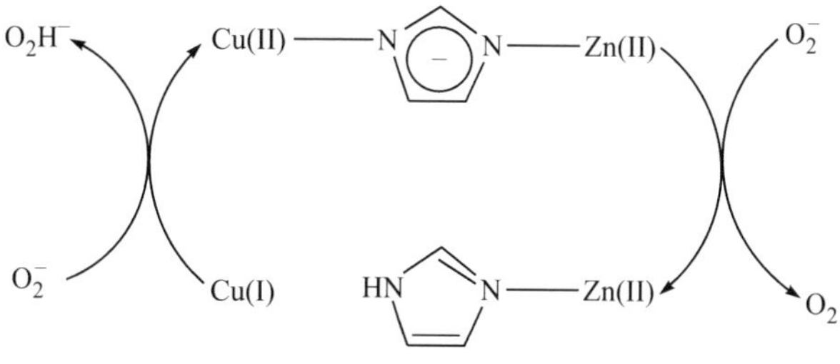

FIGURE 6.10 Mechanism of superoxide dismutase activity of Cu Zn SOD.

Various Cu(II) salts and Cu(II) complexes, which can undergo reversible redox reaction Cu(II)/Cu(I), can exhibit SOD activity in model systems. Complexes of other metal ions, which undergo reversible redox reaction, also cause dismutation of superoxide, into peroxide and molecular oxygen, in vitro systems. It is, therefore, thought why nature should evolve such a sophisticated enzyme Cu–Zn, SOD, for a reaction which can be catalysed by simple complexes. The answer could be that metal salts and complexes, exhibiting SOD activity, in vitro, may be toxic to the biological system and may not be stable in vivo conditions. SOD has evolved as an enzyme to provide a non-toxic catalyst, with SOD activity, and the enzyme is stable under the biochemical reaction conditions.

However, controversy persists about the toxicity of superoxide, as there are aerobic organisms, surviving with the possibility of having superoxide generated in their system, and have no SOD enzyme generated in their system. Thus, it is felt that though apparently this enzyme exhibits SOD activity, it may not be its only role. It may have other roles, yet to be ascertained.

Fe and Mn SOD

These are found in microorganisms. Fe SOD is dimeric and Mn–SOD is tetrameric. The subunit structure of the two is similar. The metal ion does not seem to have any structural role, because the structure of the apoprotein remains unaffected as the metalloprotein after demetallation. These enzymes serve as SODs, because the redox potentials of the metal ion site are comparable with those of O_2^-/O_2^{2-} and O_2^-/O_2 as in Cu Zn SOD.

Oxygen Transport Copper Protein – Hemocyanin

This is a copper protein, responsible for oxygen transport in invertebrate lower organisms, of the class molluscs (clams, octopus, squid), arthropods (spiders, crabs, lobsters) and annelids.

In cephalopods (Octopus), hemocyanin gets oxygenated at the gills and most of the oxygen is taken away, when the oxyhemocyanin passes through the tissues.

In the native form, hemocyanin is an exceptionally large protein, with molecular weight in the range of 4,500–9,000 K daltons. The molecule consists of large aggregates of subunits, each with molecular weight 50–70 K daltons. Each unit consists of a pair of Cu(I) bound to proteins.

Each Cu(I) in deoxyhemocyanin is bound to three amino acid residues, most probably the N atoms of the histidine residues. Sulphur coordination is ruled out, because in the electronic spectrum, bands corresponding to S → Cu(I) charge transfer are not observed. Extended X-ray absorption fine structure (EXAFS) study also does not show Cu–S bond. X-Ray and EXAFS studies show that the two Cu(I) centres are at a distance of 3.8 Å and are, most probably, bridged through the phenolate or carboxylate O^- of the protein. The geometry around Cu(I) centres is pseudotetrahedral, with the histidine nitrogens forming the trigonal base, with Cu(I) above the plane. The bridging O^- is bound at the fourth coordination site of Cu(I).

Deoxyhemocyanin is colourless. It exhibits absorption bands only in the UV region, due to the chromophoric groups of the protein. It is diamagnetic and ESR silent. This shows that the active metal centres in the protein are Cu(I). In different forms of the enzyme, the binuclear Cu(I) centre can undergo change in the oxidation state of the metal ions, and accordingly, there can be change in their ligational behaviour. On this basis, they are assigned different names (Figure 6.11).

As seen above, deoxyhemocyanin takes up one molecule of oxygen to form oxyhemocyanin. Like hemerythrin, (and unlike haemoglobin), two copper centres get bound to one oxygen molecule. Again unlike haemoglobin, the uptake of oxygen by hemocyanin is followed by one electron oxidation of each of the two Cu(I) centres and O_2 gets bound as peroxide, at the two Cu(II) centres. Oxyhemocyanin exhibits two bands at 345 and 570 nm and weak shoulders in the region 425–700 nm. The band at 570 nm corresponds to d–d transition of Cu(II), and this leads to blue colour in oxyhemocyanin. This implies that deoxyhemocyanin takes up one molecule of O_2, with consequent change in the oxidation state of Cu(I) to Cu(II). That is why, the oxygen carrying protein, in small animals such as spiders, crabs and octopus, is blue in colour, and they are said to be having blue blood.

The intense band at 345 nm in the electronic spectrum has been assigned to O_2^{2-} → Cu(II) charge transfer. As this band is absent in met hemocyanin, it cannot be due to protein → Cu(II) charge transfer.

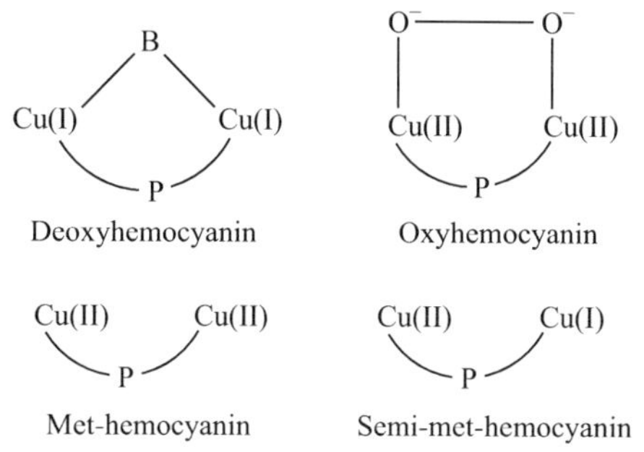

FIGURE 6.11   Oxidation state dependant ligation behaviour of copper centres in hemocyanin.

Resonance Raman spectrum of oxyhemocyanin shows $v_{o^- - o^-}$ band at 744 cm and supports the presence of m peroxo bridge. The peroxide is symmetrically bridged to the two Cu(II), in one of the following three forms shown in Figure 6.12.

Both met hemocyanin and oxyhemocyanin are diamagnetic and ESR silent, due to anti-ferromagnetic spin coupling of unpaired electrons over the two Cu(II) centres. The super exchange, in both the cases, is through the endo-bridge atom of the protein. Thus, in oxyhemocyanin, the super exchange is not through the bridging peroxide.

Like in haemoglobin, O_2 uptake in hemocyanin is cooperative. The subunits take up oxygen, simultaneously. The probable reason, for the inter subunit cooperativity, is that the pseudotetrahedral geometry, available in the protein, is favoured by Cu(I). Furthermore, the cavity has a lining of hydrophobic residues, which does not favour the formation of charged residues. However, with a bigger Cu(I) ion in the cavity, the subunits in deoxyhemocyanin are in a tense (T) state, as in haemoglobin. With the uptake of O_2, Cu(I) gets oxidised to Cu(II), and the cationic size is reduced from 0.95 Å (Cu$^+$) to 0.72 Å (Cu^{2+}). The distance between two Cu(II), in a subunit, decreases by 0.2–0.3 Å. Hence, the distance between the opposite helices changes by 0.6–0.7 Å. Thus, the T form of the protein changes to R (relaxed) form. Hence with the uptake of O_2 at one centre, other centres open up to take O_2 in a cooperative way.

Considerable amount of work on synthesis of binuclear Cu(I) model complexes, mimicking the oxygen uptake by hemocyanin, has been undertaken .Binuclear Cu(I) complex of binucleating ligand, with two tridentate coordinating sites and a bridging phenolate O$^-$, has been shown to take up one O_2 molecule (Figure 6.13).

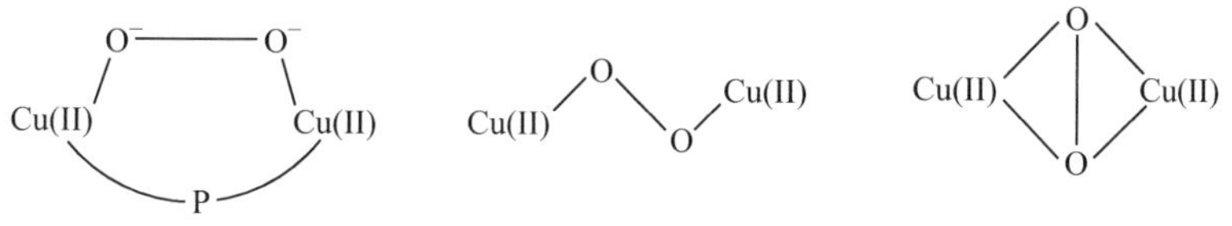

FIGURE 6.12   Possible bridging modes of peroxide in oxyhemocyanin.

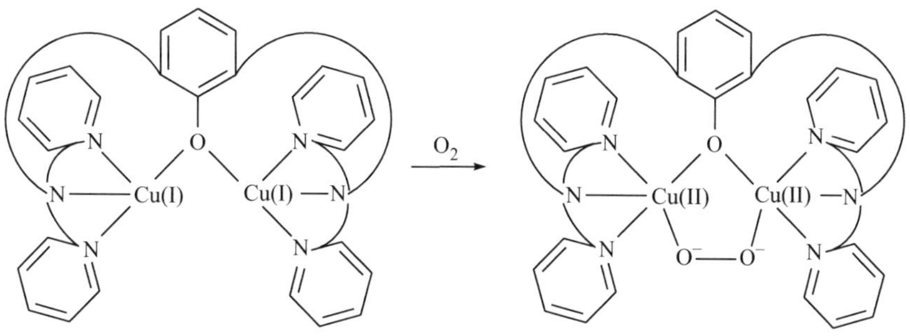

FIGURE 6.13   Oxygen uptake by binuclear Cu(I) complex as model for hemocyanin.

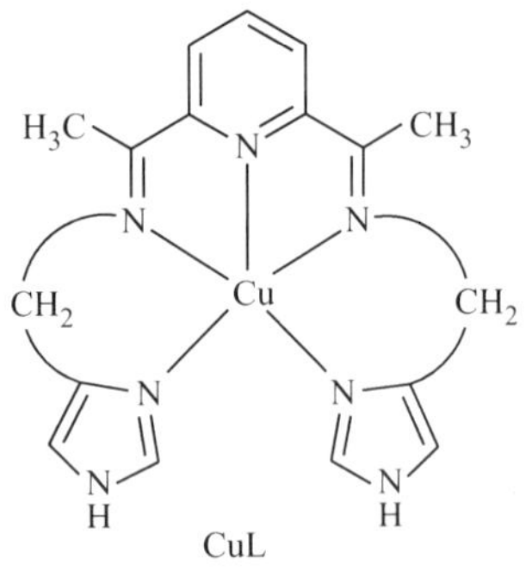

FIGURE 6.14   Mononuclear Cu(I) complex as model for hemocyanin.

$$2CuL + O_2 \longrightarrow LCu—O—O—CuL$$

FIGURE 6.15   Oxygen uptake by mononuclear Cu(I) complex as model for hemocyanin.

Another mononuclear Cu(I) complex of a pentadentate ligand (Figure 6.14) takes up one O_2 molecule, involving two molecules of the complex. The Cu(I) of the two complex molecules loses one electron each to the O_2 molecule, and the resulting O_2^{2-} gets coordinated to the resulting two Cu(II) centres (Figure 6.15).

The deoxygenated solution of the Cu(I) complex has red colour. If it is exposed to atmosphere, it takes up one oxygen molecules to form a green solution. On heating the solution gently (40°C) and passing N_2 gas through the solution, or vigorously shaking the solution, the blue colour changes to red, showing that there is a reversible uptake and release of O_2, and there is reversible change between red mononuclear Cu(I) complex and the green, peroxo-bridged, binuclear Cu(II) complex.

Copper Transport

i. **Ceruloplasmin:** This is an intensely blue-coloured copper protein, present in the blood plasma of the vertebrates; 100 mL of human plasma contains 20–40 mg of the protein. Besides blood, it is also present in spinal and joint fluids and the secretions of eyes, ear and digestive system. It is a glycoprotein and has nine to ten saccharide chains, each containing different kinds of carbohydrates. It is synthesised in the liver; 95% of copper in the liver is bound to ceruloplasmin.

The molecular weight of the protein varies in different classes of animals, human beings 140 K daltons, rabbit 142 K daltons and horse 120 K daltons. The protein has tetrameric structure and each subunit contains six to seven copper atoms. Chemical analysis, and electrochemical and magnetic studies have shown that half of the copper is present as Cu(II) of type I, type II and type III, and the other half is Cu(I).

The copper is bound very strongly with the protein and cannot be exchanged by radioactive Cu(II), added externally. The stability of the Cu–protein bond is more, due to the stability of the total structure of the protein. If the enzyme is broken into smaller units, Cu–protein bond is weakened. It breaks into dimer in guanidine solution and dissociates completely in a 0.2% mercaptoethanol solution.

Copper can be removed from the protein by the addition of diethyldithiocarbamate. The apoprotein undergoes conformational change and loses enzymatic activity. However, the activity is revived on the addition of Cu(II).

Ceruloplasmin acts as a copper transport protein. It provides copper to cytochrome C oxidase and other copper-containing oxidases. Albumin, which transports copper between intestine and liver in the portal vein, also takes copper from ceruloplasmin.

Ceruloplasmin also acts as a catalyst in the oxidation of polyphenols, polyamines, adrenalin, serotonin, etc. It also has the role of sequestering excess Cu(II) and stores it in the blood plasma, so that the delirious toxic effect of catalytic oxidative reaction by free Cu(II) is prevented, and ceruloplasmin-bound Cu(II) is made available to the Cu(II)-containing enzymes, as per their requirement.

Ceruloplasmin has also been recognised to have ferroxidase activity. As discussed in Chapter 5, iron is released in the plasma from ferritin in the form of Fe(II). It is transported by transferrin for the synthesis of haemoglobin. For binding with the protein apotransferrin to form transferrin, iron must be oxidised to Fe(III) form. It has been suggested that the oxidation of Fe(II) to Fe(III) is catalysed by ceruloplasmin, thus exhibiting ferroxidase activity. This indicates that haemoglobin synthesis depends on ceruloplasmin.

Deficiency of ceruloplasmin causes a disease, called Wilson's disease. In this genetically acquired metabolic defect, copper cannot be stored in the bound form in ceruloplasmin, as normal in the body system, and hence, copper gets deposited in liver, brain, eyes and kidney. This leads to liver and kidney failure and various neurological disorders and formation of brown or green rings in the cornea of the eyes.

It was observed that though there is deficiency of ceruloplasmin in the case of Wilson's disease, the patient does not suffer from marked deficiency of haemoglobin. This shows that there must be copper proteins, other than ceruloplasmin, performing the ferroxidase activity, necessary for the synthesis of haemoglobin.

However, it has been observed that, if there is deficiency of ceruloplasmin, there is accumulation of iron in the body. This may be because ceruloplasmin assists the oxidation of Fe(II) to Fe(III) in the transport of iron as transferrin, necessary for haem synthesis. If there is deficiency of ceruloplasmin, the synthesis of haemoglobin is retarded and the excess of Fe(II) gets deposited in various parts of the body and causes many diseases.

ii. **Metallothionein:** Thioneins are proteins, found in animals and plants. In animals, it is found in the liver and kidney, and has a strong tendency to combine with metal ions with soft acid character.

It is a small protein, with molecular weight 6–10 K daltons. Its structure has been determined in solid state by X-ray and, in solution, by NMR. The structures of thionein from different sources have certain similarity. There are seven units in the structure of thionein, each with two cysteine and other amino acid residues. The protein does not show absorption in the UV region, indicating that there is no aromatic amino acid residue present in the protein.

Thionein binds four or eight metal ions in the form of cluster, through cysteine sulphurs. So, the protein prefers soft metal ions. The metal ion is in a tetrahedral site. This has been inferred by observing that EXAFS, EPR and electronic spectra of Co(II) thionein are similar to Co(II) bound to rubredoxin, where Co(II) is bound to four cysteinyl groups, in a tetrahedral way.

When there is an excess of the essential metal ion, like Cu(II) or Zn(II), or of non-essential metal ions such as Cd, Ag, Au and Hg, thionein gets strongly bound to them, forming metallothionein. Thus, it has the role of storage of the essential metal ions and sequestration and detoxification of the non-essential metal ions.

It has been observed that whenever there is an excess of metal ion in the biological system, more thionein is produced. The metal ion activates the genes which produce thionein. The nature has provided the protein thionein to counteract the toxic effect of excess metal ions.

iii. **Role of metal ion in causing different diseases**

  a. **Wilson's disease:** In the case of deficiency of ceruloplasmin, there is leakage of copper into the blood plasma. There is excess accumulation of free copper in the liver, and Cu(II) also passes to different tissues. This excess copper deposition in liver and other organs leads to Cu(II)-catalysed oxidation of the tissues, and there is inflammation and cirrhosis of tissues. When excess copper gets deposited in the brain, it causes neurological problem. This pathological condition is called Wilson's disease.

b. **Menkes disease:** In this disease, unlike Wilson's disease, copper uptake in the liver by ceruloplasmin and other proteins is normal. But in this disease, the intestinal capacity of the patient to absorb the copper ingested in the diet is less. This is due to the weakening of a gene that encodes an ATPase enzyme that works as a pump for Cu transport through the intestine. This results in lower availability of Cu(II) to the liver, and hence, the formation of copper-containing enzymes is less. The lower availability of Cu in the brain causes neurological impairment. If the availability of Cu(II) to the brain in the foetus stage is less, the neurological problem is serious after birth and causes brain damage and death in the childhood.

c. **Alzheimer's disease:** In this disease also, there is dementia of the brain, leading to impairment in body functions. Copper has a role in the origin of this disease. Excess of copper in the brain can get bound to β-amyloid peptide, resulting in the formation of plaques. This copper deposition causes oxidative damage to the brain. Furthermore, it has been observed that due to the formation of plaques, the distribution of Cu in the brain is not uniform. There is excess Cu in parts of the brain with plaque deposits, whereas in the other parts, there is deficiency of copper in other neighbouring cells. This may cause neurological disorders and impairment of related body functions. However, work is still in progress to understand the exact role of copper in the case of Alzheimer.

d. **Cancer:** Increase in the body tissues of ceruloplasmin and free copper could be a probable cause for cancerous growth in the tissues. Excess Cu(II) can generate reactive oxygen species and free radicals, which can cause malignancy. Furthermore, it is known that copper proteins cause metabolic changes in cancer cells, and angiogenesis due to proliferation and migration of the endothelial cells in human beings.

QUESTIONS

1. On what basis are the copper proteins present in biological systems classified?
2. List major functions of copper proteins in biosystems.
3. What are blue copper proteins? Why are they called so?
4. Discuss structural features of blue copper proteins plastocyanin, azurin, and stellacyanin.
5. Why do the blue copper proteins exhibit high reduction potential at Cu(II) centre?
6. The blue oxidases are different from the blue copper proteins involved in electron transfers in biochemical reactions. Explain.
7. Describe mechanism of electron transfer reaction from substrate to O_2 mediated by blue oxidases.
8. Name the protein responsible for synthesis of pigment melanin.
9. Why is tyrosinase EPR silent?
10. Describe activity of some non-blue copper protein in biochemical processes.
11. Is it possible that copper in oxidase enzyme does not undergo redox change during its activity? Give example.
12. What is superoxidase dismutase activity? Which enzymes perform this task?
13. How is oxygen transported in invertebrate lower organisms? Give structures of oxy- and deoxy-forms of the enzyme.
14. Give the experimental evidence for μ peroxo bond in oxyhemocyanin.
15. Is ceruloplasmin classified as blue copper protein? What are its functions?

7

Cobalt in Vitamin B_{12} in Biochemical System

Vitamin B_{12} is a tetraaza macrocyclic ligand complex of Co(III). The structure is similar to myoglobin, with the difference that the ligand is corrin, containing fifteen atoms in the ring, one less than in porphyrin. One methine (=C–H) bridge between one pair of pyrrole rings is absent (Figure 7.1). The peripheral carbon atoms, of the corrin ring system, are substituted by seven amide moieties. C_2, C_7 and C_{18} are substituted by acetamide, and C_3, C_8 and C_{13} bear propionamide. C_{17} has N-substituted propionamide.

On coordination with Co(III), only one – NH of the ligand gets deprotonated, and hence, the macro cycle acquires negative charge. The Co(III) is in the plane of the macrocyclic ring and is coordinated to the four pyrrole nitrogens of the corrin ring. As the carbon atoms linking A and D pyrrole rings are sp³ hybridised, the fourteen π electrons of the ligand mono anion are delocalised over nine carbon and four nitrogen atoms of the ring and the metal ion.

The Co(III) corrin complexes, which have the group a D-ribofuranose-3-phosphato-5,6-dimethyl benzimidazole, bound to the corrin ring, through the amide group at C_{17} of the ring and coordinated to Co(III) at the axial fifth position, through benzimidazole, are called cobalamins. However, if the sugar and phosphate of the side group get removed by hydrolysis, and the fifth coordination position is occupied by any heterocyclic ligand, other than benzimidazole, the complex is called cobalamide (Figure 7.1).

In cobalamin, vitamin B_{12}, the sixth coordination position of Co(III) is occupied by cyanide ion (Figure 7.1), and hence, it is also called cyanocobalamin. It is water soluble, and is red in colour.

Only small traces of vitamin B_{12} are required for the growth of animals. In the human blood, about 2×10^{-4} mg/mL of it is present. Cobalamin, bound to blood plasma protein fractions, is carried to the tissue, where it is bound to a variety of protein receptors. Any excess is stored in the liver as vitamin B_{12} coenzyme. Out of the total 2.5 mg of cobalamin in human adults, only 1.5 mg is in the liver.

Though vitamin B_{12} is essential for higher animals, it is not required by plants. It cannot be synthesised by animals or plants, but is biosynthesised by certain bacteria. It is obtained through the food. It is absorbed through the ileum, by a mucopolysaccharide, present in the gastric juice. Pernicious anaemia is caused if there is either deficiency of vitamin B_{12} in the diet, or failure to absorb it from the food. Patients, who undergo total removal of stomach, cannot absorb cobalamin and hence develop pernicious anaemia. This indicates that the vitamin B_{12} is essential for the synthesis of haemoglobin.

The first type of corrin ring complex, cobalamin, isolated, was cyanocobalamin. It was given the name vitamin B_{12} and is still known by that name. However, this is very inert in nature and does not act as a coenzyme for any apoenzyme protein. The vitamin B_{12} derivative, which is physiologically active and can act as a cofactor for an apoenzyme protein is the one, which has 5'-deoxyadenosyl unit linked through –CH_2 carbon atom at the sixth coordination position of Co(III). With Co(III) bound to the C atom (Figure 7.1), this is an organometallic compound and the first example of occurrence of organometallic compound in biological systems. This is known as B_{12} coenzyme and combines with the apoenzyme protein, to form the biologically active enzyme.

Vitamin B_{12} coenzyme is water soluble and orange in colour. 80% of vitamin B_{12} (cobalamin) in mammals and birds is present in the form of vitamin B_{12} coenzyme. The naturally occurring B_{12} coenzyme of the above type with nucleotide bound to Co(III) at sixth position is considered complete. The others, without the nucleotide base at the sixth position, are called incomplete, like cyanocobalamin. The most complete B_{12} coenzyme has 5,6-dimethyl benzimidazole as the nucleotide base. However, if the coordination, at the sixth coordination site, of the nucleotide base is from adenine, instead of 5,6-dimethyl benzimidazole, it is called pseudovitamin B_{12}.

Both vitamin B_{12} and coenzyme are diamagnetic, as expected for low spin d⁶ Co(III) ion.

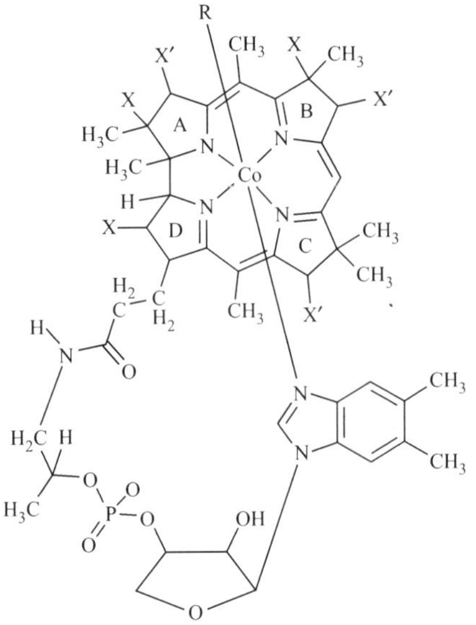

FIGURE 7.1   Vitamin B_{12}.

The other derivatives of cobalamin are vitamin B_{12}(a), aqua cobalamin, with water molecule in the sixth position and vitamin B_{12}(c), nitrocobalamin, with nitro group at the sixth position. Vitamin B_{12} can be reduced to vitamin B_{12}(r) with Co(III) converted to Co(II) and then to vitamin B_{12}(s) with cobalt in monovalent state (Figure 7.1).

- **Cobalamin:** Benzimidazole of nucleotide coordinated at the fifth position of Co(III) and the nucleotide is attached to the 17th carbon of the ring through amide linkage.
- **Cobalamide:** Benzimidazole nucleotide replaced by other heterocycle, at fifth coordination position.

Sugar and phosphate are removed due to hydrolysis.
$X = -CH_2CONH_2$ acetamide.
$X^1 = -CH_2-CH_2-CONH_2$ propionamide.
If R = CN cyanocobalamin, vitamin B_{12}.
If R = H_2O aquo cobalamin, vitamin B_{12}(a).
If R = nitro, Nitro cobalamin, vitamin B_{12}(c)

If R = 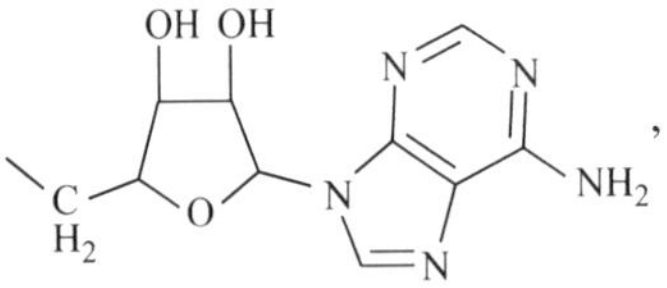   , vitamin B_{12} coenzyme

If R = CN and cobalt is Co(II) vitamin B_{12}(r)
If R = CN and cobalt is Co(I) vitamin B_{12}(s)

The electronic spectrum of vitamin B_{12} is similar to that of the free corrin ligand. This indicates that the colour in both vitamin B_{12} and coenzyme is due to $\pi \rightarrow \pi^*$ intraligand transition.

Co(II)B_{12}(r) is brown in colour and shows an additional band at 600 nm. It also shows a signal in the EPR spectrum, corresponding to one unpaired electron in the dz^2 orbital, as expected in the low spin Co(II) complex. It is like a free radical and can combine with O_2 forming a dioxygen complex.

Vitamin B_{12}(s) with Co(I) has low spin d^8 configuration and is diamagnetic.

X-Ray crystallographic studies were carried out by H C Hodgkin to establish structure of vitamin B_{12} (cyanocobalamin) and vitamin B_{12} coenzyme (5′-deoxy-5′-adenosyl) cobalamin.

In Co(II)alamin (vitamin B$_{12}$r), the ESR study shows the presence of unpaired electron at the Co(II) centre. The structure of Co(I)alamin (vitamin B$_{12}$s) is still not fully known.

X-Ray crystal study of the enzyme from *Escherichia coli* (with methionine synthetase activity) with the cofactor cobalamin and the apoprotein shows that the corrin macrocycle exists between two domains, one created by the helical amino terminal and the other by α/β carboxyl terminal of the apoenzyme protein.

It is seen that there is a change in the conformation of methyl cobalamin on getting bound with the apoprotein. The coordination at the fifth position of Co(III) from the N of the dimethyl benzimidazole in the free enzyme breaks and the ligand bound to the corrin ring moves away. Instead, a histidine of the apoprotein gets bound at the fifth position.

Redox Changes in Cobalt in B$_{12}$ Enzyme

The Co–C centre of vitamin B$_{12}$ coenzyme is not strong enough and can be broken down in three ways, during catalysis by B$_{12}$ enzyme:

i. Co–C bond undergoes homolytic split, producing the free radical R$^{•}$, and Co(III) gets converted to Co(II):

$$R \bullet \bullet Co(III)B_{12} \rightarrow R^{•} + Co(II)B_{12}$$

Thus, homolytic fission involves one electron redox process.

ii. Heterolytic fission results in the formation of a carbocation and Co(I), and involves a two electron redox process:

$$R \bullet \bullet Co(III)B_{12} \rightarrow R^{+} + Co(I)B_{12}$$

iii. Heterolytic fission can also result in the formation of Co(III) and a carboanion R^{-}:

$$R \bullet \bullet Co(III)B_{12} \rightarrow R^{-} + Co(III)B_{12}$$

Thus, under physiological conditions B$_{12}$ can exist in three different oxidation states of cobalt: III, II and I. This possibility of oxidation reduction is of great significance in the biological role of B$_{12}$ enzyme. According to Jack Halpern, the fact that the cobalt can exist in three different oxidation states makes CH$_3$–B$_{12}$ a powerful methylating agent, better than Grignard's reagent.

The number of ligands in the fifth and sixth positions depends on the oxidation state of the cobalt ion. The number of axial ligands decreases as the oxidation state goes down. Co(III) can get bound to two axial ligands as in cobalamins, whereas Co(II) and Co(I) can get bound to one axial ligand and no axial ligand, respectively.

In the vitamin B$_{12}$ coenzyme, which is complete, with the axial coordination of the nucleotide base at the sixth position and 5,6-dimethyl benzimidazole at the fifth position, the Co(III) corrin ring gets stabilised and cannot be easily reduced. Hence, reduction of alkyl Co(II) corrin takes place at more negative potential than reduction of B$_{12}$r/B$_{12}$s.

Formation of Vitamin B$_{12}$ Coenzyme from Vitamin B$_{12}$ and the Mechanism of Alkylation Reaction

In the conversion of vitamin B$_{12}$ to vitamin B$_{12}$ coenzyme, there is substitution of CN^{-} by adenosyl group. This takes place in two steps. First, a reducing enzyme causes two electron reduction of Co(III) to Co(I). The vitamin B$_{12}$(s) formed reacts with ATP in the next step to form vitamin B$_{12}$ coenzyme, as shown below:

$$Co(III) B_{12} \xrightarrow{\text{2e}} Co(I) B_{12}s \xrightarrow{\text{ATP}} Co(III) \text{ adenosyl } B_{12} \qquad (7.1)$$

This mechanism is supported by the fact that both Co(I) and Co(II) containing cobalamins react with alkyl derivatives (RI) to form Co(III) alkyls:

$$Co(II)B_{12}(r) + RI \rightarrow R-Co(III)B_{12} - I \tag{7.2}$$

$$Co(I)B_{12}(s) + RI \rightarrow R-Co(III)B_{12} - I^- \tag{7.3}$$

Due to low oxidation state and square planar structure, vitamin $B_{12}(r)$ and vitamin $B_{12}(s)$ act as nucleophiles with RI.

The above alkylated vitamin B_{12} compounds are examples of organometallic compounds, present in biochemical systems, other than vitamin B_{12} coenzyme.

The Co(III)–C bond has a dissociation energy of 100 KJ/mol and is a weak bond. Hence, the alkyl group can be easily broken thermally or photochemically. The methyl group can also be transferred easily to another nucleophile, causing its alkylation:

$$CH_3 - Co(III)B_{12} + X \longrightarrow CH_3X + Co(II)B_{12}(r) \tag{7.4}$$

However, if this is the mechanism, the above alkylation reaction should be slow, as in the non-enzymatic model systems (Figure 7.4). The faster alkylation in enzymatic system shows that it must be involving another pathway. Recently, it has been suggested that the first step of the reaction involves homolytic fission of the Co–R bond, as shown in the following:

$$R - Co(III)B_{12} \longrightarrow \dot{R} + Co(II)B_{12}r$$
$$\dot{R} + SH \rightarrow \dot{S} + RH \longrightarrow R - S \tag{7.5}$$

The free radical $\dot{R}$ combines with the substrate SH to form substrate free radical $\dot{S}$ and RH. Then, there is a rearrangement reaction, resulting in the alkylated substrate R–S. The mechanism of the last stage is not established yet.

Viewed as an anion, the alkyl group is strongly electron donating, and hence, the electron density on Co(III) increases in R–Co(III)B_{12}. The corrin ligand, with p conjugated system, reduces the electron density on the cobalt, by accommodating the extra electron density in the lower energy vacant π* orbitals.

However, excess π delocalisation in the ligand is not favourable for the optimum stability of the M–C bond in R–Co(III)B_{12}. Corrin ring has extensive, but not excessive, π conjugation, which is suitable for the alkylation reaction, affected by R–Co(III)B_{12}.

The Co–C bond strength is also dependent, on the presence of the protein part of the enzyme. In the model alkylated Co(III) system, the alkylation reactions can be affected by generating free radicals, using photolysis or thermolysis. However, in the enzymatic system R–Co(III)B_{12}, the metal-alkyl bond undergoes self-homolytic fission, during alkylation of the substrate, showing that the protein has a major role in inducing homolytic fission of R–Co(III) bond. The coenzyme part is in the resting state, and the protein is strained. The protein part modulates the strength of Co–C bond. In the model systems, it has been shown that on changing the base at the sixth position, there is conformational distortion of the equatorial bonds, inducing Co–C bond weakening and lengthening.

Studies in the model systems have revealed that if the bulk of the group around the carbon bound to Co(III) is increased, there is a labilisation of the Co–C bond. This is because of the steric hindrance of the alkyl group with the protein part on the complex, and the strain on the complex part is increased, resulting in weakening of Co–C bond. It has, therefore, been suggested that on the approach of the substrate near to the enzyme, the system, with strained protein and resting coenzyme, is changed to strained coenzyme and relaxed protein. This conformational change in the coenzyme determines the strength of the Co–C bond and can switch the alkylation reactions, on or off. Thus, for facilitating alkylation reaction, the macrocyclic ligand has to be flexible. For this reason, the nature has selected 15-membered corrin, rather than 16-membered porphyrin, as ligand, in vitamin B_{12}.

Roles of Vitamin B$_{12}$

i. **Alkylation of organic compounds and metals:** As stated above, Co(III)B$_{12}$–R has a weak Co–C bond, and hence, the alkyl group can easily be transferred to a substrate, causing its alkylation. First, Co(I)B$_{12}$(s) gets alkylated by R–X. The resulting Co(III)B$_{12}$–R causes alkylation of the substrate. The catalytic cyclic reaction is shown in Figure 7.2.

The alkyl Co(III) corrins are resistant to proteolytic cleavage of Co–C bond under normal oxidations. However, in the protein bound B$_{12}$ coenzyme, in presence of visible light, the breaking down of Co(III)–R bond is facilitated. This may be because of the change in the structure of Co(III)–B$_{12}$, due to the binding of the apoprotein, as discussed earlier.

Reaction involving the transfer of methyl group are of significance in reactions in biochemical system, of the type, anaerobic acetogenesis, methanogenesis and breaking down of acetic acid into methane and CO$_2$. There are several compounds which act as source for alkyl groups, such as methyl alcohol, methyl amines and phenyl methyl ethers.

Following are the two examples of methylation reaction:

a. The enzyme methionine synthetase in E coli, involving vitamin B$_{12}$(r) cofactor, causes methylation of homocysteine, converting it to methionine, using the methylating agent CH$_3$–THF (N^5 ethyl tetrahydrofolate) (Eq. 7.6).

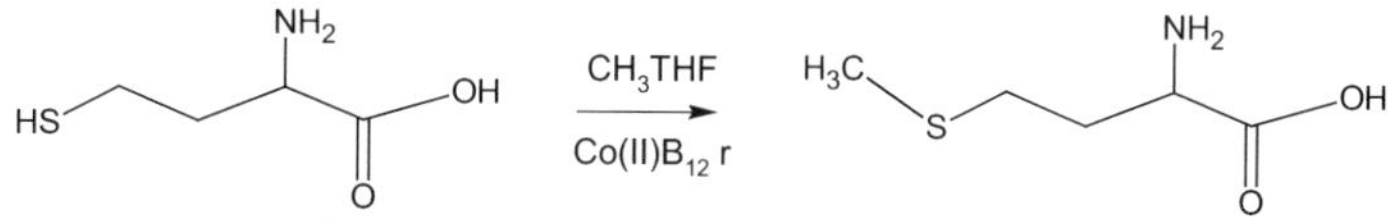

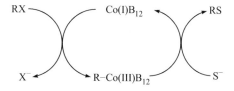

$$\text{(7.6)}$$

Homocysteine　　　　　　　　　　　　　　　　Methionine

b. Methylated cobalamin, CH$_3$–Co(III)B$_{12}$, can transfer methyl group to Hg(II) forming methyl mercury, (CH$_3$ – Hg$^+$) and dimethyl mercury, (CH$_3$)$_2$Hg. Thereby, CH$_3$–Co(III)B$_{12}$, is converted to Co(II)B$_{12}$r.

In nature, metal alkyls are biosynthesised, due to the response of the microbes to the toxicity, generated by the metal ions. In case of toxicity due to metals such as lead, tin, platinum, gold, selenium and arsenic, there is biomethylation of the metals, assisted by CH$_3$–Co(III) B$_{12}$. However, toxicity is not completely eliminated, as the methylated metals are also toxic in nature.

ii. **Generation of methane:** Methyl–Co(III)B$_{12}$ is also found to react with molecular hydrogen to form methane (Eqs. 7.8 and 7.9). It has been suggested that the methyl group, on cobalt, is not directly reduced and displaced, but is transferred to a coenzyme with Ni (II)-containing prosthetic group (to be discussed in Chapter 9), to give methyl coenzyme M. This is followed by reductive cleavage of CH$_3$ coenzyme M to form methane.

When the substrate gets bound with the B$_{12}$ coenzyme, the Co(III)–C bond breaks and this is followed by the formation of 5′-deoxy-5-adenosyl radical and Co(II)B$_{12}$. The radical abstracts H atom from the substrate, and a new radical is formed. This new radical reorganises to form the product. There is little role of the enzyme in this reaction.

FIGURE 7.2 Catalytic cycle for alkylation of substrate by Co(III)B$_{12}$.

In this homolytic fission reaction, the formation of 5′-deoxy-5′-adenosyl radical and $Co(II)B_{12}$ is supported by the ESR spectral study, which shows two signals corresponding to the radical electron and the unpaired electron in $Co(II)$.

The isomerisation reaction can be illustrated by the following reactions:

$$CH_3 - Co(III)B_{12} + CoenzM \rightarrow Co(II) - B_{12}r + CH_3 - CoenzM \tag{7.7}$$

$$CH_3CoenzM + H_2 \rightarrow CoenzM - H + CH_4 \tag{7.8}$$

The cell free extracts of bacteria, *Methanosarcina barkeri* and *Methanosarcina Omelianski*, containing $Co(III)B_{12}$, form methane in the presence of CO_2, formate, pyruvate and other compounds, containing carbon and nitrogen.

iii. **Isomerase activity:** Many enzymes, with $Co(III)B_{12}$ cofactor, can cause exchange of groups on vicinal carbon atoms of the substrate. For example, dialdehydrase enzyme, which requires vitamin B_{12} coenzyme, catalyses the conversion of diol to aldehyde. (Eq. 7.9). In the intermediate stage of this reaction, there is 1, 2 interchange. This is followed by dehydration (Eq. 7.10).

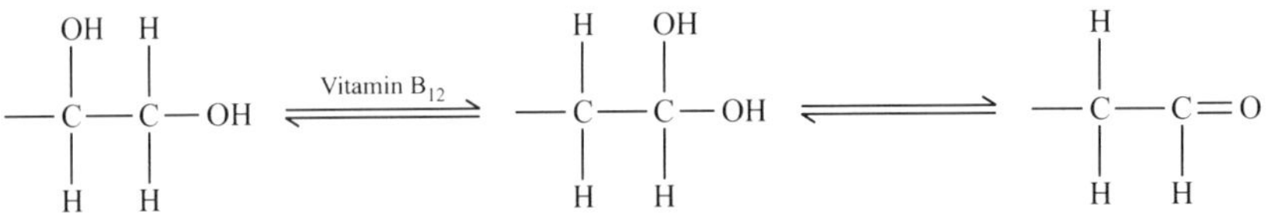

$$\tag{7.9}$$

Isomerisation reaction, like conversion of glutamic acid to β-methyl aspartic acid, is assisted by the enzyme glutamate mutase, involving vitamin B_{12} as coenzyme (Eq. 7.10). This is the first step in the metabolism of glutamate in the biochemical systems.

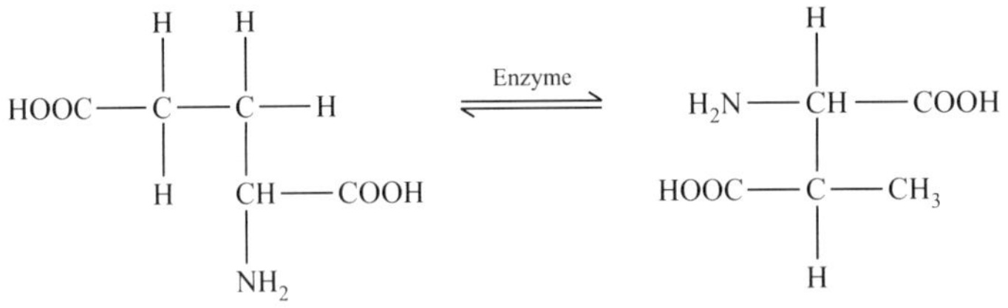

Glutamic acid   β-Methyl aspartic acid   (7.10)

Another vitamin B_{12}-containing enzyme is methyl malonyl CoA mutase, which catalyses the conversion of methylmalonyl–CoA into succinyl CoA (Eq. 7.11).

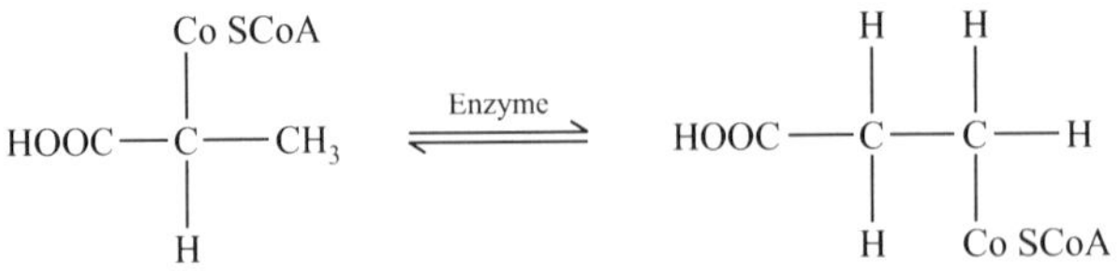

Methyl malonyl CoA   Succinyl CoA   (7.11)

In case of pernicious anaemia, because of deficiency of vitamin B_{12}, there is an increased excretion of methyl malonic acid in urine. This supports the requirement of vitamin B_{12} in the conversion of methyl malonic acid to succinic acid.

iv. **Deamination reaction:** Vitamin B_{12}-containing enzymes can cause deamination reactions, converting amino alcohols into aldehyde. The reaction is shown below:

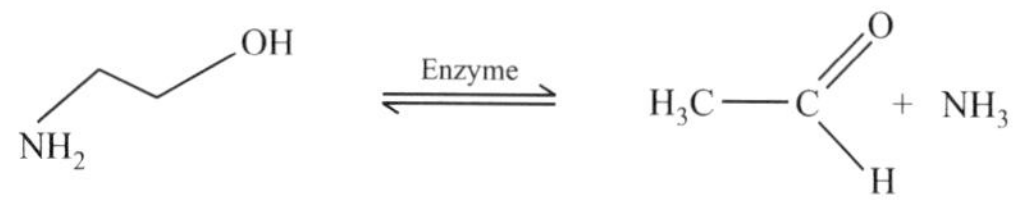

$$(7.12)$$

v. **Reduction reactions:** Co(I) cobalamin is a strong reducing agent and can reduce O_2 to H_2O. Co(I) cobalamin also assists the reduction of CHOH group of the ribose of ribonucleotide triphosphate to $-CH_2$, catalysed by the enzyme ribonucleotide reductase (Eq. 7.12).

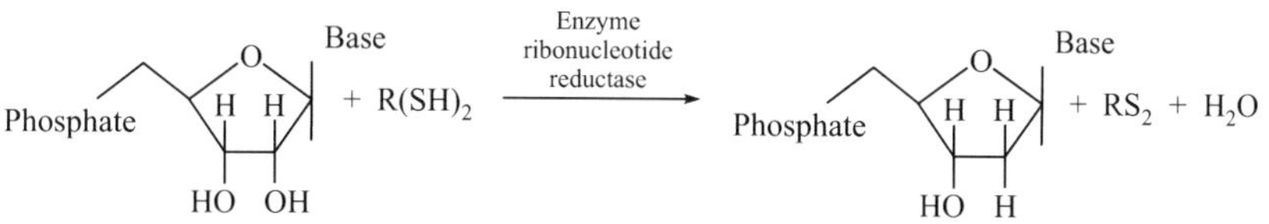

$$(7.13)$$

vi. **Reaction of vitamin B$_{12}$ with nucleotides:** Recent studies have revealed that coenzyme B$_{12}$ can combine with the RNA directly and cause genetic changes. Thus, B$_{12}$ coenzyme has control over genetic processes and this is termed B$_{12}$ responsive genetic switches. Laboratory studies have also shown that RNA compounds can be synthesised, which bind with vitamin B$_{12}$ coenzyme. This shows that the B$_{12}$ compounds have some role in the biochemical processes involving RNA.

vii. **Role in synthesis of blood:** Vitamin B$_{12}$ is known to have a role in the synthesis of blood in association with folic acid. Exactly how it works is a matter of further study.

Blood synthesis takes place in three steps:

i. Transport of iron from bone marrow to mitochondria in the form of Fe(III) transferrin.
ii. Synthesis of the macrocyclic compound protoporphyrin in mitochondria and cytosol.
   The above two reactions have been discussed in Chapter 5, "Synthesis of Haem". However, in both these processes, there is no role of vitamin B$_{12}$.
iii. In the third step, Fe(III) of transferrin is reduced to Fe(II) and is liberated to combine with protoporphyrin and form haem.

In the reduction of Fe(III) of transferrin, it gets bound to the cell forming a vesicle (endosome), which is a piece of membrane with transferrin bound to it. At low pH, there is demetallation and the liberated Fe(II) is taken up by protoporphyrin in the immature blood cells (reticulates or retics) which are very active in iron uptake and form haemoglobin. The reduction of Fe(III) of transferrin to Fe(II) depends on the binding sites of transferrin to the erythrocytes. Hence, formation of healthy erythrocytes is essential for the formation of healthy RBC. Vitamin B$_{12}$, folate and iron are essential for the production of erythrocytes. Thus, at this stage, vitamin B$_{12}$ has role in the synthesis of blood.

Erythrocytes are produced by the process of erythropoiesis by hematopoietic tissue of the bone marrow. The process begins with the hematopoietic stem cells and continues to erythrocytes formation in several steps. Normal human erythrocyte has an average life span of 120 days. In normal human adult, about 1% of the total erythrocytes die and new ones are produced daily. This approximates to about 200 billion erythrocytes. In case of loss of RBCs due to bleeding or some other cause, rate of formation of erythrocytes increases and can reach about one trillion. A glycoprotein hormone, viz., erythropoietin (EPO), regulates the erythropoiesis process. EPO is produced in renal cortex in kidney. There is decreased response of hematopoietic tissue of the bone marrow to EPO when there is deficiency of any of the three nutrients – iron, vitamin B$_{12}$ or folate.

Plant and animal tissues contain folate, generally as 5-methyl-tetrahydrofolate (5-CH$_3$-THF). One of the functions of vitamin B$_{12}$ is to transfer a methyl group from 5-methyl-THF to homocysteine via methylcobalamin, thereby regenerating methionine (Figure 7.3). This reaction represents the link between folate and vitamin B$_{12}$ coenzymes and accounts for the requirement of both vitamins in

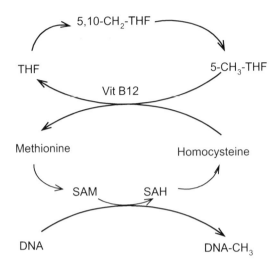

FIGURE 7.3 Abbreviations: THF, tetrahydrofolate; 5,10-CH$_2$-THF, methylenetetrahydrofolate; SAM – s-adenosyl methionine; SAH – s-adenosyl homocysteine.

normal erythropoiesis. Methionine-mediated methylation of DNA is considered an important step in the process of erythropoiesis. Methionine is activated to s-adenosyl methionine (SAM) which serves as methyl group donor, leaving behind s-adenosyl homocysteine (SAH) and methylated DNA. SAH is then cleaved to adenosine and homocysteine. The regeneration of methionine from homocysteine requires vitamin B$_{12}$ in the form of methyl cobalamin. With deficiency of vitamin B$_{12}$ or folate, the DNA synthesis as well as repair is affected, hence insufficient number of reticulocytes are formed, eventually causing anaemia.

The overall process is shown in the following scheme.

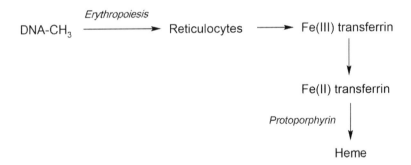

The whole process of the synthesis of RBCs takes place in the bone marrow. Bone marrow is the spongy tissue, located in the medullary cavity of the bone. There are microscopic canals in the hard part of bone, which connect the marrow with outer circulation of blood. The bone marrow has large, thin-walled, permeable spaces or modified capillaries called sinusoids. Blood cells can pass through the walls of the sinusoids into the blood stream. Reticulocytes are released into the blood stream, where they mature into erythrocytes in about a day or two. In the process, size of the cell decreases, cytoplasm volume increases, nucleus is lost and all these steps make room for more haemoglobin.

Model Compounds

Synthesis of Co(III) complexes of tetradentate Schiff bases, with coordination from four N atoms, has been reported. These compounds mimic vitamin B$_{12}$ and can form alkyl derivatives with R groups bound

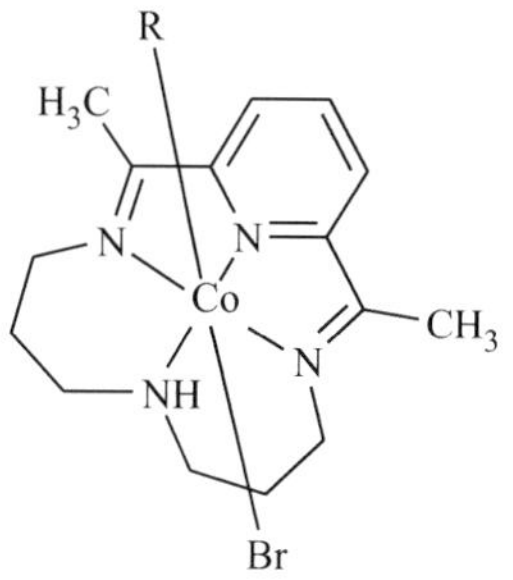

FIGURE 7.4 Co(III) complex of Schiff's base obtained by condensation of dipropylene triamine and diacetyl pyridine.

at sixth position. Alkylation reaction of organic molecules using the alkyl derivatives of Co(III) Schiff base complexes has been attempted. These compounds serve as models for vitamin B$_{12}$. One of the model complexes is shown in Figure 7.4.

QUESTIONS

1. How is the macrocyclic ring in vitamin B$_{12}$ different from that in haemoglobin?
2. What is cyanocobalamin?
3. Which organisms produce vitamin B$_{12}$ in nature?
4. What happens in humans in case of deficiency of vitamin B$_{12}$?
5. List various derivatives of vitamin B$_{12}$.
6. What are various roles of vitamin B$_{12}$?
7. Cite an example where methylation of a metal carried out by cobalamin leads to more toxicity.

8

Molybdenum in Nitrogen Fixation in Plants

Gaseous form of nitrogen cannot be absorbed by plants. Nitrogen fixation is the process of conversion of nitrogen to ammonia and subsequently to soluble nitrogen compounds. The enzyme nitrogenase, in plants, plays the vital role of assisting fixation of nitrogen.

Nitrogenase is obtained in the symbiotic bacteria, living in the root pods of leguminous plants, like pea and bean. This group of bacteria is called Rhizobium. The enzyme is also found in the nonsymbiotic bacteria, *Azotobacter* (aerobic) and *Clostridium pasteurianum* (anaerobic), photosynthetic bacteria and some blue green algae and fungi.

Ammonia formed, due to nitrogen fixation, is converted to soluble nitrogen compounds, which can be taken up by the plants for the synthesis of DNA, RNA and proteins. These nitrogenous biomolecules are reconverted to ammonia in due course.

The excess ammonia, not used for the synthesis of biomolecules, is oxidised to hydroxyl amine and further to nitrate by the nitrification process. This nitrate goes to the soil. Soil nitrate is also formed by the decay of nitrogenous organic compounds. A variety of microorganisms, and higher plants convert the soil nitrate into ammonia. The soil nitrate is first reduced to nitrite, using the enzyme nitrate reductase, and nitrite is reduced to ammonia, assisted by the enzyme nitrite reductase. In microorganisms, nitrite may alternatively be reduced to N_2O and to N_2 in the denitrification process. The nitrogen, liberated, goes back to the atmosphere and is reconverted to ammonia by the enzyme nitrogenase.

This cycle of fixation of atmospheric N_2 into ammonia, its incorporation into biological system, due to the formation of nitrogenous biomolecules, their decay into nitrate and the return of N_2 to the atmosphere, due to the denitrification of nitrate, is called the nitrogen cycle (Figure 8.1).

Mechanism of Nitrogen Fixation

Though nitrogen molecule is kinetically inert, the reduction of N_2 by H_2 to NH_3 is thermodynamically favourable, as shown in the following:

$$N_2 + 3H_2 \longrightarrow 3NH_3 \quad \Delta G = -3.97 \, kcal/mol \tag{8.1}$$

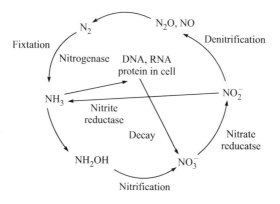

FIGURE 8.1 The nitrogen cycle.

However, there has to be concerted reaction of six electrons and six H⁺ with N_2 (Eq. 8.2). This implies that formation of NH_3 is difficult. Hence, the reaction may proceed through the formation of the intermediates N_2H_2 and N_2H_4 (Eqs. 8.3 and 8.4):

$$N_2 + 6H^+ + 6e \longrightarrow 2NH_3 \tag{8.2}$$

$$N_2 + 2H^+ + 2e \longrightarrow N_2H_2 \tag{8.3}$$

$$N_2 + 4H^+ + 4e \longrightarrow N_2H_4 \tag{8.4}$$

However, the formation of intermediates is less probable, as they are thermodynamically less stable. The intermediates are higher in energy than the reactants $N_2 + H_2$ and also the product NH_3 (Figure 8.2).

In the nitrogenase-assisted nitrogen fixation, the formation of the intermediates may be facilitated, as they are stabilised due to coordination with the metal centres. The formation of the intermediates in the nitrogenase activity may also be assisted by the liberation of energy, due to the hydrolysis of ATP.

In biological systems, the formation of NH_3 is always associated with the production of H_2 (Eq. 8.5). In the absence of N_2, the active nitrogenase reduces H_3O^+ ion into H_2 gas. This is suggested to be due to the presence of metal hydride in the systems, which reacts with H_3O^+ to form H_2:

$$N_2 + 8H^+ + 8e \longrightarrow 2NH_3 + H_2 \tag{8.5}$$

Nitrogen-fixing process requires a source of electrons and a source of energy for the activation of electrons and N_2. Nitrogenase receives the electrons formed by the oxidation of pyruvates (generated from carbohydrates in the respiration process) or other sources. These electrons are carried to the electron receiving site of nitrogenase by the Fe–S protein, ferredoxin.

The energy required for the activation of the electrons is provided by the hydrolysis of ATP to ADP, as shown in the following:

$$ATP + H_2O \longrightarrow ADP + PO_4^{3-} + Energy \tag{8.6}$$

The reaction (Eq. 8.6) uses up water and hence maintains anhydrous condition, required for nitrogen fixation. Twelve to fourteen molecules of ATP are required to reduce one molecule of N_2.

Furthermore, it has been observed that in the red root pods of the leguminous plants, there is a Fe(II) haem protein called leghemoglobin. This binds with O_2 and thus provides anaerobic environment required for the catalytic activity of nitrogenase. The oxygen in oxyleghemoglobin also assists the respiration process of the microorganisms and liberates additional energy for the excitation of electrons.

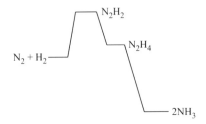

FIGURE 8.2   Relative energies for intermediates in reduction of nitrogen to ammonia.

Structure of Nitrogenase

The enzyme nitrogenase consists of two proteins. They are low molecular weight (50,000 daltons), iron protein and high molecular weight (270,000 daltons), molybdenum iron protein. Individually these are catalytically inactive, but when present together, they can assist nitrogen fixation.

a. **Iron protein:** It is made up of two identical subunits of molecular weight 30,000 daltons. The protein contains four iron and four sulphide ions, which can be extruded as one $[Fe_4S_4]^{2+}$ cluster. This shows that there is one $[Fe_4S_4]^{2+}$ centre, bound between the two subunits of the protein. This has been confirmed by X-ray studies also (Figure 8.3).

Fe_4S_4 is diamagnetic, due to antiferromagnetic interaction between the pairs of Fe(II) and Fe(III) centres, through sulphide bridges. It is thus similar to $[Fe_4S_4]^{2+}$ in ferredoxin with two Fe(III) and two Fe(II).

The Fe_4S_4 centres of iron protein receive electron through ferredoxin chain and undergo one electron reduction. The reduced form shows EPR activity. However, the EPR spectrum of the reduced iron protein shows two spin states $S = 1/2$ and $S = 3/2$, raising the controversy earlier, that there are more than one type of Fe_4S_4 sites. However, it has now been established that the reduced state of same types of Fe_4S_4 site may exist in different spin states. It has also been suggested that one of the coupled Fe_2S_2 sites in the reduced Fe_4S_4 [1Fe(III), 3Fe(II)] is paramagnetic. The spin coupling due to antiferromagnetic interaction may lead to $S = 1/2$ or $S = 3/2$. However, there is no difference in the redox behaviours of the Fe_4S_4 with different spin states.

During the enzyme turnover, reduced Fe_4S_4 centres of the iron protein transfer electrons to Fe–Mo–protein, in single step.

The Fe protein is bound to two molecules of Mg–ATP, at the cleft between the two subunits of the protein. For transfer of each electron to the Fe–Mo–protein, two Mg–ATP molecules must undergo hydrolysis. Thus, hydrolysis of 12 molecules of Mg–ATP is required for the reduction of one molecule of N_2, which requires transfer of six electrons. Fe protein receives activated electrons and transfers to Fe–Mo–protein, where N_2 gets reduced. Thus, the iron protein reduces Fe–Mo–protein to its catalytically active form.

Though the presence of Fe protein is essential for the transfer of electron, and thus for nitrogenase activity, only Fe–Mo–protein is responsible for the reduction of N_2.

b. **Fe–Mo–protein:** This protein receives the electron from the iron protein, and its reduced form acts as catalyst for reduction of nitrogen. Fe–Mo–protein contains two types of subunits and has a molecular weight in the range of 220,000–245,000 daltons. The two types of subunits are

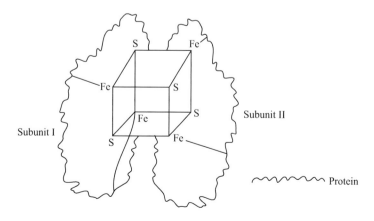

FIGURE 8.3 The iron protein in nitrogenase.

called P cluster and M cluster. Both types of centres have specific properties, even after being extracted out of the protein. Their structures and roles are discussed as follows:

i. **P cluster:** The P cluster consists of four Fe_8S_8 units as pairs of Fe_4S_4. These Fe_4S_4 differ from Fe_4S_4 found in ferredoxin, as revealed by their electronic spectra and Mössbauer (M.B.) spectra. EPR shows that the P clusters are paramagnetic with $S = 7/2$. This indicates that there is incomplete antiferromagnetic coupling between pairs of Fe(II). M.B. spectra reveal more than one isomer shift, showing that all four irons in Fe_4S_4 are not equivalent. This indicates that the Fe_4S_4 units are highly distorted, unlike the cubane structure in ferredoxin.

The spectroscopic studies of P clusters after extrusion (though with uncertainties about structural rearrangement during extrusion) reveal that the four clusters are not equivalent. They seem to be present in the form of two subsets.

The X-ray crystallographic studies show that there are four Fe_8S_8 clusters, very close together, each in the form of a pair of Fe_4S_4. Each pair has a doubly bridged (through cysteine S) double cubane structure. Each cubane has two intracubane bridged Fe(II). The third and fourth Fe(II), of one Fe_4S_4, are bound to cysteine S at the fourth coordination site. In the other Fe_4S_4 third iron is bound to cysteine S and the fourth iron is bound to one cysteine sulphur and one serine sulphur. Thus, this iron is penta-coordinated. The intercubane bridging may be responsible for the unusual properties of the P clusters.

The P clusters are considered to be reservoirs for accepting the high-energy (low-potential) electrons from Fe protein, to be passed on to the M cluster, where reduction of nitrogen takes place.

c. Structure of P cluster is shown in Figure 8.4.

ii. **M cluster:** The spectroscopic and redox studies have revealed the presence of iron and molybdenum in the M cluster, and hence, it is also known as Fe–Mo–cofactor (Fe–Mo–co). The detailed characterisation of Fe–Mo–co has been carried out by the studies of the unit in vivo in the protein and also in the extracted form, in various organic solvents.

The composition of the M cluster is $Fe_{6-8}S_{7-10}$. The structure of the cluster is not definitely known. Each, multiple iron-sulphur cluster, contains one molybdenum.

The M.B. spectral studies, in the presence of magnetic field, and also Electron nuclear double resonance spectroscopy (ENDOR) studies confirm the presence of four types of iron, one each at A_1, A_2 and A_3 sites and three equivalent irons at the B site. Thus, the number of iron atoms in each cluster must be minimally six, and it can be more.

The iron EXAFS shows that iron is coordinated to sulphur at a distance of about 2.2 Å. There are also distant Fe–Fe interactions at about 2.6 Å.

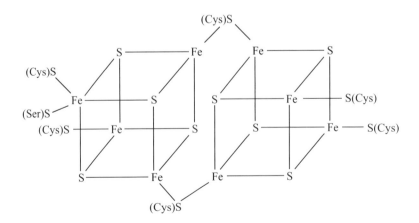

FIGURE 8.4   Structure of P-cluster of Fe-Mo protein in nitrogenase.

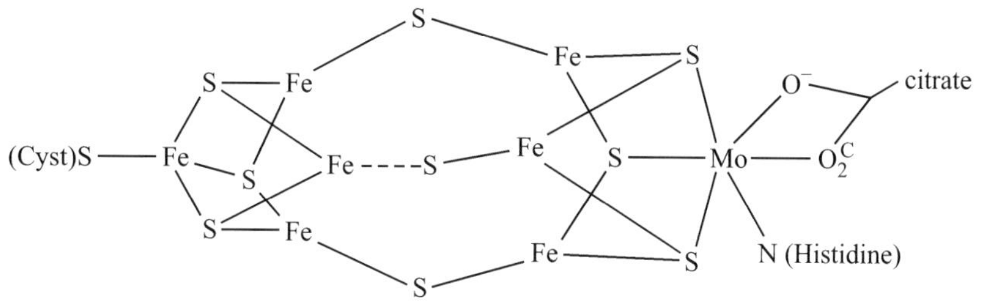

FIGURE 8.5 Structure of Fe-Mo co-factor.

EXAFS study on the molybdenum K edge indicates that Mo is coordinated to three or four sulphurs at 2.4 Å, one to three oxygens or nitrogens at 2.2 Å, and to three irons at 2.7 Å. However, the evidence of O and N coordination by EXFAS study is subject to question. On the basis of the above studies and the recent X-ray study, it has been concluded that the M cluster is $MoFe_7S_8$. Fe_7S_8 has two halves, bridged through three nonprotein sulphides. Fe_7S_8 unit is bound to molybdenum through three sulphides. Mo is six coordinated, bound to two oxygens of bidentate citrate and nitrogen of histidine. The structure of the Fe–Mo cofactor is shown in Figure 8.5.

Mechanism of Nitrogen Reduction and Ammonia Formation

As discussed earlier, iron protein receives the electron generated from the oxidation of pyruvates, through electron transfer protein ferredoxin. The reduced iron protein gets bound to the two molecules of MgATP. The resulting (Mg ATP)$_2$ Fe protein gets bound with Fe–Mo–protein. There is hydrolysis of two molecules of ATP, providing energy for the excitation of electrons of Fe protein, to be transferred to the P cluster of Fe–Mo protein and subsequently to the Fe–Mo–cofactor, to which is bound the substrate nitrogen. A simplified picture of electron transfer is shown in Figure 8.6.

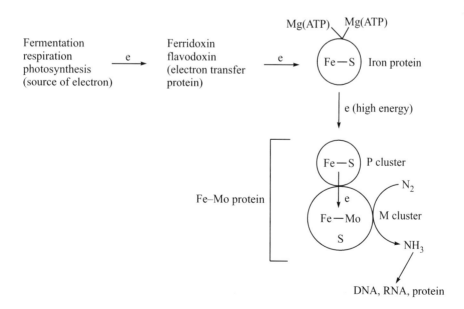

FIGURE 8.6 Electron transfer in the mechanism of nitrogen reduction.

Electrons are successively transferred to N_2, resulting in steps, in the formation of N_2^{6-} ion. These negatively charged nitrogen species combine with H^+, giving a sequence of intermediates, leading to the formation of ammonia, as shown in Eq. (8.7).

$$Mo-N\equiv N \xrightarrow[2H^+]{2e} Mo-NH=NH \xrightarrow[2H^+]{2e} Mo-NH_2-NH_2 \xrightarrow[2H^+]{2e} Mo+NH_3 \qquad (8.7)$$

$$2NH_3 + 2H^+ \rightarrow 2NH_4^+ \qquad (8.8)$$

As stated earlier, reduction of N_2 to ammonia is thermodynamically favourable. However, hydrolysis of ATP, to provide energy for the excitation of electrons, is a kinetic requirement. The two components of nitrogenase, the iron protein and iron molybdenum protein, catalyse the reduction of one molecule N_2 to $2NH_4^+$ (Eqs. 8.7 and 8.8). Along with the formation of NH_3, there is also evolution of H_2. Hence, the limiting stoichiometry is shown in the following equation:

$$N_2 + 10H^+ + 8e \rightarrow 2NH_4^+ + H_2 \qquad (8.9)$$

The electrons of the nitrogenase reduce H^+ also, resulting in the formation of molybdenum hydride sites on the enzyme. The hydride ion combines with H^+ of water to form H_2. If N_2 is not present, all the electrons of the enzyme are used for hydrogen evolution.

If carbon monooxide is present, it inhibits reduction of nitrogen and, hence, NH_3 formation. But it does not affect H_2 formation. This shows that there are different sites present in the enzyme for N_2 and H^+ reduction.

It is observed that there is formation of symmetrical intermediate products during the reduction of nitrogen. The formation of di–imide NH=NH, and hydrazine H_2N–NH_2, shows that two nitrogen atoms are reduced simultaneously. This indicates that N_2 is bound to the two metal centres, through two nitrogen atoms. Thus, a binuclear Fe–Mo protein is suitable for N_2 binding. On this basis, it has been suggested that the reduction may be in steps, as shown in Eq (8.10):

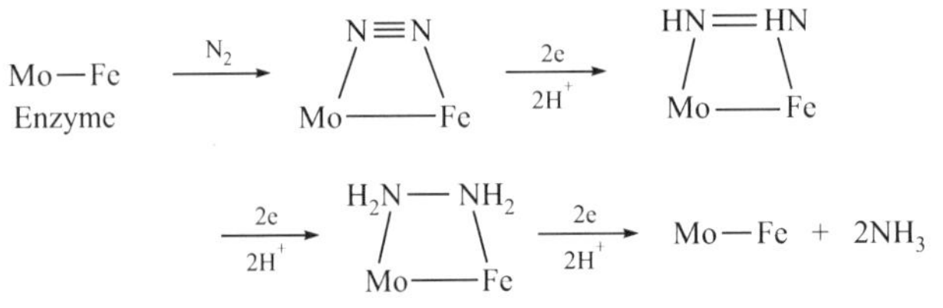

$$(8.10)$$

However, more recent X-ray crystallographic studies suggest that the Mo is coordinatively saturated and so may not be the site of nitrogen binding. N_2 molecule may be bound inside the cluster. Thus, the mode of binding of N_2 is not known with certainty.

Since, the iron protein and Fe–Mo protein sites are not solvent exposed, electron transfer, substrate binding and product release should involve structural change in the enzyme, exposing the active sites during the reduction process, similar to the gating process, discussed in Chapter 3.

Nitrogenase also reduces other substrates with triple bond, like acetylene, cyanides and isocyanides (Eqs. 8.11–8.15):

$$C_2H_2 + 2e + 2H^+ \longrightarrow C_2H_4 \qquad (8.11)$$

$$HCN + 6e + 6H^+ \longrightarrow CH_4 + NH_3 \qquad (8.12)$$

$$N_3H + 6e + 6H^+ \longrightarrow NH_3 + N_2H_4 \qquad (8.13)$$

$$RNC + 6e + 6H^+ \longrightarrow RNH_2 + CH_4 \qquad (8.14)$$

$$RCN + 6e + 6H^+ \longrightarrow RCH_3 + NH_3 \qquad (8.15)$$

It has been shown recently that the Fe–Mo–protein part of the nitrogenase can also cause hydrogenation of the dyes, in the presence of molecular hydrogen, and thus can act as a hydrogenase. This is the only example, where only one component of nitrogenase shows catalytic activity, independently.

Other Forms of Nitrogenase

Vanadium Nitrogenase

Though it has been emphasised since 1930 that molybdenum is an essential component of nitrogenase, in 1971 nitrogenase, extracted from *Azotobacter vinelandii*, was found to contain vanadium. This enzyme has small amount of molybdenum also and hence has lower activity than normal nitrogenase. However, because of the presence of vanadium, the enzyme shows selective activity for specific substrates. This shows that it may not have structure similar to normal nitrogenase with Fe–protein and Fe–Mo protein units. Because of being associated with some vanadium, it may have different structure.

The essentiality of molybdenum in nitrogenase was ruled out in 1980, when a molybdenum-free nitrogen-fixing enzyme was identified in the organism *A. vinelandii*, on not being supplied molybdenum for some time. This new form of nitrogenase, without molybdenum, was also isolated from Azotobacter. It was demonstrated that this alternative nitrogenase has two units: one containing vanadium and the other, very similar to iron protein in normal nitrogenase, with Fe_4S_4 cluster. None of the units contain molybdenum.

The vanadium enzyme has selectivity for the substrate and the products formed are also different. Fe–V enzyme is much less reactive to acetylene than Fe–Mo enzyme. Fe–Mo enzyme reduces acetylene to ethylene, whereas Fe–V enzyme reduces acetylene to ethane by four electron transfer process.

Nitrogenase with Iron Only

A nitrogenase was extracted from a mutant of *A – vinelandii*, which contained only iron protein and was able to fix nitrogen. This should be called Fe–Fe protein and shows least activity in nitrogen fixation, the order being Fe–MO > Fe–V > Fe–Fe. Fe–Fe protein also reduces ethylene to ethane.

The identification of all iron nitrogenase indicates that metals, other than iron, are not the primary requirement for nitrogen fixation. This casts further doubt on the mechanism of direct binding of nitrogen with molybdenum during its hydrogenation process, as suggested in Eq. (8.10).

Nitrogenase Models

There is no satisfactory Fe–S cluster model for the P site of Fe–Mo protein. However, many FeMoS clusters have been synthesised, mimicking FeMoCo site (M centre). However, reduction of nitrogen, using the FeMoS clusters, has not been possible, because of the nonavailability of suitable reductant.

Other Pathways of Ammonia Formation

As discussed earlier, besides nitrogen fixation process, ammonia can also be produced by reduction of nitrate and nitrite.

Nitrate Reductase

Nitrate is reduced to nitrite by two-electron reduction. This is catalysed by the enzyme nitrate reductase. This protein has FAD, haem b_{557} and Mo pterin complex, as prosthetic groups. Nitrate is bound at the

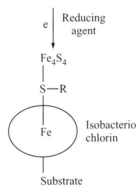

FIGURE 8.7   Mechanism of reduction of nitrate to nitrite by nitrate reductase.

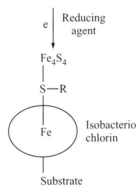

FIGURE 8.8   Binding mode of substrate with iron centre in isobacteriochlorin.

molybdenum (VI) centre, replacing a ligand, possibly H_2O. The electron received from NADPH, through FAD and b_{557}, reduces Mo(VI) to Mo(IV). Two electrons are transferred to the bound NO_3^-, reducing it to NO_2^-, and Mo(VI) is regenerated for repeating the catalytic cycle (Figure 8.7).

Nitrite Reductase (Sulphite Reductase)

Reduction of nitrite to ammonia (or sulphite to S^{2-}) involves six electron reduction. It is catalysed by the enzyme nitrite or sulphite reductase (SR). This enzyme protein has a Fe_4S_4 – siroheme prosthetic centre.

Siroheme is the iron complex of the macrocyclic isobacteriochlorin ligand, which has less unsaturation, and hence is more electron rich than porphyrin. Fe_4S_4 unit is bridged to the iron of siroheme, through S–R group in the axial direction.

The substrate NO_2^- or SO_3^- is linked at the vacant sixth coordination site of iron of siroheme (Figure 8.8). The Fe_4S_4 unit receives the electron from the natural reducing agent and transfers it to the iron centre of the siroheme. The reduced iron centre transfers six electrons to the substrate, and there is a reductive cleavage of N–O or S–O bond of the substrate.

Resulting N^{3-} and S^{2-} combine with H^+ to form NH_3 or H_2S as shown in the following equations:

$$NO_2^- \xrightarrow[7H^+]{6e} NH_3 + 2H_2O \tag{8.16}$$

$$SO_3^{2-} \xrightarrow[7H^+]{6e} HS^- + 3H_2O \tag{8.17}$$

QUESTIONS

1. Name the metals present in the enzyme that performs nitrogen fixation.
2. Describe the nitrogen cycle.
3. Give all the chemical reactions involved in nitrogen fixation in terms of mechanistic aspects.
4. Discuss structure and magnetic properties of iron protein in nitrogenase.
5. What is the experimental evidence to show that the Fe_4S_4 units in P cluster and in ferredoxin are different?
6. Apart from fixing nitrogen, which other reactions does the nitrogenase perform?
7. Are there any other forms of nitrogenase as compared to the iron–molybdenum proteins?
8. What is function of nitrate reductase?

9

Magnesium and Manganese in Photosynthesis in Plants

Photosynthesis is a bioenergetic process. It proceeds in two stages. In the first stage, occurring in the sun light, called light reaction, there is photo-induced splitting of water molecules. O^{2-} loses electron and is converted to molecular oxygen, which is liberated in the air. The light energy is converted into chemical energy and is stored in the form of ATP. The electrons liberated in the formation of O_2 reduce $NADP^+$ to NADPH.

In the second stage, a series of enzymatic reactions lead to the reduction of CO_2 to glucose, by NADPH, which gets converted back to $NADP^+$. The energy for this reaction is provided by the breakdown of ATP into ADP. This reaction is called dark reaction, that is the reaction which does not need light, and can occur in the dark. The name should not wrongly imply that the reaction occurs in dark conditions only.

The two steps of the photosynthetic reactions can be shown as follows (Eqs. 9.1–9.3). The gross reaction is shown in Eq. (9.4):

$$2H_2O \xrightarrow{h\nu} O_2 + 4H \tag{9.1}$$

$$2NADP^+ + 4H \longrightarrow 2NADPH + 2H^+ \tag{9.2}$$

$$6CO_2 + 24NADPH \longrightarrow C_6H_{12}O_6 + 24NADP^+ + 6H_2O \tag{9.3}$$

$$6CO_2 + 6H_2O \longrightarrow C_6H_{12}O_6 + 6O_2 \tag{9.4}$$

The fact that O_2, liberated in photosynthesis, is from H_2O has been confirmed by using ^{18}O labelled water. It is observed that only $^{18}O_2$ with molecular weight 36 is obtained. As seen above (Eq. 9.2), in the normal photosynthetic process, the H used for the reduction of $NADP^+$ to NADPH is also provided by water.

In photosynthetic process, in anaerobic bacteria, other sources of hydrogen are used and hence O_2 is not liberated in these reactions. For example, H_2S could be the source of hydrogen. In that case, sulphur is liberated instead of O_2 (Eq. 9.5):

$$2H_2S + 2NAD^+ \rightarrow 2NADH + 2S + 2H^+ \tag{9.5}$$

The oxidation potential of O^{2-} being more positive than that of S^{2-}; liberation of electrons from O^{2-} utilizes solar energy, and hence special photo system, is required for "light harvesting". In other words, for the capture of solar energy. The photo systems in green plants and algaes, have organic molecules, like chlorophyll and carotenoids, in large number of clusters, which absorb light, and thus function as molecular antenna, for capture of solar energy.

Chlorophylls are Mg(II) complexes of tetra pyrrole ligands of porphyrin family, with different substitutions at positions 2, 3, and 10 (Figure 9.1). There is a cyclopentanone ring E, between pyrrole ring C and D of the macrocycle, and it can exist in keto or enol form. The pyrrole ring D is in the reduced form. There is a long phytol chain attached to the carbon atom at the seventh position of the D ring. This helps chlorophyll to bind with the cell membrane.

If X at position 3 in the pyrrole ring B (Figure 9.1) is CH_3, it is called chlorophyll a, and if X = CHO, it is called chlorophyll b.

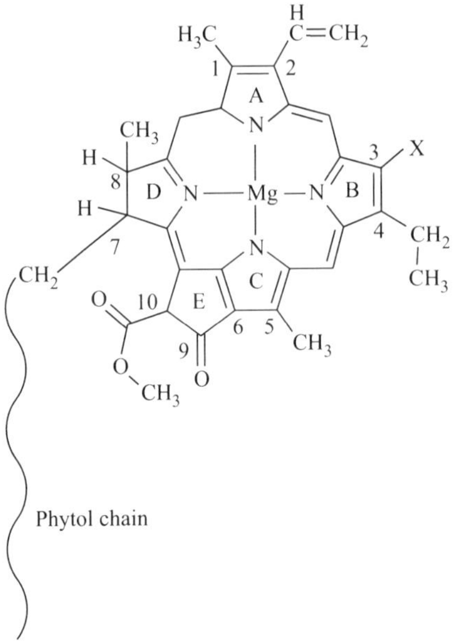

FIGURE 9.1   Chlorophyll.

Mg^{2+} is bound to two N and two N^- of the porphyrin ring and is 0.3–0.5 Å above the plane of the ring (Figure 9.1). Mg^{2+} in chlorophyll can be replaced by $2H^+$ or metals like Cu^{2+}, Ni^{2+}, Co^{2+} or Fe^{2+} ions, but the unique photocatalytic property, of the Mg(II) chlorophyll, is lost.

Bacteriochlorophyll has structure similar to chlorophyll, but in it, the pyrrole ring B is reduced, and there is an acetyl group, instead of vinyl, at position 2 in the ring A.

Chlorophylls absorb at ~700 nm, due to the $\pi_b \rightarrow \pi^*$ transition, in the porphyrin ring. The energy required, for the electronic transition, from the π_b HOMO to π^* LUMO of chlorophyll is less, and hence, the absorption maxima appears in the visible range.

There is a small change in the band position, with change in substitution over the porphyrin ring. The photosynthetic systems also contain additional pigments, like carotenoids (β-carotene) or blue or green phycoerythrobilins. These pigments have a broad range of absorption wavelengths and thus absorb the entire range of visible spectrum from the sunlight. The absorbed light is not only useful for the photo activation of the chlorophyll, but this process of absorption of the sunlight by chlorophyll, also protects the biochemical system of the plant from photochemical damage.

Chlorophyll is insoluble in water and forms surfactants in non-polar solvents, due to the presence of the polar head, Mg(II) macrocycle site and the non-polar phytol chain (Figure 9.2a). Antenna chlorophyll, in plants, is now believed to consist of $(chl)_n$ chain. In the formation of the chain, the keto group at C_9 carbon of the ring E, of one chlorophyll coordinates with Mg(II) of the second chlorophyll (Figure 9.2b).

Light Reaction

There are two different sites, where light reaction occurs and light energy is converted to chemical energy. These reaction sites are called photo system I (PS I) and photo system II (PS II). They contain the green pigments chlorophyll a and chlorophyll b in different proportions. They also differ in accessory chemicals for processing the tapped energy of the photon.

Let us first consider the PS II. It has a characteristic absorption at wavelength 680 nm and is termed P680 (chl). It absorbs a quantum of light, followed by transfer of one electron from the π bonding to the π* antibonding molecular orbital of the porphyrin ring. The molecule P680 chl* in the excited state is

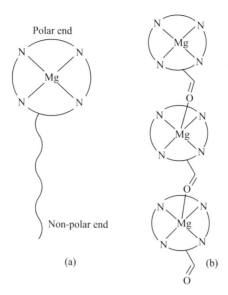

Polar end

Non-polar end

(a)

(b)

FIGURE 9.2 (a) polar and non-polar parts in chlorophyll and (b) surfactant formation of chlorophyll.

a strong reducing agent. It transfers the electron to an acceptor molecule C of which the composition is not definitely known. The composition of C_{550} has now been suggested to be an iron–quinone complex, associated with a series of quinones. On reduction, C_{550} shows a characteristic absorption at 550 nm, hence the name C_{550}.

The oxidised form of PS II, P680 chl+, can accept an electron from a suitable donor. The redox potential of P680 chl+ is in the range of 0.8–1.0 V, which is comparable with the potential for the oxidation of O^{2-} of water to O. Hence, two electrons from O^{2-} are transferred in steps to two P680 chl+, and O^{2-} is converted to O, which combines with another O and gets liberated as O_2.

The electron is not directly transferred from O^{2-} to P680 chl+, but is transferred through a manganese protein, called oxygen evolving protein or water-oxidising complex (WOC).

The WOC comprises of four Mn ions, and one or two calcium ions bound to a protein. An unspecified number of chloride ions are linked to the metal ions. The four Mn exist in an electronically coupled tetramer.

No other metal ion can replace manganese. Calcium plays a structural role, in stabilising the manganese site. Chloride acts as a bridging ligand between manganese and might promote intracluster electron transfer.

The oxidation state of manganese (Mn^{n+}) is not definitely known. It may be IV or III. During electron transfer process, manganese shuttles between two oxidation states, IV and III or III and II. A probable mechanism is that two H_2O molecules get deprotonated, and the resulting OH^- get bound to the two of the four Mn^{n+} centres. Four electrons are transferred from the two OH^- to the four Mn^{n+} centres, reducing them to $Mn^{(n-1)+}$. OH^- ions are oxidised and liberate $2H^+$ and one O_2 molecule. A probable mechanism is shown below:

$$4Mn^{n+}\text{protein} \xrightarrow{2H_2O} 4Mn^{n+}\text{protein(OH)}_2 + 2H^+ \rightarrow 4Mn^{(n-1)+} + O_2 + 4H^+$$

Oxidation of O^{2-}, by the metal ion, does not depend only on the thermodynamic requirement of matching of the redox potentials, but kinetically, it depends on the activation barrier of mutual approach of oxygen atoms to form O_2 molecule. The binding of O with the manganese ion, in the catalyst, lowers down the activation energy of the formation of O_2 molecule. The metal catalyst also lowers down the reorganisation energy of the electron transfer, by allowing stepwise electron transfer, rather than simultaneous transfer of four electrons from $2O^{2-}$ to P 680 chl+. Thus, a 2e/2e pathway or a 2e/1e/1e pathway, provided by the WOC, which meets both thermodynamic and kinetic requirements, and is preferred over 4e transfer pathway.

Reduced form of WOC transfers the electrons to P680 chl$^+$ and comes back to the original oxidation state. The P680 chl$^+$ is then reduced back to the original form P680 chl, absorbs light to get into the excited state and transfer electrons to C_{550}, to continue the process, as stated earlier.

Thus, effectively, the electron from O^{2-}, of water, is transferred to C_{550}.

PS II is, therefore, the site where O_2 is liberated. Accordingly, photosynthetic bacteria which do not evolve O_2 do not have PS II, but have only PS I.

The electron from reduced C_{550} is transferred to the oxidised form of PS I. The electron transfer is not direct, but takes place through a series of electron carriers, cytochrome b$_3$, plastoquinone (PQ) Riseke's Fe–S protein, cytochrome f and plastocyanin, arranged in the order of their increasing oxidation potential, or increasing tendency to accept electrons.

The electron transfer is stepwise because there is a large difference in the redox potentials of reduced C_{550} and oxidised form of PSI. The large amount of energy generated, due to the electron transfer, is not liberated in one step, but is liberated in small amounts in each step. The transfer of a pair of electrons, from O^{2-} to PS I$^+$, liberates sufficient energy for the synthesis of one molecule of ATP. There is also release of protons in the oxidation of water. There is transmembrane gradient of these protons, releasing energy, and this also assists the synthesis of ATP.

The photosystem I has a characteristic absorption at 700 nm, that is at a lower energy region than PS II, and is called P700 chl. It has an oxidation potential of 0.4–0.5 V and hence does not have tendency to lose electrons as such. On absorption of energy at 700 nm, one electron from π bonding molecular orbital of the porphyrin ring is excited to the π^* antibonding molecular orbital. In the excited state, P700 chl* transfers an electron to the electron acceptor molecule P430 and results in the formation of P700 Chl$^+$ or PSI$^+$. P430 is an Fe–S protein, different from ferredoxin. It has an oxidation potential −4 V. In the reduced form, it absorbs at 430 nm, hence the name P430. The electron from the reduced form of P430 is transferred to NADP$^+$.

Since the oxidation potential of NADP$^+$ is more positive (has strong tendency to accept electrons), the transfer of electron to it, from reduced P430, is a downhill process, liberating a lot of energy. Hence, the electron transfer is carried out in steps by a number of electron carriers like ferredoxin, 2Fe–2S protein in plants and 4Fe–4S protein in bacteria, and a flavoprotein (FP). The electron carriers are arranged in the increasing order of their oxidation potentials.

The reduction of NADP$^+$ to NADPH, by the transfer of two electrons from P700 chl*, through P430 and ferredoxin (Fd), can be shown as follows:

$$2Fd(III) \xrightarrow{2e} 2Fd(II) \tag{9.6}$$

$$2Fd(II) + NADP^+ + H^+ \longrightarrow NADPH + 2Fd(III) \tag{9.7}$$

NADP$^+$ requires two electrons to get converted to NADPH. The first electron reduces NADP$^+$ to NADP. The second electron reduces NADP to NADP$^-$. The H$^+$ of H_2O, brought by transmembrane proton gradient, to the site where NADP$^-$ has been formed, combines with it to form NADPH (Eq. 9.8):

$$NADP^+ \xrightarrow{+e} NADP \xrightarrow{+e} NADP^- \xrightarrow{+H^+} NADPH \tag{9.8}$$

After losing electron to P430, the oxidised form of PS I, that is, P700 chl$^+$ is ready to accept one more electron from reduced C_{550} and the process of transfer of electrons to NADP$^+$ continues.

Thus, in the light reaction, two electrons from O^{2-} of H_2O are ultimately transferred to NADP$^+$. As the redox potential of NADP$^+$ is much more negative than O^{2-}, the transfer of electrons from O^{2-} to NADP$^+$ is an uphill process and cannot take place directly. In the light reaction of photosynthesis, the PS II and PS I act as boosters. They absorb light energy and utilise it to push the electron from low energy (positive potential) O^{2-} to high energy (negative potential) NADP$^+$.

The light reaction and dark reaction can be represented, as shown in Figure 9.3.

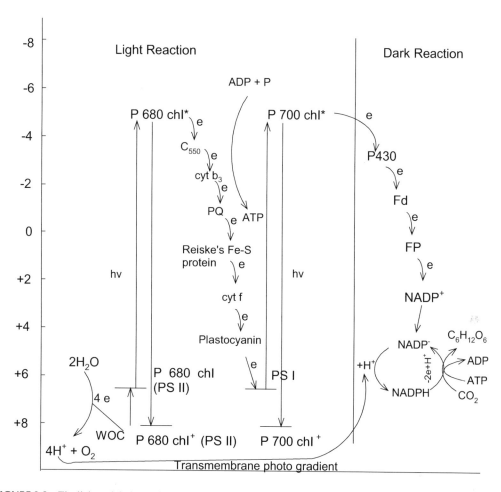

FIGURE 9.3 The light and dark reactions in photosynthesis.

Dark Reaction

In the dark reaction, the electrons are transferred from NADPH to carbon dioxide. As a result, it gets reduced to form glucose, as shown in Eq. (9.3). The required energy, in this dark phase, is provided by the breaking down of ATP to ADP. The dark phase of the reaction is not known to involve any metalloenzyme.

Model Studies of WOC and Photo Systems

Model mono- and binuclear Mn(III) and Mn(IV) Schiff base complexes and mono- and binuclear Fe(III) and Ru(III) bipyridyl complexes have contributed significantly to our present understanding of the role of the WOC and the photosystems. However, well-characterised functional models are rare. Such model complexes have direct application as catalysts in artificial photosynthesis, for splitting of water into O_2 and H_2. Hydrogen can be used in future fuel cells, for the generation of electricity. Such systems are commercially appealing, in view of the depleting energy sources and also the environmental pollution due to the combustion of coal, oil or fuel gases.

An interesting example is of a binuclear Mn (III) complex, as WOC and $Ru(bipy)_3^{3+}$ as acceptor of electrons. Mn (III) centres transfer electrons to $Ru(bipy)_3^{3+}$, get oxidised to Mn(V) and accept electrons from O^{2-} of water, oxidising it to molecular O_2. $Ru(bipy)_3^{3+}$ is reduced to $Ru(bipy)_3^{2+}$. On photoexcitation, $Ru(bipy)_3^{2+}$ gets into the excited state $Ru(bipy)_3^{2+*}$ and transfers the electron to an oxidative quencher Mn(IV) pyrophosphate, reducing it irreversibly into Mn(III). Thus, $Ru(bipy)_3^{2+*}$ acts as a photosensitiser. After the transfer of electrons, $Ru(bipy)_3^{3+}$ is reformed.

Similar $Ru(bipy)_3^{3+}$ and analogous complexes, whose oxidised form can act as oxidant and the reduced form, $[Ru(bipy)_3]^{2+}$, can act as a photosensitiser, are being tried as photo catalysts for water splitting.

Structure, Photochemical Property and Water Splitting Property of Tris-bipyridyl Ru (II) and Analogous Complexes

In the crystal field model of $[Ru(bipy)_3]^{2+}$, with octahedral geometry and D_3 symmetry, the d orbitals get split up into t_{2g} and e_g orbitals. Besides that, there are π_b and π^* orbitals of the ligand. In the case of strong field ligands like bipyridyl (bipy), the ground-state electronic configuration is $t_{2g}^6 e_g^0$. On excitation, there are four possible electron transitions, $t_{2g} \longrightarrow eg\left(t_{2g}^6 e_g^0 \longrightarrow t_{2g}^5 e_g^1\right)$,

$$t_{2g} \longrightarrow \pi^* \left(t_{2g}^6 e_g^0 \to t_{2g}^5 e_g^0 \pi^{*1}\right), \quad \pi_b \longrightarrow e_g \left(\pi^n t_{2g}^6 e_g^0 \longrightarrow \pi^{n-1} t_{2g}^6 e_g^1\right) \text{ and } \pi_b \longrightarrow \pi^*.$$

The first transition is a forbidden low-intensity d–d transition. Second and third are charge transfer transitions, metal ligand charge transfer (MLCT) and ligand metal charge transfer (LMCT), respectively (Figure 9.4). These are allowed transitions and are of high intensity. The fourth is an intraligand transition and is also an allowed transition, of high energy and intensity.

This consideration of metal- and ligand-centred electronic transitions presumes that the metal and the ligand are point charges, with no covalent interaction. However, in a complex like $[Ru(bipy)_3]^{2+}$, there will be significant covalent interaction and a quantitative interpretation, of the electron spectral properties, has to be based on molecular orbital model.

However, the crystal field model can adequately explain the absorption spectrum and luminescent properties of the complex. The complex shows three high-intensity bands at 185, 208 and 285 nm. These have been assigned to intraligand $\pi \to \pi^*$ transition, by analogy with the absorption spectrum of protonated bipyridyl.

The low-intensity bands at 323, 385, 238 and 285 nm are assigned to $t_{2g} \to e_g$ transitions. The high-intensity band in the visible region at 452 nm has been attributed to MLCT. The spin forbidden component of MLCT occurs at 620 nm.

The emission spectrum of $[Ru(bipy)_3]^{2+}$ shows an emission band at 579.6 nm. This has been assigned to $\pi^* \to \pi$ charge transfer fluorescence. The π^* orbital can be of two types: (i) antibonding MO of one of the bipyridyl ligands, or (ii) a multi-ring orbital, extended over all the ligands, instead of a single chelate orbital.

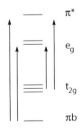

FIGURE 9.4 Electronic transitions in tris-bipyridyl Ru(II) and analogous complexes.

The direct evidence, for the first possibility of π^* being a single-ring antibonding orbital, is that the Raman spectrum, of the excited state of $[\text{Ru}(\text{bipy})_3]^{2+}$, is comparable with the Raman spectrum of reduced bipyridyl (bipy$^-$). Furthermore, the EPR of $[\text{Ru}(\text{bipy})_3]^{2+}$ shows that the received electron is located on a single ring, rather than being delocalised over all the three rings.

The most interesting is the redox chemistry of the photo excited state $[\text{Ru}(\text{bipy})_3]^{2+*}$, $(t_{2g}^5 e_g^0 \pi^{*1})$. It acts both as a strong reductant, because the electron, being located in the high-energy ligand π^* orbital, can be easily transferred, and also as a good oxidant, because the metal centre with t_{2g}^5 configuration is electron-deficient and can accept an electron easily. From MO consideration, oxidation is metal-centred and reduction is ligand-centred.

The excited state can be deactivated to the ground state, by interaction with another molecule, called quencher. This can happen in three ways:

i. The energy from the excited state is transferred to the quencher, raising it to the excited state. $\left[\text{Ru}(\text{bipy})_3\right]^{2+*}$ comes back to the original ground state $\left(t_{2g}^6 e_g^0\right)$.

$$[\text{Ru}(\text{bipy})_3]^{2+*} + Q \rightarrow [\text{Ru}(\text{bipy})_3]^{2+} + Q^* \qquad (9.8)$$

ii. Electron is transferred from the excited state to the quencher. Thus, the quencher is reduced, and the complex is oxidised to $\text{Ru}(\text{bipy})_3^{3+}$. The process is called oxidative quenching.

$$[\text{Ru}(\text{bipy})_3]^{2+*} + Q \rightarrow [\text{Ru}(\text{bipy})_3]^{3+} + Q^- \qquad (9.9)$$

iii. Electron is transferred from the quencher to the excited state. Thus, the quencher is oxidised, and $\left[\text{Ru}(\text{bipy})_3\right]^{2+}$ is reduced. This is called reductive quenching.

$$[\text{Ru}(\text{bipy})_3]^{2+*} + Q \rightarrow [\text{Ru}(\text{bipy})_3]^{+} + Q^+ \qquad (9.10)$$

This property of $\left[\text{Ru}(\text{bipy})_3\right]^{2+}$, to get oxidised or reduced by interaction with different quenchers, supports that in $t_{2g}^5 e_g^0 \pi^{*1}$ state, the electron is localised on one of the bipy, and hence, charge distribution in the excited state can be shown as $[\text{Ru}^{3+}(\text{bipy})_2 \text{ bipy}^-]$.

Mechanism of Photo Decomposition of Water

It can take place in the following two ways.

i. Reduction of H^+ to H_2 by receiving an electron from the donor A^-.

$$H_2O \xrightarrow{+e} OH^- + \tfrac{1}{2}H_2 \qquad (9.11)$$
$$A^- \quad\quad A$$

ii. Oxidation of O^{2-} to O_2, by transferring two electrons to the acceptor A^{2+}.

$$H_2O \xrightarrow{-2e} 2H^+ + \tfrac{1}{2}O_2 \qquad (9.12)$$
$$A^{2+} \quad\quad A$$

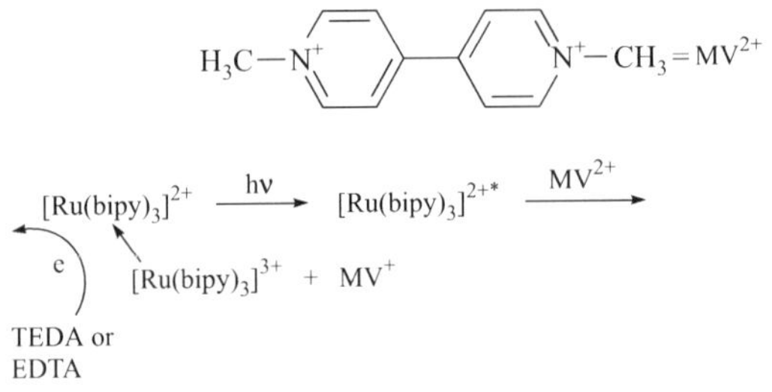

$$H_3C-N^+ \hspace{2cm} N^+-CH_3 = MV^{2+}$$

$$[Ru(bipy)_3]^{2+} \xrightarrow{h\nu} [Ru(bipy)_3]^{2+*} \xrightarrow{MV^{2+}}$$

$$e \begin{matrix} \\ \end{matrix} [Ru(bipy)_3]^{3+} + MV^+$$

TEDA or
EDTA

FIGURE 9.5   Photoreduction of water to hydrogen by ruthenium tris-bipy complex.

i. For the first process of photoreduction of water to form H_2, $[Ru(dipy)_3]^{3+}$ takes up one electron from a donor to get reduced to $[Ru(dipy)_3]^{2+}$. EDTA or triethanol amine acts as electron donor. On photo irradiation, $[Ru(dipy)_3]^{2+}$ gets into the excited state $[Ru(dipy)_3]^{2+*}$. This can directly transfer electron to H^+ of water. However, the electron transfer is brought about through an electron acceptor relay, which acts as an oxidative quencher. Dimethyl pyridine salt also called methyl viologen (MV^{2+}), acts as an oxidative quencher. Upon visible light excitation, $[Ru(dipy)_3]^{2+}$ reduces MV^{2+} to MV. This transfers electron to H of water, reducing it to H_2 (Figure 9.5).

   However, this electron transfer from MV^+ to H^+ is not efficient and needs a redox catalyst. In this photo reduction of water, platinum acts as a catalyst, wherein platinum particles act as electrodes, polarised by MV^+, as shown in the following:

$$2MV^+ + 2H_2O \xrightarrow{Pt-cat} 2MV^{2+} + 2OH^- + H_2 \tag{9.13}$$

   The yield of hydrogen is dependent on the pH. The optimum pH is 4.5, when EDTA is used as an electron donor. The maximum yield attained is 15%. Attempt is being made to improve the quantum yield by using different viologens as quenchers and Ruthenium complexes of bipyridyl and phenanthroline with substituents as photosensitisers. Further, polymeric systems, with pendant viologen groups and surfactant derivatives of $[Ru(bipy)_3]^{2+}$, are being used as quenchers and photosensitisers, respectively.

ii. In the second process of photo oxidation of water to form O_2, the electron from O^{2-} of water is transferred to $[Ru(bipy)_3]^{3+}$, through a redox catalyst. Thus, O^{2-} is oxidised to O_2 and the $[Ru(bipy)_3]^{2+}$, formed, is subjected to photoexcitation and is oxidatively quenched with a suitable acceptor molecule, so that $[Ru(bipy)_3]^{3+}$ is reformed to accept electron from O^{2-}. The sequence of reactions can be shown as follows:

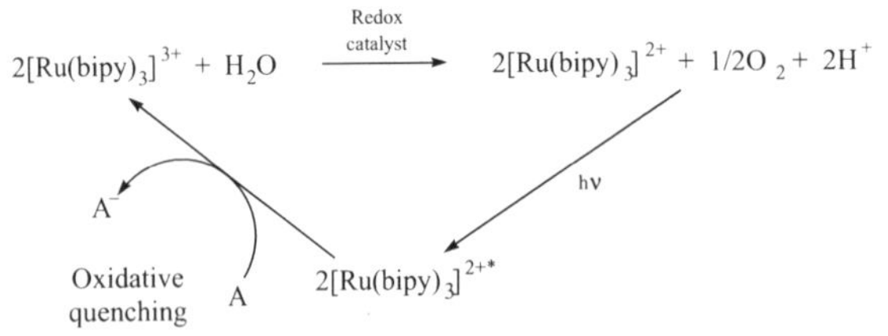

$$2[Ru(bipy)_3]^{3+} + H_2O \xrightarrow{\text{Redox catalyst}} 2[Ru(bipy)_3]^{2+} + 1/2 O_2 + 2H^+$$

A   Oxidative quenching   A $2[Ru(bipy)_3]^{2+*}$ $h\nu$ \hfill (9.14)

The oxidative quencher A should be such that it can be irreversibly reduced to A^-, on accepting electron from $[Ru(bipy)_3]^{2+*}$, so that it does not react back with the $[Ru(bipy)_3]^{3+}$ formed. Some of the electron acceptors (oxidative quenchers) used are $[Co(NH_3)_6Cl]^{2+}$, $[Co(C_2O_4)_3]^{3-}$, S_2O_8, Tl^{3+} and Ag^+.

The redox catalysts used in the electron transfer process from H_2O to $[Ru(bipy)_3]^{3+}$ are homogeneous, which use Fe^{3+}, Co^{2+}, Mn^{2+} ions, or heterogeneous, that use RuO_2 or IrO_2.

It is thus observed that in both the processes, a redox catalyst is necessary in the electron transfer process. This redox catalyst is analogous to WOC in the natural photosynthetic system.

A model system, incorporating the role of both the electron transfer catalyst and the photosensitiser of the oxygen evolution system PS II, has been suggested.

Katakis et al. found a system, wherein tungsten dithiolene complex acts both as a catalyst and photosensitiser and methyl viologen brings about electron shuffle. The complex catalyst accepts electrons from O^{2-} of water, oxidising it to molecular oxygen. On irradiation with light > 350 nm, the reduced catalyst is excited to higher energy state. It transfers the electrons, through methyl viologen to $2H^+$ of water to form H_2. Using D_2O, it has been proved that the H_2 is from water. Tungsten dithiolene complex is a robust catalyst and remains active even after thousand cycles.

QUESTIONS

1. What is photosynthesis?
2. Write the chemical equations showing photosynthesis process.
3. Which experimental evidence proves that O_2 produced in photosynthesis is from H_2O?
4. Why solar energy is required for photosynthesis by green plants? How is this achieved?
5. How chlorophyll is structurally different from haemoglobin?
6. Though chlorophyll is present in plants for absorption of solar radiation for photosynthesis, there are other pigments also necessary in the system. Which are the pigments and why are they needed?
7. What are photosystems?
8. Differentiate between photosystem I and photosystem II.
9. Which part of photosynthesis takes place as dark reaction? Does this part involve any metalloenzyme?
10. Which metal complexes have been studied as models for WOCs? What is importance of such studies?
11. Give mechanism of photodecomposition of water catalysed by metal complexes.

10

Less Common Trace Metal Ions in Biochemical Systems

Besides the roles of the trace metals zinc, copper, iron, cobalt and molybdenum in biochemical systems, discussed in previous chapters, the presence of other trace metals, nickel, aluminium, vanadium and chromium, has also invited the attention of bioinorganic chemists. The probable roles of these metal ions, in biochemical systems, are being discussed in this chapter.

Role of Nickel

Nickel is present in some of the enzymes as follows.

Hydrogenase

The enzymes, which catalyse the evolution of hydrogen from a substrate or the uptake of hydrogen by the substrate, are called hydrogenases. In the first case, the substrate HS gets deprotonated and the proton acts as an oxidant, accepts electron from S^- and oxidises it to S. H^+ is reduced to atomic H and is liberated as H_2. The substrates, oxidised by the dehydrogenation process, are normally the compounds, which get decomposed during the oxidative degradation of sugar in the respiratory process.

In the second process of hydrogenation of the substrate, the enzyme activates the molecular hydrogen, which causes hydrogenation of the substrate; that is, the enzyme methanogenase takes up hydrogen and transfers it to CO_2, for its reduction to CH_4.

Hydrogenases occur as iron hydrogenases or nickel–iron hydrogenases.

To understand the role of nickel–iron hydrogenase, the iron hydrogenases are being discussed first. These iron enzymes are being discussed in this chapter for comparison with nickel–iron hydrogenase, though they could have been discussed in Chapter 5, along with iron-containing biomolecules.

The iron hydrogenases are derived from the bacteria. These enzymes, derived from different genera of bacteria, have similar structure and properties.

Iron is present in this enzyme as iron clusters, Fe_4S_4. They are of two types:

a. **F cluster:** It has a thiocubane structure $Fe_4S_4^+$, like reduced ferredoxin (1Fe III, 3Fe II (Chapter 5), with S = ½, as revealed by ESR spectrum.

b. **H cluster:** This site is called H cluster, because it is here that the hydrogen is activated. This is also an iron sulphur cluster, but differs from ferredoxin or high ionisation potential iron protein. It shows three EPR signals.

In the hydrogen liberation process, the F cluster accepts the electron from the substrate X^-, intermolecularly, through the electron carriers and oxidises it to X. The accepted electron is transferred by the F cluster to the H cluster intramolecularly. It is here that H^+ gets attached and accepts the electron to get reduced to H and gets liberated as H_2.

In the hydrogenation process, atomic hydrogen transfers the electron to the H cluster forming H^+. The electron from the H cluster, in turn, is transferred to the F cluster, and then to the substrate X, attached at the F cluster. The substrate is reduced to X^- and binds with H^+, generated at the H cluster, to form XH, leading to the hydrogenation of the substrate.

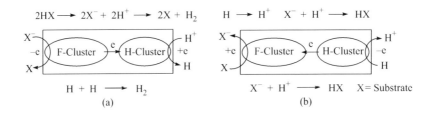

FIGURE 10.1 Hydrogen liberation and hydrogen uptake reactions catalysed by iron hydrogenases. (a) Hydrogen liberation and (b) Uptake of hydrogen.

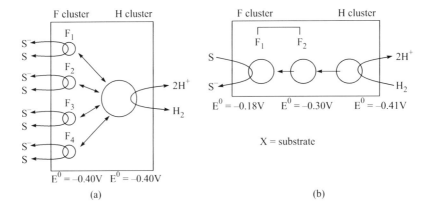

FIGURE 10.2 Hydrogen liberation and hydrogen uptake reactions catalysed by two hydrogenases present in bacteria *Clostridium pasteurianum*. (a) Hydrogenase I (b) Hydrogenase II.

The hydrogen liberation and hydrogen uptake reactions are shown in Figure 10.1.

The bacteria *Clostridium pasteurianum* contains two different hydrogenases. They differ in the number of iron atoms and the number of F and H clusters, as revealed by ESR studies.

Hydrogenase I – contains 20 iron atoms, in four F clusters (16 Fe) and one H cluster (4 Fe). Hydrogenase II – contains 14 iron atoms, in two F clusters (8 Fe) and one H cluster (6 Fe atoms).

These two hydrogenases differ in their roles also. Hydrogenase I catalyses both H_2 uptake and liberation, whereas hydrogenase II catalyses only hydrogen uptake.

This difference in roles arises because, in hydrogenase I, both F and H centres have reduction potential (−0.40 V at pH 8), similar to that of hydrogen. Hence, the electron can move from H to the H cluster or vice versa, thus catalysing both the uptake and liberation of hydrogen.

In hydrogenase II, the F_1 cluster has $E^0 = -0.18$ V, F_2 cluster has $E^0 = -0.30$ V and H cluster has $E^0 = -0.41$ V. Hence, the electron can move only one way, from hydrogen to H cluster, leading to its reduction, and uptake of the formed H^+ by the reduced substrate X^-, to form XH. Hence, only hydrogen uptake is catalysed by this enzyme.

The reactions are shown in Figure 10.2.

Nickel–Iron Hydrogenase

It was observed that some of the hydrogenases contain some other metal ions, besides iron. It was initially mistaken as impurity. But the ESR study showed that the additional signals, other than that of iron, were due to the presence of nickel. Such enzymes, catalysing hydrogen uptake or liberation of hydrogen, are called nickel–iron hydrogenase.

These enzymes originate from bacteria and the Ni-Fe hydrogenases from different bacteria have similarities in the protein structure. Some of these enzymes contain selenium in the form of selenocysteinyl residue in the protein structure. The structure of hydrogenase from bacteria *Dusulfovibrio gigas*, which reduces sulphate, has been extensively studied.

It has two protein subunits, with molecular weights of 30,000 and 60,000 daltons. There is one nickel present in the latter protein units. There are two Fe_4S_4 clusters and one Fe_3S_4 cluster present in the protein. Nickel is the active site for the activation of hydrogen.

Nickel may be present in the enzyme in three different forms, as revealed by the ESR studies. Ni(A) and Ni(B) sites exhibit three g values and correspond to low spin Ni(III). Ni(C) site shows one g value, corresponding to Ni(I).

X-Ray absorption spectroscopic study shows that the nickel ions are bound to the S atom of the cysteine residue of the protein. In the case, where the protein contains selenocystenyl residue, there exists a Ni–Se bond.

Electron spin echo envelop modulation (ESEEM) spectroscopy shows that in Ni(A) and Ni(C), there is a hyperfine splitting of the ESR signal, due to the coupling of electron spin with the nuclear spin of nitrogen. This indicates that the nickel ions are coordinated to the N of the histidyl residue of the protein. Furthermore, Ni(C) shows electron spin coupling with a proton. Therefore, it was initially thought that Ni(C) could be Ni(III)–H. However, the proton coupling constant is too small to suggest nickel hydride structure. Hence, it has been concluded that Ni(C) is Ni(I) with proton close to the nickel. The presence of proton may have some role in the mechanism of the enzymatic activity.

The different forms of Nickel are obtained by the oxidation and subsequent reduction of another form. The enzyme with Ni(C) site is most active. On further reduction, it gives a fully reduced ESR silent form. This has been assigned to Ni(0) or Ni(II).

Presuming that the active form of the enzyme has nickel as Ni(II), the mechanism shown in Figure 10.3 can be suggested for the enzyme activity.

Study of the model complexes has shown that molecular hydrogen can get bound to the metal centre. The σ bonding orbital of H–H acts as the donor, and there is a weak back donation from the metal dπ orbital to the vacant σ* orbital of H_2 (Figure 10.4). The formation of dihydrogen complexes require that the electronegativity at the metal centre should be such that there is only optimum back bonding from $M \rightarrow H_2$.

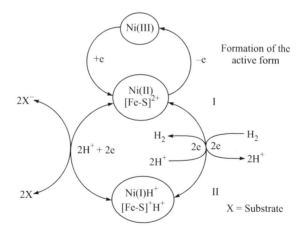

FIGURE 10.3 Catalytic cycle showing mechanism of action of nickel–iron hydrogenase.

FIGURE 10.4 Dihydrogen metal complex in model complexes depicting activity of nickel–iron hydrogenase.

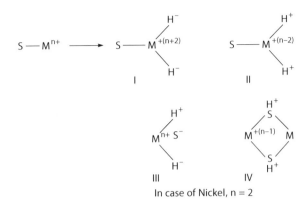

FIGURE 10.5   Different possibilities of hydrogen binding in the activity of nickel–iron hydrogenase.

If H–H to M, σ bond is strong (dependent on the electronegativity of the metal centre and the nature of ligand bound), there is a shift of electron density to the metal ion forming two H^+, which are bound to the reduced metal ion, as suggested in the case of iron hydrogenase activity (Figure 10.1b).

If the metal to H_2 π bonding is strong, there is a shift of electron density to hydrogen, leading to the formation of dihydride. In the synthetic model complexes, like Wilkinson's catalyst $[Rh(PR_3)_3Cl]$ for hydrogenation of olefins, there is initial formation of $Rh(III) - (R_3P)_2 - Cl(H_2)$ and the activated hydrogen is transferred to the olefin.

Since, in the nickel–iron hydrogenases, Ni(II) is bound to sulphur, an alternative mechanism has been suggested. It has been observed that in the reaction of hydrogen with metal sulphide, the H_2 can get cleaved by binding with sulphide, without binding to the metal ion. The resulting thiol group bridges the two metal ions.

Thus, it is seen that hydrogen can get bound to the M–S site in different ways, either directly to the metal or to the sulphur. Hence, all the following possibilities should be considered, to suggest the mechanism of hydrogenase activity (Figure 10.5).

Nickel in Methanogenic Bacteria

Nickel is present in factor F_{430}, which is a cocatalyst of coenzyme M, which catalyses the reduction of methyl into methane. Coenzyme M ($HS \cdot CH_2 - CH_2 \cdot SO_3^-$) is found in the methane generating bacteria (methanogens). Coenzyme M first forms methyl coenzyme M, by reacting with methyl cobalamin (discussed under vitamin B_{12} in Chapter 7). Methyl coenzyme M is reduced by molecular hydrogen into methane, as shown in Eq. (10.1).

Factor F_{430} has a role, as a catalytic cofactor, in the above reaction, but the mechanism of how it catalyses the methane formation reaction is not established.

$$Me - B_{12} + HSCH_2\text{–}CH_2\text{–}SO_3^- \longrightarrow B_{12} + CH_3\text{–}S\text{–}CH_2\text{–}CH_2\text{–}SO_3^-$$

Methyl cobalamin   Methyl coenzyme

$$\text{Factor } F_{430} \bigg\downarrow H_2$$ (10.1)

$$CH_4 + HS\text{–}CH_2\text{–}CH_2\text{–}SO_3^-$$

Factor F_{430} is a tetrapyrrole complex of Ni(II). The coordinated macrocyclic ring has the characteristics of both porphyrin and corrin, and is sometimes said to have corphin structure. The ring has more reduced sites and hence is less conjugated. The macrocyclic ligand has an additional carbocyclic ring (Figure 10.6).

Factor F_{430} has a molecular weight of ~1,000 daltons and exhibits an absorption maximum at 430 nm, hence the name F_{430}. It has yellow colour and is diamagnetic in non-coordinating solvents, indicating a square planar structure of the Ni(II) complex. It has, however, been indicated that the macrocyclic ring

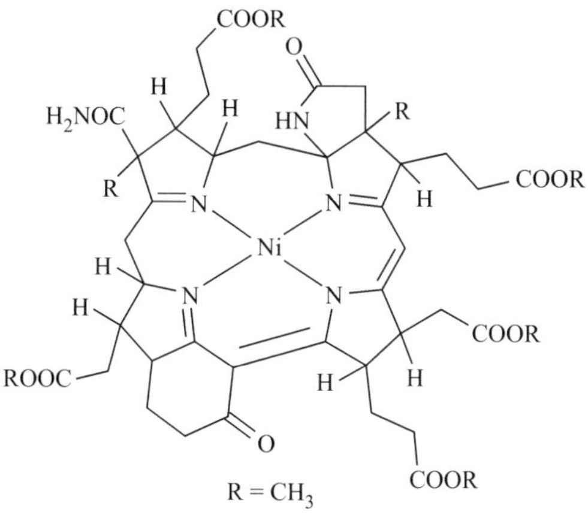

FIGURE 10.6 Factor F_{430} – tetra pyrrole complex of Ni(II).

in F_{430} is substantially nonplanar and buckled. In polar solvents (water methanol mixture), it is paramagnetic, indicating axial coordination of the solvent molecules to Ni(II), leading to octahedral structure.

In the bacteria, F_{430} may occur in two forms, free or bound to protein, probably at both the axial coordination sites of nickel. In the protein-bound form, the composition is F_{430}, two proteins. However, both the free and the protein-bound forms of F_{430} have identical role. Probably, in case, where more of F_{430} is produced, there may not be sufficient protein to bind with it, and it may probably be left in the free form.

The probable role of F_{430}, in methanogenic activity, may be that in the protein-free form, it gets incorporated with the methyl bound form of the coenzyme M, and assists in catalysing the hydrogenation of $-CH_3$ to methane, as shown in Eq. (10.1).

Role of Aluminium in Biochemistry

It has been recently known that aluminium in very small quantity is essential for life processes. However, in slight excess, the element has toxic effects.

Though aluminium is the third most abundant element in the earth, it is never found in the free elemental form. It occurs in the form of insoluble salts such as hydroxides, phosphates and silicates which remain bound to the soil. Thus, in nature, free aluminium occurs only in traces. Hence, in the initial stages of evolution of the biological species, there was no excess of soluble aluminium salts to enter the biological system. However, with the increase in the chemical activity in the nature, there was release of acids lowering the pH of water and soil. This led to the dissolution of the insoluble Al salts into soluble forms like chloride and nitrate, leading to abundance of free Al^{3+} ions in food and water, which in turn entered the biological system through the food chain.

Toxicity of aluminium

Inspite of abundance of Al in the environment, it is present in vegetable and animal systems only in traces. Any excess of Al has toxic effects, leading to stunted growth of the roots of the plants and their destruction. In human beings, disorders such as Alzheimer's and Parkinson's diseases are attributed to the toxic effects of excess aluminium. Fishes are more sensitive to excess Al, and it can have fatal effects on fishes, even in very small excess. The reason for high toxicity of aluminium for animals is that the Al^{3+} ion, with three charges, is highly acidic in nature and has strong tendency to bind with hard base ligands such as amino acids, polypeptides, proteins, DNA and ATP, present in the biological systems. In the gills of the fishes, there are several anionic sites. Al^{3+} binds with these sites, stimulating the production of large excess of mucus in the gills. This causes blocking of the respiration and leads to the death of the fish.

Extensive studies of reactions of Al^{3+} with organic molecules, mimicking biomolecules, have been carried out. The various species of complexes formed in solution have been studied, and their formation constants have been determined, by using precise computation programmes. This has helped in locating the Al^{3+} binding sites of the biomolecules. In such studies, the various hydrolytic species of Al^{3+} in aqueous solution are also considered. Thus, an overall picture of speciation involving Al – biomimicking ligand (L) complex, Al(OH)x complexes and Al(OH)xLy complexes is obtained.

For example, the study of interaction of Al with amino acids and phosphate, which are the building blocks of biomolecules, can reveal the sites of binding of Al^{3+} with the biomolecules and also the strength of binding of the metal ion with the biomolecules. Comparison of the binding constants of the Al^{3+} complexes of biomolecules with the corresponding complexes of other metal ions can reveal how binding of Al^{3+}, with the biomolecules, can affect its bioavailability and also possibility of Al^{3+} completely knocking out essential metal ions from the metalloenzymes thereby deactivating them. This throws light on the toxic effects of Al^{3+}. For example, Mg^{2+} can be displaced by Al^{3+}, and thus, Mg^{2+}-dependent enzymes such as alkaline phosphatases, adenylate cyclase and acetylcholinesterase are deactivated or their activity is altered by Al^{3+}.

Al^{3+} prefers to combine with ligand molecules, with carboxylate groups like in oxalate and citrate. Citrate, with three negatively charged oxygen donor sites, forms very strong chelates with Al^{3+} and thus acts as the best carrier of Al^{3+} in the blood.

It has been seen that Al^{3+} binds with the high-molecular-weight proteins at the negatively charged carboxy site of the amino acid residue. Various species of Al protein complexes that are formed in the blood have been studied. Detailed studies of Al binding, with the protein transferrin, the best carrier of Al^{3+} in the blood, have been carried out.

Study of Al^{3+} binding, with β-amyloids (proteins in the brain, excess of these causes Alzheimer's disease), also indicates the formation of stable complexes. This has been suggested as a possible reason for Al^{3+} causing mental dementia.

Thus, the knowledge of the speciation of Al complexes, with biomolecules, in the blood serum can help in understanding the toxic effects of Al^{3+}. However, this kind of the study of all the Al-biomolecules complex species in blood serum is difficult, even by using precise computation technique, because of the low concentration of Al in the blood.

Restriction on Al^{3+} Poisoning

Though there is large amount of Al all around, its concentration in biological system is low and thus the poisoning effect is lower. The toxicity of Al^{3+} has been restricted in the evolutionary process by the presence of silicon. It is known that the oxides of Al and Si have strong affinity, because of the similarity of the structure of silicate $(SiO_4)^{4-}$ and aluminate $(AlO_4)^{5-}$ ions. When Al and Si are together in water at pH, higher than 4.5, they combine to form highly insoluble hydroxy aluminosilicate (HAS) and thus the availability of free Al^{3+} is inhibited. It has been shown that the fishes can survive, even at higher concentration of Al at pH 5, if there is excess of silicon available. This is because HAS is formed which does not affect the biological processes. Similarly, in the plants, it has been shown that the presence of excess of silicon reduces the root damage, due to excess aluminium. The requirement of silicon for the inhibition of Al poisoning, is more at lower pH, because HAS is more soluble at lower pH and can liberate free Al^{3+} that can cause poisoning. In the presence of excess silicon, the dissolution process of HAS is reversed.

The formation of HAS also reduces the intake of Al^{3+} in the human intestine. Further, the presence of silicic acid in human system also helps in excreting Al deposits from body tissues and bones.

Silicon and aluminium have also been found to exist together in some parts of plants, like roots of sorghum and tabasheer of bamboo plant. However, in these plants Al and Si do not exist as HAS, but Al^{3+} is associated with the silica deposits. Such co-location of Al and Si has also been found in human beings in the plaques found in the brain, causing senility in patients of Alzheimer's and other mental diseases. In normal persons, also, such deposits are formed in the brain, with advancing age. High-resolution NMR studies of such deposits, in the brain, have shown that the deposits are in the form of aluminosilicate (HAS). However, there is no direct evidence of the Al deposits in the brain causing mental dementia in Alzheimer's or other mental diseases.

Transport of Al in Biosystems

The following are the two types of biomolecules which bind with Al^{3+}:

   i. Low-molecular-mass (LMM) citrates
   ii. High-molecular-mass (HMM) protein, apotransferrin.

It can, therefore, be considered that these two types of ligands are, largely, responsible for transporting Al^{3+}, inside the cell. It has also been suggested that Al^{3+} may also be bound to many other ligands in the biomolecules.

Initially, it was believed that Al^{3+} in blood is mainly carried by apotransferrin, the protein, which is known to carry Fe^{3+} in the biological system in the form of the metalloprotein transferrin (as discussed in Chapter 5 on iron). Apotransferrin consists of a protein chain, folded into two globular lobes, connected by a short protein linkage. Each lobe has a metal binding site, where Al^{3+} gets bound to six coordination sites. X-Ray crystal study shows that the protein has two different orientations: (i) the open form, when not bound to the metal ion, and (ii) the closed form, when it gets bound to two Al^{3+}.

By analogy with Fe^{3+} transport by apotransferrin, it was suggested that Al^{3+} is also transferred by the same pathway. However, unlike Fe^{3+}-apotransferrin (transferrin), which has great affinity for the transferrin receptor site in the cell, where it gets bound, Al^{3+}-bound apotransferrin does not have affinity for the transferrin receptor site. Hence, it was suggested that apotransferrin may not be the carrier of Al^{3+}.

Furthermore, it was observed that the fluids present in the brain and spinal cord contain a large amount of Al^{3+}, bound to citrate. Hence, it was suggested that citrate may be binding with Al^{3+} and enters the biological system, mainly in the brain cell.

Citrate has three –COOH and one –OH sites. At low pH, citrate forms a $[Al(cit^{2-})]^+$ type of complexes in which a terminal carboxylate, the central carboxylate and the –OH group are bound to Al^{3+}. The coordinated –OH and the third uncoordinated –COOH remain undissociated. On increasing the pH, however, the coordinated –OH gets deprotonated and the species $[Al(cit^{3-})]^0$ neutral species is formed. The uncoordinated –COOH remains undissociated. At still higher pH, uncoordinated –COOH also gets deprotonated, and the species $[Al(cit^{4-})]^-$ is formed. Thus, on coordination with Al^{3+}, –OH gets deprotonated earlier than uncoordinated –COOH, though in free citric acid, three COOH groups first get deprotonated and then deprotonation of –OH takes place. The species $[Al(cit^{3-})]^0$ is of significance as it is neutral and could pass through the brain cell membrane, thus transporting Al^{3+} inside the biological system (Figure 10.7).

$$[Al(cit^{2-})]^+ \longrightarrow [Al(cit^{3-})]^0 \longrightarrow [Al(cit^{4-})]^- \qquad (10.2)$$
$$\text{At low pH} \qquad\qquad \text{At moderate pH} \qquad\qquad \text{At moderate pH}$$

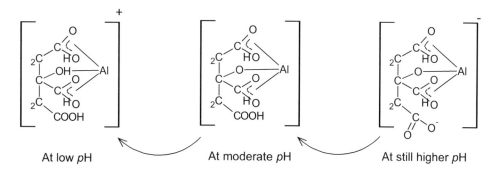

    At low *pH*   At moderate *pH*   At still higher *pH*

FIGURE 10.7   Different species formed between Al and citrate ion at various pH values.

However, the mechanism of Al^{3+}, transport in biochemical system, remains undecided, as the above two mechanisms of transport by apotransferrin or citrate, are only probable suggestions. Further studies are in progress to establish the mechanism.

Role of Aluminium in Promotive Oxidation

Aluminium, being a non-transition metal, forms only trivalent ion. It has no variable oxidation state and is redox inactive. However, it was shown by Fridovick in 1974 that Al^{3+} exhibits pro-oxidative property and thus causes oxidative stress in the biological system. This unusual activity of Al^{3+} was explained by Exley, arguing that Al^{3+} stabilises the superoxide formed in the biological system, due to various processes, by forming the superoxo complex species $[Al^{3+} - O_2^-]^{2+}$. Recently, the stability of such superoxo species has been determined by ESR studies, which show the presence of unpaired electron of O_2^-. It has been shown that the $M^{2+} - O_2^-$ bond is mainly electrostatic in nature and is stable because of the smaller size and high (3+) positive charge of the metal ion, leading to strong attraction between Al^{3+} and O_2^-. The species $[Al^{3+} - O_2^-]^{2+}$ can promote oxidation by either of the following two pathways:

i. $[Al^{3+} - O_2^-]^{2+}$ can combine with H^+ forming the free radical O–O–H, and Al^{3+} is liberated. The O–O–H free radical can generate various oxy species and can thus promote oxidation.

ii. Alternatively, $[AlO_2^-]^{2+}$ can be considered to combine with Fe^{3+} in biological systems reducing it to Fe^{2+}. O_2^- in turn gets oxidised to molecular oxygen, and Al^{3+} is liberated, as shown in the following equation:

$$\left[AlO_2^-\right]^{2+} + Fe^{3+} \rightarrow Fe^{2+} + O_2 + Al^{3+} \tag{10.3}$$

The Fe^{2+} formed in the reaction can generate various oxygen containing radicals by reacting with H_2O_2, generated in the biological system. Fe^{2+} gets oxidised to Fe^{3+}, and there is a formation of OH$^\bullet$ free radical and OH$^-$ anion, as shown in Eq. (10.4) (Fenton's reaction).

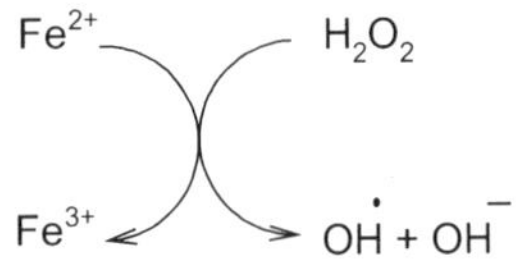

$$\tag{10.4}$$

There may be many such reactions generating active oxygen species, which promote oxidation.

The Al^{3+} regenerated in both the probable pathways is ready for further activation of superoxide. Thus, though Al^{3+} promotes oxidation, it does not take part directly in oxidation.

The active oxygen species generated in the above reactions of $(AlO_2^-)^{2+}$ can cause degradation of various biochemicals. Thus, Al^{3+} by activating O_2^- causes oxidative stress in the biochemical system. The authors of this book have to suggest that the Al^{3+}-assisted activation of O_2^- species in human brain may probably be responsible for the oxidative stress on brain tissues, causing mental dementia, like Alzheimer's disease in older people.

Role of Trace Metal Ions in Control of Diabetes

As discussed earlier, zinc is the constituent of insulin, which is a zinc protein, secreted by pancreas and controls diabetes. Besides that, two other trace metal ions have been shown to have role in control of diabetes. In the following section, role of vanadium and chromium in control of diabetes and in other biochemical reactions has been discussed.

Vanadium

Vanadium is the 22nd most abundant element in the earth's crust. This is present in small amounts in edible plants. However, V enters the animal and the human biochemical system mainly through the food. Many cereals, fruits and fresh vegetables and sea food (like parsley, black pepper, mushroom, shellfish) contain up to 40 mg/L g V as $VOSO_4$, $NaVO_3$, Na_3VO_4 and V_2O_5. Absorption of V compounds in human system also depends on the solubility of the vanadium compounds in the gastric juice and blood serum. Soluble V compounds, inhaled and collected in lungs, are quickly absorbed through the lungs. It has been shown that the persons working in vanadium processing factories, who are exposed to vanadium dust, show high vanadium content in the urine. There is also dermal absorption of V in animals, whereas skin absorption of V in men is insignificant.

Heaz observed in 1917 that the blood cells of ascidians (squirts) contain vanadium, as high as 100 mM, which is hundred times that of V in the sea water. Similarly, organisms like tunicates have very high amount of vanadium in their bodies. This shows that these organisms collect vanadium in their bodies. The purpose for which this collection takes place was not known earlier. It has now been suggested that the accumulated vanadium is helpful for storage of energy, transport of signals and protecting the organisms from predators.

In the 18th century, vanadium was being used as a drug for rheumatism, diabetes, anaemia and even tuberculosis. But not much study on the effectivity of vanadium and its adverse effects had been carried out till 1978, when Carley observed that vanadium can inhibit the activity of the enzyme Na–K ATPase, which in turn can affect various biochemical processes.

Vanadium occurs in nature in various oxidation states. In food, it is found as the $H_2VO_4^-$ (V being in oxidation state V^{5+}). This is absorbed in the duodenum. $H_2VO_4^-$ gets reduced to V^{4+} and exists as oxycation VO^{2+} in the stomach. This has strong affinity for fibre in food. Hence, most of VO^{2+} gets bound to the fibre and is excreted out.

In the biochemical system, O_2 can oxidise V^{4+} back to VO_4^{3-} (with V^{5+}) and VO_4^{3-} can be reduced back to V^{4+}. Thus, both V^{4+} and V^{5+} exist together. Some of the vanadium in anionic form $H_2VO_4^-$ is transported inside the tissues. The transport of the anion is five times more than VO^{2+}, through anion transfer mechanism. The cation vanadium cannot be directly transferred through the membrane of the tissue. Based on the fact that vanadium metabolises similar to iron, it has been suggested that the protein apotransferrin, which transports iron, can also act as a carrier of vanadium in the cationic form VO^{2+}. The cation can get coordinated to the protein apotransferrin or albumin in the blood serum, and vanadium gets transported to liver, kidney, etc.

Finally, there is a possibility that vanadium is stored in the bones like iron, bound to the storage protein apoferritin.

A simple outline of circulation of vanadium in biological system is shown in Figure 10.8.

Role of Vanadium in Biochemicals

Deficiency of V has been shown to have some adverse effects. For example, the pregnant goats having very low intake of V exhibit tendency to higher abortion rate. Similarly, deficiency of V in rats and chickens leads to impaired reproduction.

Deficiency of V also causes rise in glucose level in the blood, lowering of the activity of thyroids, and inhibition of the activity of Na/K ATPase in the brain, kidney and heart. However, the effects are seen more in lower animals. There is no established evidence of the adverse effects of deficiency of V in human beings.

It has been shown that bacteria, which help in the fixation of nitrogen, have vanadium in the constitution of the enzyme nitrogenase, which catalyse N_2 fixation. Vanadium substitutes molybdenum in the structure of nitrogenase. The resulting vanadium nitrogenase has improved properties, and can fix nitrogen at lower temperature and acidic condition.

Vanadium also gets substituted at the active site of another enzyme haloperoxidase. As discussed in Chapter 5 "Catalase and peroxidase activity", haloperoxidases are iron proteins, which assist the oxidation of Cl^- by H_2O_2 to ClO^-, which in turn, oxidises other molecules. The presence of vanadium in the

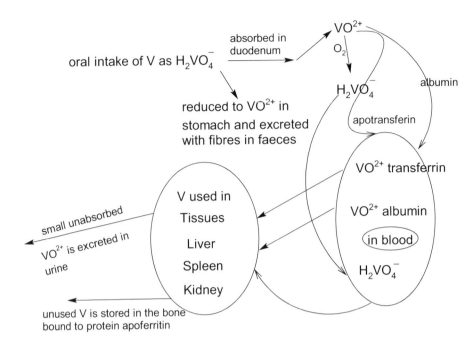

FIGURE 10.8   Circulation of V in biological system.

haloperoxidase enzyme in algae has been shown. X-Ray crystal study of the chloroperoxidase enzyme in the fungus curvularia inaequalis confirms that there is substitution of V in the iron protein enzyme.

It is seen that vanadium is in 5+ state in a five coordinated trigonal pyramidal structure with V^{5+} coordinated to an azide N, three non-protein oxygens and a histidine N atom of the protein. The probable role of V^{5+} in the haloperoxide could be as a Lewis acid centre, assisting the role of haloperoxide in H_2O_2 decomposition. However, the exact structure of active V^{5+} site and how vanadium acts as a better catalytic site is not well established. Besides being a constituent of enzymes and activating them, vanadium can be involved in redox reactions in biomolecules. V^{5+} assists oxidation of the reducing compound NADH in cells, which in turn oxidises other molecules. Vanadium in state V^{4+} gets oxidised to V^{5+} using molecular oxygen and in turn transfers oxygen to other biomolecules causing their oxygenation and comes back to V^{4+}.

V^{5+} also exists as the anion VO_4^{3-}. This has a tetrahedral structure similar to the PO_4^{3-} ion. Hence, VO_4^{3-} can substitute the PO_4^{3-} in the biochemically active phosphates and affect their bioactivity.

Role of V in Sugar Metabolism

Since 1995, clinical trials have been carried out to see the efficacy of V as a drug for diabetes. Detailed study of the mechanism of how V compounds lower the level of sugar in the blood is being carried out. A probable mechanism is the interference of V with the phosphate signal cascade (Chapter 4).

It has been observed in recent years that V salts act like insulin and can regulate the phosphate transfer process. The antidiabetic effect of V is not exactly known, but it has been suggested that like insulin, V may be activating IRK for phosphate transfer and phosphorylation of molecule and deactivating protein tyrosine phosphatase (PTP) molecules to lower the dephosphorylation process.

However, there is an alternative mechanism suggested. It has been proposed that vanadium may not interact with the cell membrane IRK and therefore may not affect the phosphate signal cascade. But V may be active only after entering the cell and deactivating PTP to lower the dephosphorylation process. Thus, phosphorylation process of IRK is not suppressed and phosphate signal cascade process remains activated and helps decomposition of glucose.

It has been shown that administration of vanadium salts, to diabetic rats, reduces their blood sugar level. The antidiabetic effect has been mainly studied using VO_4^{3-}, VO^{2+} salts and hydroxyl amine or maltol complexes of VO^{2+}. The complexes of vanadium are found to be more effective in antidiabetic activity than the free ion. This may be because ligands bound to the vanadyl ion may neutralise the charge on the metal ion and provide a hydrophobic environment around the metal ion and thus facilitate the transport of the charged vanadyl ion through the hydrophobic membrane of the cell. The metal ion or the complex binds with the enzyme PTP and affects its dephosphorylation activity. In case of the complexes being used as drugs, the ligands around the vanadyl ion affect the binding tendency of VO^{2+} with PTP and may regulate the phosphatase activity of PTP.

One feature of the protein of all members of PTP family is a conserved amino acid sequence, and there is importance of all the amino acid residues in the sequence. The presence of cystine residue in the enzyme protein is a must because this is the site for bonding with the substrate phosphate through covalent bonding and affects dephosphorylation. The cystine S–H bond is more dissociable due to adjacent histidine residue. However, the X-ray study of the enzyme protein structure reveals that the histidine residues are situated at a large distance from cystine residue. Hence, there is less possibility of direct bonding between hist and cystine residues. The greater dissociation of S–H bond may be due to network of H bonding between the hist and cyst residues.

The phosphatase activity of PTP has also been shown to depend on the orientation of the protein. Hence, there is an alternative suggestion for the role of V in reducing the dephosphorylation activity of PTP. Binding of the vanadyl ion with phosphatase may change its orientation and thus affect its phosphatase activity. In case of complexes of VO^{2+}, being used as drugs, the ligand bound with the metal ion may affect the orientation of the PTP more and hence lowers its phosphatase activity.

Detailed study of the mechanism of the role of vanadium in lowering the sugar level in the blood of animals is being carried out. Clinical trials are being carried out to see the possibility of using V compounds and complexes as drugs for the treatment of diabetes.

It has been suggested that the property of modulating the carbohydrate metabolism and controlling the diabetic condition may also enable vanadium compounds to affect the pathogenesis of the cancerous glands, by glycogenetic response, like insulin. Vanadyl complexes of 1,10-phenanthroline have been shown to have strong anti-tumour activity. However, the exact mechanism of the role of V as drug for cancer is not known.

Toxic Effects of Vanadium Compounds

The anionic form of V(5+), that is VO_4^{3-}, is like PO_4^{3-}, but in cationic form, VO^{2+} and VO^{3+}, vanadium behaves like a metal ion. It can bind with proteins at cationic sites similar to the non-transition metal ions like Ca^{2+}, Mg^{2+} and d^{10} metal ion Zn^{2+}.

V oxy cations can also combine with the coordinating parts of the biomolecules, like transition metals, and thus can inhibit their activity. It competes with other metal ions in coordinating with biomolecules and thus affects their activity.

It was observed that V compounds deactivate various enzymes such as ATPase, protein kinase and phosphatase by binding with them. V can also deactivate enzyme proteins by the oxidation of the –SH group of the cystine residue present in them. Vanadium compounds activate or deactivate DNA or RNA producing mutagenic and genotoxic effects.

Vanadium administration also leads to the formation and growth of tumours in human beings. However, as stated earlier, now it is known that vanadyl complexes have proved to be anticancer drugs.

Thus, though there is a probability that intake of large quantity of vanadium may cause serious toxic effects, actually normal vanadium toxicity only results in minor irritation in eyes and respiratory system. There is no systemic effect due to vanadium toxicity. Even patients with diabetes tolerated administration of 50 mg of $VOSO_4$ twice daily for a long time, and a lowering of sugar level in blood was observed.

The reason for low toxicity effect of vanadium is due to small absorption of vanadium in the animal system, as mentioned earlier. The small amount of V, which is absorbed, is stored in the organs and is released to the system only when required. Thus, the amount of V in the cells is very small, and it remains bound to fat. Hence, V cannot bring any serious toxic effects. However, the lower toxicity of V is yet to be established. As per the present knowledge, vanadium does have toxic effect and hence should be administered with caution.

Chromium

Chromium is found in large amount in earth's crust and also in water of oceans and seas. It is a transition metal with [Ar] $3d^4 4s^2$ configuration and can exist in oxidation states 3+ and 6+. Cr^{3+} is essential for life processes and has minimum toxic effect. However, Cr^{6+} formed by the oxidation of Cr^{3+} is highly toxic.

Chromium is found in various foods such as vegetables, green beans and broccoli, whole grains, nuts, egg yolk, brewer's yeast, coffee, and some alcoholic drinks. It was observed that brewer's yeast controls diabetes in animals and this was attributed to the presence of a glucose tolerance factor (GTF).

Chromium has been recognised since 1950 as a nutrient for improving glucose metabolism in animals, though, the mechanism of chromium affecting glucose metabolism was not known. Moreover, there is a difference of opinion about the role of chromium, supplied as a food supplement on the human health. However, it is now recognised that GTF, present in yeast, is a biologically active compound, containing Cr^{3+}. As a result, Cr^{3+} compounds were tried for the treatment of diabetes. It was shown that on administration of Cr^{3+} compound, the sugar level of patients was brought down and the dose of insulin required for external administration could be lowered.

Cr^{3+} is the most stable form in biological systems, but it cannot penetrate through the cell membrane easily. However, it has a strong tendency to bind with organic ligands to form neutral coordination compounds or Cr^{3+} complexes with hydrophobic ligand environment. In this form, Cr^{3+} complex is easily absorbed.

Studies have now established that Cr^{3+} is essential for human health and therefore every adult individual is recommended to take a dose of 50–200 µg/day of Cr^{3+}. However, a very small part (0.4%) of the administered Cr^{3+} is taken up by the human system. The extent of absorption depends on the nature of complex form of Cr^{3+} administered. Though free Cr^{3+} ion is absorbed only 0.4%, the neutral picolinate complex of Cr^{3+} is absorbed up to 2.8%. After absorption, Cr^{3+} gets distributed in the different parts of the body and the maximum amount is in the liver, kidney, spleen, bone and muscles. The unused Cr is mainly excreted in urine, faeces, hair and sweat.

Biochemical Role of Chromium

Earlier it was believed in the beginning that Cr^{3+} present in the GTF of brewer's yeast gets bound to insulin and increases its activity, and thus makes it more effective. However, later it was realised later that Cr(III) interacts with insulin and also the insulin receptor kinase (IRK). This indicates that Cr(III) first interacts with insulin and boosts the mechanism of sugar entering the cell. In the second step, it activates IRK and assists in the binding of the insulin on the cell surfaces with IRK.

Cr^{3+} absorbed as the free ion Cr^{3+} or in the form of neutral complex gets distributed in various parts of body. The absorption of Cr^{3+} is facilitated by certain amino acids on the carrier protein like histidine which forms stable chelates with Cr^{3+}, stable at high pH. Hence, Cr^{3+} does not get precipitated as insoluble hydroxide in the basic environment in the intestine. Absorption of Cr^{3+} is also affected by the presence of other metal ions such as Zn, V and Fe. This may be because the other metals may compete with Cr^{3+} and block the coordination sites on the stored carrier proteins where Cr^{3+} is supposed to get bound. Hence, absorption of Cr^{3+} is reduced.

The absorbed Cr^{3+} gets bound to the protein apotransferrin (ATf) and is transported in the extracellular region. ATf has two binding sites A and B, combining with two Fe^{3+} at different pH forming transferrin. However, ATf binds with Cr^{3+} at site B only. Cr^{3+} and Fe^{3+} compete to bind with ATf. Thus, Cr^{3+} intake may interfere with iron metabolism.

Inside the insulin receptor cells, which are initially not sensitive to insulin, there are transferrin (Tf) receptor sites, attached to the vesicles and also a low-molecular-weight protein called apochromodulin. On interaction of the cell with insulin, there is insulin-dependent movement of the Tf receptor from inside the cell to the membrane of the cell. ATf bound to Cr^{3+} gets bound to the transferrin receptor site on the membrane. Subsequently, by a process called endocytosis, the Cr-ATf along with the transferrin receptor site goes back inside the cell and gets bound to the vesicle. This results in lowering of pH of the vesicle by the entry of H^+ in the cell with the help of ATP-driven proton pumps (like Na^+ and Ca^{2+} pumps discussed in earlier chapters). Due to the lowering of pH, Cr-ATf bond breaks and Cr^{3+} is released. This gets bound to the low-molecular-weight oligopeptide apochromodulin to form the compound chromodulin. The oligopeptide bound to Cr^{3+} has a molecular weight of 1,500 daltons and is constituted of glycine, cysteine, glutamic acid and aspartic acid residues.

Apochromodulin exists in inactive form inside the cells, which are insulin receptive. On combination with Cr^{3+}, the resulting chromodulin gets activated and attains halo form. The halochromodulin gets bound to the IRK site, already bound to insulin. Binding of halochromodulin with IRK increases its tyrosine kinase activity and also increases the activity of insulin bound to IRK, causing greater phosphorylation of IRK. Thus, there is overall increase of phosphorylation of IRK. The overall process is shown in Figure 10.9.

Furthermore, Cr^{3+} also binds with the enzyme protein tyrosine phosphatase (PTP), which removes the phosphate from the IRK, and deactivates it. It has been suggested that there is a redox process involved. Cr^{6+} reacts with PTP and gets reduced to Cr^{3+} which binds with PTP, so due to the redox reaction and Cr^{3+} binding, PTP is deactivated, and thus, the dephosphorylation of IRK is inhibited. Thus, chromodulin increases the phosphorylation of the insulin receptor in two ways. The balance between these two processes and the optimum phosphorylation of the insulin receptor facilitates the role of insulin in assisting rapid access of glucose in the insulin sensitive cells and get decomposed there, assisted by the phosphate transfer cascade.

The three catalytic processes, the activation of insulin, greater phosphorylation of IRK and the deactivation of phosphatase enzyme, are affected only by Cr^{3+} bound apochromodulin, Apochromodulin, without Cr^{3+} or apochromodulin bound to any other metal ion, does not affect the activation of insulin.

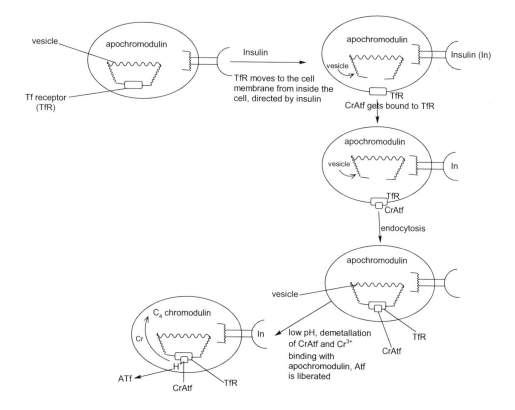

FIGURE 10.9 Increase of phosphorylation of IRK due to chromium intervention.

Chromodulin also increases the activity of the β-cells of the pancreas, so that more insulin is released and more number of insulin get bound to the IRK. Thus, Cr(III) first acts in generation of more insulin, helps in the entry of more sugar into the cell and then facilitates the interaction of insulin with the receptor IRK on the cell membrane, and thus assists the metabolism of sugar.

It is observed that of the above suggested roles of Cr^{3+} in increasing the activity of insulin, consider the process to be intracellular. An alternative extracellular mechanism of chromodulin activity has now been suggested. It has been indicated that Cr^{3+} first gets bound to the insulin receptor site (IRK). Cr(III) also gets coordinated to insulin, resulting in the formation of the ternary complex IRK–Cr–Insulin. This activates the binding of insulin with IRK, and there is an increase in antidiabetic activity. Attempts are being made to study the model systems of ternary complexes of Cr(III) mimicking IRK–Cr–insulin and to understand the mechanism of Cr^{3+} activity in small molecules like Cr picolinate, which are being used as antidiabetic drugs. The authors of this book have a suggestion that there may be intramolecular interligand interaction between insulin and IRK, leading to the stabilisation of the ternary complex IRK–Cr–insulin and increase in activity of insulin on IRK.

Besides the antidiabetic role, Cr^{3+} has the following other roles in biochemical system:

i. Intake of excess Cr^{3+} in food by human beings results in reduction of fat, and the muscles become stronger. Thus, chromodulin helps in controlling the lipid levels. Thus, it has been suggested that there is a relationship between the chromium content in blood and the extent of glucose and lipid metabolism. This, in turn, affects the cardiac health. If there is a deficiency of Cr in food, there is a loss of chromium during pregnancy in women, or during old age, it is followed by impaired metabolism of lipid and sugar. It is also found that in diabetic persons, the level of Cr is less, emphasising that Cr is essential for the metabolism of sugar, through the control of insulin activity.

ii. Chromium also protects the animals against lipid peroxidation and lowers down the effects of the free radicals in the diabetic patients.

iii. Binding of Cr^{3+} with nucleic acid has been found to simulate the DNA-dependent RNA synthesis.

iv. Cr^{3+} is similar to Fe^{3+} in charge and size, and therefore, there can be replacement of Fe^{3+} in transferrin and ferritin by Cr^{3+}, and this can be helpful in treatment of some disease.

v. Cr(III) is also known to reduce the body weight in diabetic patients. Obesity causes impaired sugar metabolism and associated heart disease. Thus, Cr^{3+} by reducing obesity controls the occurrence of diabetes and associated heart disease.

vi. Cr(III) supplemented diet also improves lipid profile in diabetic patients. However, the studies of the effect of Cr^{3+} on obesity, lipid profile and heart disease are preliminary and need further confirmation.

Model Cr(III) Complexes as Drugs

As it is clear that chromodulin involved in the insulin signalling pathway is a low-molecular-weight chromium (LMWCr) peptide, an attempt was made to synthesise model chromium LMWCr peptides and to compare with the naturally occurring ones. A penta-peptide with sequence Glu-Glu-Cys-Gly-asp was synthesised. Two molecules of these peptides were oxidised so that they get bridged through the S atoms of the cysteine residues of the two peptides. This dimer peptide was reacted with Cr^{3+} salt to prepare the Cr^{3+} complex. It was found that the electronic spectrum of the complex was comparable with LMWCr(III) complexes in biological systems, showing antidiabetic character like chromodulin. EPR spectrum of the synthesised Cr(III) complex showed the presence of clusters of three Cr(III) and is comparable with the EPR spectrum of naturally occurring chromodulin. The structure of the synthesised complex is shown in Figure 10.10.

The antidiabetic properties of such LMWCr proteins have not been studied in detail. However, the fact that in the LMWCr peptides, there is a regular pentameric sequence provides direction for the study

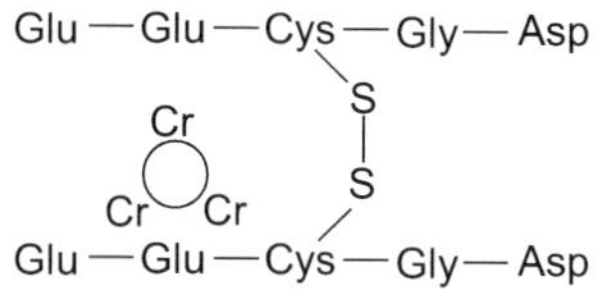

Glu—Glu—Cys—Gly—Asp

Cr

Cr Cr

Glu—Glu—Cys—Gly—Asp

FIGURE 10.10 Synthetic low molecular weight chromium peptide.

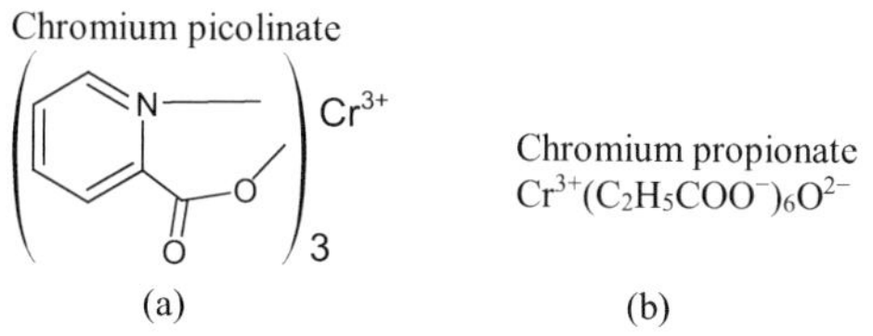

Chromium picolinate

Cr^{3+}

Chromium propionate
$Cr^{3+}(C_2H_5COO^-)_6O^{2-}$

(a)   (b)

FIGURE 10.11 LMWCr models (a) chromium picolinate and (b) chromium propionate.

of the biocoordination chemistry of their Cr^{3+} complexes and their role in activating the signal pathway of insulin.

Chromium has been mainly administered as brewer's yeast, containing Cr^{3+}, $CrCl_3$ and chromium picolinate and chromium propionate (Figure 10.11a and b) to diabetic patients in different amounts. It is observed that neutral chromium picolinate is better absorbed in the biochemical system and better antidiabetic effect is observed, than in the case of $CrCl_3$, with charged Cr^{3+} ion.

However, some studies have pointed out that the administration of Cr^{3+} or its complexes as food supplements has no significant effect on the blood sugar levels and lipid profiles of patients with type 2 diabetes and also of persons with no diabetes. Hence, they have suggested that more studies should be carried out to arrive at the efficacy of administration of $CrCl_3$ or Cr picolinate as antidiabetic drug. However, another study, on the efficacy of chromium picolinate, in the treatment of diabetes, has shown that the Cr(III) complex does have antidiabetic effect and intake of chromium up to 1000 µg per day, in the drug, for up to five years, has no toxic effects.

Carcinogenicity of Cr(VI) Compounds

Cr(VI) finds access to the biological system in the form of anions chromates (CrO_4^{2-}) and dichromates ($Cr_2O_7^{2-}$), through the anion channels like for PO_4^{3-}. Though Cr^{3+} is essential in limited quantities and chromium(III) complexes have been used as antidiabetic drugs, Cr(VI) shows harmful effects in animal system and is toxic.

The toxic character has been attributed to the redox processes, and Cr(VI) gets reduced to Cr(III) interacting with biochemicals and consequently oxidises the biochemicals. Proteins and nucleic acids present in cell reduce Cr(VI) to Cr(III). There may be formation of some active oxygen compounds, formed during the reduction of Cr(VI) which can cause oxidative degradation of the biochemicals and it gets damaged. The presence of active oxygen species can make the cell cancerous. It has been observed that the incidence of cancer among workers in companies producing chromate and dichromate is abundant, indicating that Cr(VI) causes malignancy, by reaction with genetic matter DNA and RNA.

Besides the Cr(VI) entering the animal system, externally as anion, there is also generation of Cr(VI) by the oxidation of Cr^{3+} ingested in the animal system, as food. This is caused by the oxidants superoxide and peroxide present in the biological system in the extracellular region. The anionic CrO_4^{2-} formed enters the cell and exhibits carcinogenic effect, as discussed above. The steps of oxidation and reduction are shown diagrammatically in Figure 10.12. This is another reason that Cr(III) complexes should be used as antidiabetic medicines, with great caution.

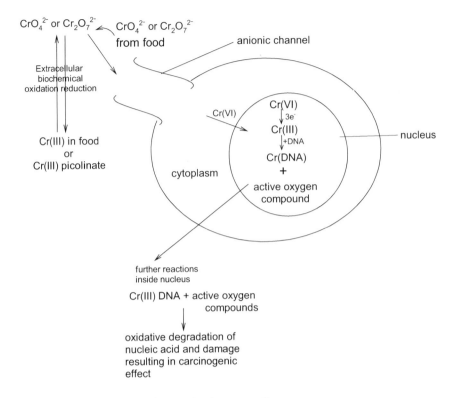

FIGURE 10.12   Redox reactions of chromium causing damage to cell.

Role of Vanadium

Vanadium has been detected in very small amounts in biological systems. Though its role is not definitely known, small amount of vanadium is required by living system for normal health.

If it exists in oxidation state (V) as anion VO_4^{3-}, it is analogous to phosphate and may mimic phosphate in the biochemical system. Hence, it deactivates the enzyme like Na–K–ATPase, which catalyses the breaking down of ATP to ADP, through the binding of the triphosphate at the alkali metal ion, and subsequent hydrolysis, as shown in Eq. (10.5):

$$\text{ATP} \xrightarrow[\text{(NaK protein)}]{\text{NaKATPase}} \text{ATP} - \text{Na-K-protein} \longrightarrow \text{ADP} + PO_4^{3-} + \text{Na} \cdot \text{K-protein} \qquad (10.5)$$

The $H_2VO_4^-$ gets bound to the Na–K–ATPase, at the site where phosphate group of ATP binds for phosphorylation to take place. Thus, Na–K–ATPase is deactivated, and hence, ATP decomposition is inhibited in the presence of $H_2VO_4^-$.

If vanadium has oxidation state V^{4+}, it may be present as the cation VO^{2+}. This also deactivates the enzyme Na–K–ATPase.

Thus, mainly vanadium has the role of deactivating the enzyme in the biological systems. However, there are few cases where vanadium activates some enzymes.

The enzyme peroxidases are essentially Fe(III) protoporphyrin proteins (as discussed in Chapter 5) and catalyse the oxidation of substrates by H_2O_2. Alternatively, the oxidised form of the enzyme may assist the oxidation of a redox bromide, which in turn oxidises the substrate. Such enzymes are called bromoperoxidase.

Bromoperoxidase reaction is shown in the following equation:

$$Bromoperxidase + H_2O_2 + Br^- \longrightarrow BrO^- + H_2O$$

$$BrO^- + AH_2 \longrightarrow A + H_2O + Br^-$$

(10.6)

In some algae, the enzyme bromoperoxidase contains vanadium, and it acts as a cofactor.

Vanadium may replace molybdenum in nitrogenase as discussed in Chapter 8. However, it is not definitely known, if vanadium independently catalyses the reaction between N_2 and H_2 or it stabilises Fe–Mo, present in the enzyme, which actually catalyses nitrogen fixation.

Vanadium also has a probable role of oxygen transport. It is present in the blood cells of ascidians (sea squirt), a primitive vertebrate. A preliminary study showed that there was small reversible uptake of oxygen by the vanadium containing biochemical in the ascidians. The compound was therefore named hemovanadin by analogy with haemoglobin, though it is not a porphyrin complex. The vanadium was initially supposed to be present in the ascidians in the form of a protein complex.

However, it was later realised that the blood of ascidians is highly acidic and hence the vanadium protein complex may not be stable at that high acidity.

Vanadium in the form VO^{3-} enters thorough the anionic channel, gets reduced to V^{4+} or V^{3+} and is entrapped in the cellular vacuoles of the ascidians.

X-Ray absorption spectroscopic study of the blood cells of ascidians shows that 10% of vanadium is present as VO^{2+} and the rest is simply V(III). It is still not known how vanadium in the free form can transport oxygen.

It has now been recognised that vanadium has some role in sugar metabolism and this is an active field of research in diabetes therapy.

Chromium

Chromium also occurs in biochemical systems in very small amounts and its role is not definitely known.

It has been suggested that chromium is involved in the action of the zinc protein, insulin. As deficiency of chromium affects sugar metabolism, it is evident that it is associated with the GTF in the body. However, the exact role of chromium and its mode of binding with the GTF are not known. Chromium also acts as a cofactor, in the activity of other enzymes, and may be involved in the stabilisation of nucleic acid.

Lanthanides in Biological Systems

Lanthanides (Ln) are the 14 elements in which 4f orbital is progressively filled. The general electronic configuration can be given as Kr $4f^{1-14}5d^16s^2$. As the f orbitals are filled up, these elements are called f block elements. The 4f block elements are called lanthanides, and 5f block elements are called Actinides [Ref. Threshold of Inorganic Chemistry]. The total of the first three ionisation potentials of the lanthanide elements is low, and hence, the lanthanides tend to lose the three electrons easily and form the trivalent ions. The earlier trivalent lanthanide ions have bigger size and are basic in nature. The ionisation potential increases, and the consequent basicity goes down, as the ionic size decreases with increasing atomic number, that is from cerium to lutetium.

The lanthanides are more reactive and thus behave like the alkali and alkaline earth metal ions. The heat of atomisation of the lanthanides is less, like the alkali and alkaline earth metals. Hence, they are much softer, like s and p block elements, and differ from harder transition metals. Like alkali and alkaline earth metals, the lanthanides liberate H_2 on reaction with water. Thus, lanthanides are like alkaline earth metals (group 2), and were historically considered rare in terms of their natural abundance. Hence, they are also called rare earths.

They primarily occur in nature as phosphate or carbonate minerals, such as bastnaesite, monazite and xenotime. Due to similarity in their properties, many of the members occur together in their minerals. Lanthanides are also present at significant concentrations in coal ash and acid mine drainage, along with other metals ions such as Fe^{3+}, Mn^{2+}, Cu^{2+} and Al^{3+}, which are present at higher concentrations. Lanthanides find several commercial applications. For example, europium is important in flat screen displays found in computer monitors, cell phones and televisions, as phosphors; in laptop hard drives; in catalytic processes that convert crude oil into gasoline; in dental and medical equipment; in hybrid vehicle engines and the magnets used in wind turbines; and in lasers. The lanthanides also have a variety of nuclear applications. Because they absorb neutrons, they have been employed in control rods used to regulate nuclear reactors. They have also been used as shielding materials and as structural components in reactors.

Because of this, there is need to use lanthanides in sustainable manner. Also focus is on recovery of many of the used metals. Methods of recovering the metals from mixtures are quite tedious and expensive. Hence, scientists have also focussed on using microorganism to recover them. These efforts and other independent studies indicated that nature has several roles for lanthanides in biochemical processes.

Following are the reasons for nature making optimum use of lanthanides in biological systems.

1. Ability to achieve high coordination number (CN), typically 8–12 and flexible coordination geometries, making binding with proteins easy.
2. Ionic radii similar to the more abundant metal ions like Ca^{2+} {for a CN of 8, ionic radii span from 1.16 Å (La^{3+}) to 0.98 Å (Lu^{3+})}. The ionic radius of Pr^{3+} is similar to that of Ca^{2+}, 1.12 Å (for CN = 8). Lanthanide contraction is responsible for the ionic radii pattern of the lanthanides.
3. Because 4f electrons are shielded, they do not influence the bonding; therefore, Ln^{3+} complexes are largely ionic in nature. They therefore prefer hard (oxygen containing) carboxylate ligands, abundantly present in biological systems.
4. High charge-to-size ratio makes them good Lewis acids, resulting into their hydroxide and phosphate salts to be less soluble in water.
5. Fair natural abundance (thulium, the rarest naturally occurring lanthanide, is more common in the earth's crust than silver ($4.5 \times 10^{-5}\%$ versus $0.79 \times 10^{-5}\%$ by mass)).

The lanthanide biochemistry has been broadly considered in terms of acquisition (by various proteins/enzymes), transport, utilisation and storage in biological systems, and these processes are comparable to similar processes occurring in nature for Ca^{2+} and Fe^{3+}.

Though not considered important earlier, lanthanides have been studied in biological systems due to the possible exposure of humans to radioactive lanthanides. Also, medical uses of the elements and their compounds were being explored which also needed studying lanthanides in biological systems. Current heightened interest, however, is due to demonstration of a specific biological role of lanthanides with genetic and biochemical characterisation of XoxF as a lanthanide- and PQQ-dependent methanol dehydrogenase (Ln-MDH) in several methylotrophs including methylorubrum (formerly methylobacterium) extorquens AM1 (a type of methylotrophic bacterium). PQQ is pyrroloquinoline quinine cofactor.

Methylotrophs – bacteria that utilise methanol as their carbon source, converting it to formaldehyde. They contain enzyme MDH – methanol dehydrogenase – that oxidises methanol to formaldehyde. This reaction is considered to be a formal hydride transfer from the substrate to the metal – ligated C5 carbonyl of the PQQ cofactor, facilitated by the Lewis acidity of the metal ion. This is followed by single electron transfer from reduced PQQ to a c-type cytochrome or XoxG in the case of Ln-MDH. Methylotrophs are found in soil, water and plants. They are considered to have important role in carbon cycle. In the carbon cycle, there are several microorganisms that convert methane to carbon dioxide via methanol, formaldehyde and formate.

Methanotrophs – bacteria that utilise methane as the carbon source, converting it to formaldehyde.

Gram-negative bacteria possess an MDH with PQQ as its catalytic centre. This MDH belongs to the broad class of quinoproteins, which comprise a range of other alcohol and aldehyde dehydrogenases. The Ca^{2+}- and PQQ-dependent enzymes were well known before the discovery of Ln-XoxF, lanthanide- and PQQ-dependent enzyme. After it was discovered that the enzyme action of Ln-XoxF depends on lanthanides, several more such Ln-XoxF- and PQQ-dependent enzymes have been found in methylotrophs and even non-methylotrophs during last ten years. These are called lanthanoenzymes. It has been shown that these enzymes prefer early lanthanides, viz., lanthanum to neodymium. The set of proteins and other biomolecules involved in lanthanide utilisation are called "Lanthanome". The oxidation of methanol catalysed by these enzymes is shown in Figure 10.13.

The figure shows reaction catalysed by MDHs with c-type cytochrome redox partners of Me Ca-MDH (MxaG) and Ln-MDH (XoxG), and the reactive C5 site of PQQ is denoted. (Ref.: Joseph A Cotruvo Jr, *ACS Cent. Sci.* 2019, 5, 1496–1506.)

Apart from the PQQ-dependent ADHs, other lanthanide proteins have also been discovered. In 2018, Lanmodulin (LanM) was discovered and characterised. LanM has amino acid sequence containing 4 EF hand motifs. EF hands are 12-residue Ca^{2+}-binding motifs, are widespread in biology and are usually found in pairs for cooperative binding motifs. The EF hands bind Ln^{3+} with slightly higher affinity than Ca^{2+} (about 100 folds more). Greater selectivity of these proteins for Ln^{3+} over Ca^{2+} is also in terms of LanM responding to picomolar concentrations of free Ln^{3+} compared to the response for Ca^{2+} being millimolar concentration. Studies suggest that the proline residue at the second position in EF contributes to this selectivity by uncoupling initial Ca^{2+} binding from a conformational change. Several models are being studied looking at the mechanism of these enzymes and using them for various applications. Many more proteins containing lanthanides are being discovered and characterised, and it is expected that some novel enzyme may be discovered in the process.

Lanthanides in Plants

Since last few years, research has focussed on role of rare earths in soil and in agriculture. Because of the abundance of these elements in earth's crust, it is expected that they may have some role in growth of plants. In fact, some studies have been carried out with synthetic fertilisers containing rare earth salts as additional micronutrients. In one such study, rare earth phosphate fertiliser (REPF) as base fertiliser

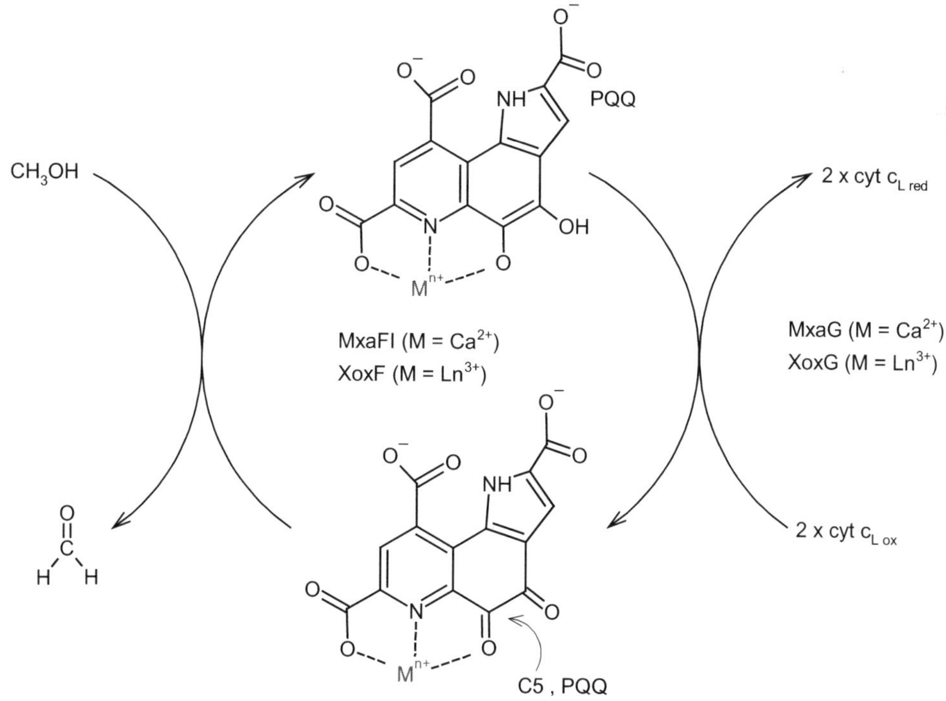

FIGURE 10.13 Oxidation of methanol catalysed by lanthanoenzymes.

(750 kg/hm^2) was applied to crops, and the following results were obtained: compared with calcium superphosphate (CK), REPF increased crop yields for maize by 17.0%, for rice by 10.5%, for wheat by 24.2%, for potato by 18.5%, for cabbage by 16.3%, for Chinese cabbage by 16.4%, for beet by 6.5%; decreased the diseased plant rate for common smut of maize by 1.0%, for maize stalk rot by 1.2%, for wheat take-all disease by 7.8%, for wheat root rot by 3.2%, for potato blackleg disease by 1.4% and for potato late blight by 6.6%; and increased the sugar content of beet by 0.9%. However, more detailed studies are being undertaken to understand influence of rare earths in plant life, their accumulation and selectivity of uptake.

Lanthanides in Medicine

Due to similarities in calcium and lanthanides, it is expected that lanthanides may share biological activities of calcium. In fact, literature on medicinal use of lanthanides is increasing. Lanthanide-based small molecules and nanomaterials have been investigated as cytotoxic agents and inhibitors, in photodynamic therapy, radiation therapy, drug/gene delivery, biosensing and bioimaging. Lanthanide radioisotopes like ^{177}Lu have been used in cancer imaging and therapy. Other forms of lanthanides, such as lanthanide oxide nanoparticles, nanodrums and nanocrystals, are being studied as imaging agents and potential anticancer drugs.

Cytotoxicity of lanthanide salts and compounds has been a subject of intense research. A variety of ligands have been reported as anticancer agents using lanthanides as metal ions. For example, complexes or nanoparticles of La, Gd, Sm, Eu, Tb, Dy, Pr and Nd with the ligands warfarin, hymecromone, coumarin-3-carboxylic acid and dihalo-8-quinolinoline (Figure 10.14) have been studied for their anti-cancer properties. In general, cytotoxicity of these compounds is attributed to their interaction with DNA, though, in some cases, other mechanisms have also been proposed. For example, Pr^{3+}, La^{3+} and Nd^{3+} were found to inhibit calcium transport in mitochondria (owing to similar ionic radii).

The USFDA approved lanthanum carbonate as drug in 2004. Hydrated lanthanum carbonate, $La_2(CO_3)_3 \cdot xH_2O$ (x = 3–5 moles), is used to decrease phosphate levels in human body. This is necessary in patients with advanced stage of kidney disease. There can be other reasons for increased levels of phosphate in blood, apart from decreased excretion due to kidney disease. Increase in phosphate level in blood is called hyperphosphatemia. Phosphate hinders absorption of calcium, causing serious complications. Lanthanum carbonate is given to patients as oral drug; in stomach, La^{3+} is released and it traps dietary phosphate as insoluble lanthanum phosphate in gut, thereby inhibiting phosphate absorption.

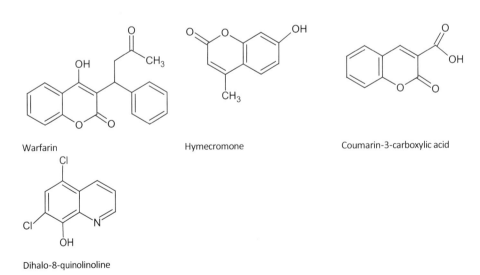

FIGURE 10.14 Ligands studied as anticancer agents along with lanthanide metal ions.

Lanthanides as Imaging Agents in Diagnostics

Imaging techniques such as MRI (magnetic resonance imaging) and PET (positron emission tomography) are used for diagnosis of diseases or to monitor treatment in internal organs, cellular processes and tumours, as well as normal tissues of human body. To improve the quality of information, certain reagents are used in the process, which are called imaging agents. Some lanthanide complexes have been successfully used for the purpose.

The most common imaging or contrast agents used in MRI use a polydentate ligand with a coordinated paramagnetic metal ion – Gd^{3+}, $Fe^{2+/3+}$, $Mn^{2+/3+}$. After administration of the contrast agent, a short radiofrequency pulse is applied to the person placed in external magnetic field (the scanner). Due to this, nuclear spins of the surrounding water protons get excited, and then, they relax to the equilibrium of the external magnetic field. The rate of relaxation depends on the specific chemical environment of the water molecules. Variation of the relaxation rates is the basis of signal intensity in MRI. Paramagnetic metal ions, containing unpaired electrons, generate magnetic moments which interact with neighbouring water molecules and increase the proton relaxation rates, improving resolution of contrast image. Gadolinium compounds as contrast agents have certain advantages:

- Gd^{3+} (S = 7/2) has the largest number of unpaired electrons for a stable aqueous ion.
- It can achieve high coordination numbers (6–10).

Gadolinium is administered as strongly coordinated complex, in order to avoid toxicity of the free metal ion. Preferred ligands used for the purpose are diethylenetriaminepentaacetic acid (DTPA) and 1,4,7,10-tetraazacyclododecane-1,4,7,10-tetraacetic acid (DOTA), of which former is linear while later is macrocyclic ligand. There is some possibility of leaching of Gd^{3+} from the linear ligand; hence, macrocyclic ligand is presently the ligand of choice. Figure 10.15 shows structures of the two ligands.

The complex of gadolinium with DTPA is usually administered as salt, with chemical formula $A_2[Gd]$ in which the cation A is protonated form of the amino sugar meglumine and the salt is called gadopentetate dimeglumine. Meglumine is an amino sugar derived from glucose, and its structure is shown below.

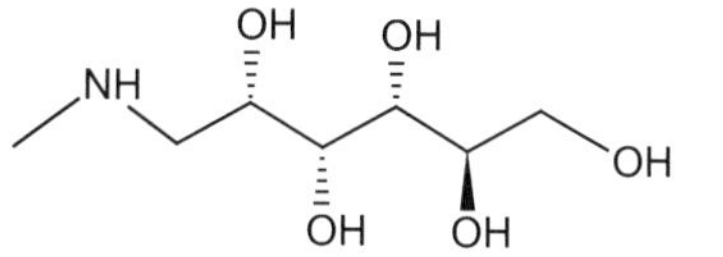

Gadoteric acid is the DOTA complex of gadolinium, used in the form of meglumine salt.

Most of the current research is focussed on targeted delivery of therapeutics, including contrast agents. Some derivatives of DTPA and DOTA have been designed yielding metal complex targeting conjugates.

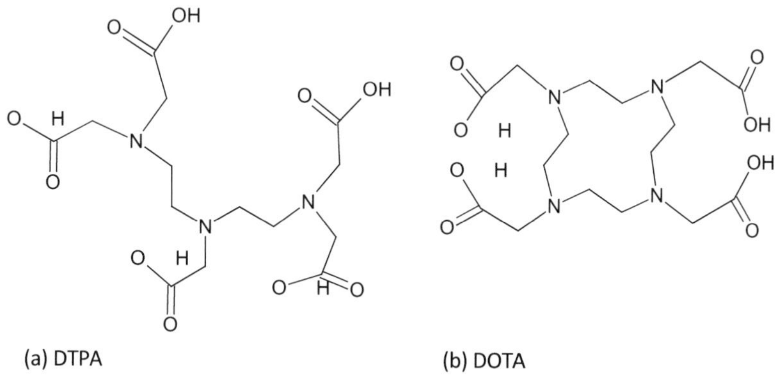

(a) DTPA

(b) DOTA

FIGURE 10.15 Gadolinium complexes of ligands (a) DTPA and (b) DOTA used in imaging techniques.

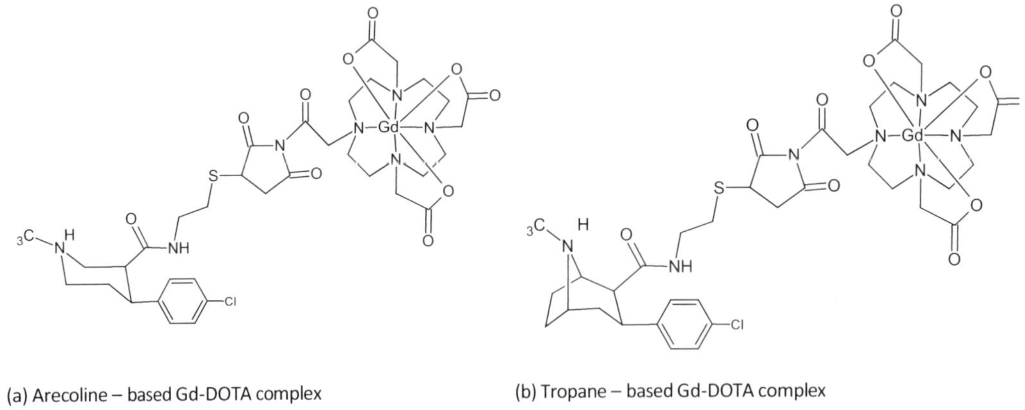

(a) Arecoline – based Gd-DOTA complex   (b) Tropane – based Gd-DOTA complex

FIGURE 10.16   Functional imaging using (a) arecoline- and (b) tropane-based Gd–DOTA metal complexes.

Thus, arecoline- and tropane-based Gd–DOTA metal complexes (Figure 10.16) have been used for the selective MRI of dopamine transporter, DAT, while studying CNS disorders. This is also called functional imaging (fMRI).

Lanthanide compounds are also used as probes for luminescence-based imaging methods either alone or along with PET or MRI techniques. Luminescence is defined as the emission of light from a molecule or substance (called luminophore) on excitation with radiation of a certain frequency. The electronically excited state of the molecule being unstable, when it comes back to ground state, it emits light of wavelength higher than that used for excitation. In luminescent optical imaging, a substance can be targeted to the organ/cell and then irradiated with light, monitoring of the emission light can then be used to ascertain the location and concentration of the luminophore within the biological sample. Such optical imaging agents therefore find applications in direct quantification of specific biomarkers involved in physiological or pathological processes at cellular and subcellular levels. But the penetration of light through tissue has limitation, and hence, such studies have been restricted to small animals only.

QUESTIONS

1. Name the Ni(II)-containing enzymes that participate in biochemical reactions. Also give brief description of its activity.
2. What is the of vanadium in biochemical reactions?
3. Does chromium has some role in biochemical processes?
4. Despite being present in abundance on earth, there is very little of aluminium in biological systems. Why?
5. Aluminium is generally considered to be toxic to biological systems. Give reason.
6. Why is aluminium present in environment in abundance presently?
7. Does nature have some system to tackle toxicity of aluminium in biological systems?
8. Write various chemical forms of vanadium that are important in biochemical processes.
9. Is vanadium important in diabetes control? Explain.
10. How does chromium help in sugar metabolism of diabetes patients?

11

Metal Ion Toxicity in Biochemical Systems

As discussed in previous chapters, metal ions, in limited quantity, are essential for life processes. The concentration of the essential metal ions, inside the cell, is controlled by an enzyme-regulated transport mechanism, by which excess, unused, metal ion is ejected, outside the cell. However, normally the biological systems do not have transport mechanism for nonessential (impurity) metal ions, which find access inside the biological system, though they have no biological role and do not normally occur in biological systems. The excess of the impurity metal ion can, thus, be created, and they can cause disorder or poisoning effect. Such metal ions are considered to be toxic.

The toxic metal ions can be classified into following two types:

i. **Essential metal ions in excess:** A metal ion, which is essential in trace amount, for the normal activity of the enzymes, may also have toxic effect, when present in excess amount. For example, magnesium and manganese, essential for the activity of various enzymes and chlorophyll, are moderately toxic in large excess. Chromium, which is essential as a sugar tolerance factor, is highly toxic as Cr(VI) and moderately toxic as Cr(III). Zn(II), which is an essential component of various enzymes, is moderately toxic in excess. Iron, which is essential for all organisms, is slightly toxic in excess. Copper, cobalt and nickel, which have significant role, as cofactors in enzymes, are highly toxic, in excess, to plants and invertebrates and moderately toxic to mammals.

ii. **Nonessential metal ions:** These metal ions are not essential for body system and have toxic effect, if ingested. The examples are the heavy metal ions of the third transition series and f-block elements. These are not essential for biochemical systems, but have strong tendency to bind with biochemical molecules and can thus have toxic effect, when introduced in the body.

The toxic metal ions can exist outside or inside the cell. The metal ion can bind to the negatively charged sites of the lipid membrane, or the carboxylate groups of the peptides of the outer membrane of the cell, and thus get collected in large amount in the biological system.

The toxic metal ions can find passage inside the cell, by the same transport mechanism, as the chemically related essential metal ions. For example, arsenate or arsenite follow the path of phosphate. Tl^+ enters using the transport mechanism of K^+, and Cd^{2+} avails the mechanism of Mn^{2+} transport.

The metal ions can cause toxic effects in following ways:

i. The impurity metal ion can get bound to the coordinating site of the enzyme proteins, which are the binding sites for the essential metal ions, as shown in Figure 11.1.

Such binding will block the essential biological functional group of the enzyme and it gets deactivated. Thus, it cannot perform its normal biochemical functions.

The toxic metal ion can also bind with the ion channels, membranes and polysaccharides, which have coordinating sites. This also interferes with the normal biochemical process.

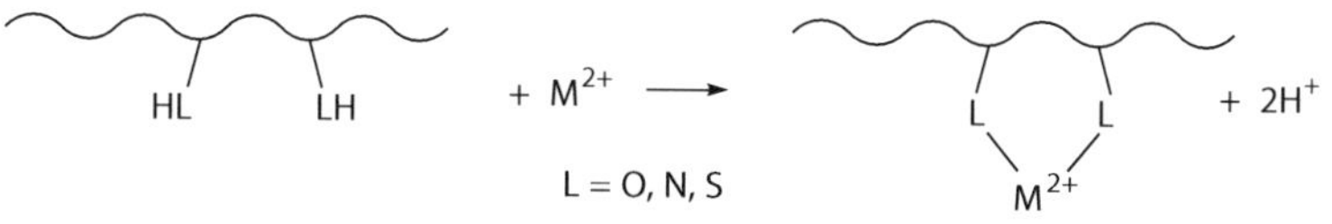

FIGURE 11.1   Binding of metal ions to enzyme proteins.

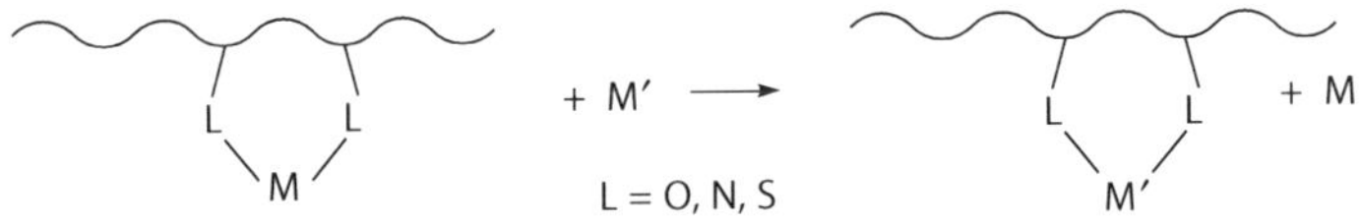

FIGURE 11.2 Displacement of metal ion in enzyme proteins.

ii. The deactivation of the enzyme can also be due to the displacement of the metal ion, present in the enzyme and essential for its catalytic activity, by the external toxic metal ion. Heavy metal ions, such as Pb^{2+}, Cd^{2+} and Hg^{2+}, can replace the essential metal ion Zn^{2+}, from the enzyme.

Normally, the essential metal ion is replaced by an external metal ion, having the same chemistry. But, since the external metal ion is biochemically inactive, it cannot catalyse the activity of the enzyme and thus exhibits toxic effect. For example, cadmium and zinc belong to the same periodic group and have similar chemical properties, and hence, Cd(II) replaces Zn(II) in the enzymes. However, Cd(II) cannot act as a cofactor for the Zn(II)-specific enzyme, and hence, its introduction deactivates these enzymes and brings metabolic disorder. Thus, Cd(II) shows toxic effect.

iii. The entry of the new metal ion in the biomolecule can change its conformation, and consequently the biological activity.

iv. The toxic metal ion can get bound to the DNA and can stimulate its replication, thus resulting in uncontrolled cell growth. Thus, the toxic metal ion can be carcinogenic, leading to the formation of malignant tumours.

v. The binding of the metal ion, with the DNA, can cause change in the base sequence. This mutagenic effect results in transmission of wrong genetic information from DNA, leading to the production of faulty proteins and enzymes. Toxic metal, binding with DNA, can cause birth defects also.

The toxic effect of a metal ion depends on its chemical form. For example, Sn^{2+} is not very toxic, but the methylated form $[Sn(CH_3)_3]^{2+}$ is more toxic. This indicates that toxicity is dependent on the whole molecule and not only on the metal ion.

As stated above, Cr(VI) is more toxic than Cr(III), probably, because the former is a stronger oxidising ion and may cause oxidative degradation of the biomolecule.

The solubility of the metal salt also affects its toxicity. For example, nitrates and halides of metal ions are more water soluble, whereas sulphates, phosphates, carbonates, oxides and hydroxides are less soluble in water. The availability of a toxic metal ion to the living system is more, if it is present in the form of a soluble salt. Hence, soluble salts of the metals are more toxic. For example, soluble barium chloride is moderately toxic, whereas insoluble barium sulphate is non-toxic and is used internally for getting X-ray photographs of the digestive system.

Detoxification Mechanism

Though the biological systems do not have mechanism for the expulsion of the toxic metal ions, they have mechanism to inhibit the toxic effect. For example, in certain bacteria there are chemical oxidation or reduction processes, which change the toxic metal ion to less toxic form, for example Cr(VI) to Cr(III) form.

There may also be sequestering of the metal ion by binding with strongly coordinating biomolecules, like proteins or even cell membrane. As discussed earlier (Chapter 5), the organisms contain a strongly coordinating biomolecule metallothionein, which has sulphur coordinating, cysteine sites. It has special role in copper metabolism. But for other toxic metal ions, it can act as a sequestering agent, masking the metal ion, and rendering it non-toxic. The metal ions are taken up in the order of their tendency to bind with metallothionein. The monovalent metal ions Cu(I) and Ag(I), bivalent Hg(II) and trivalent Bi(III) are taken up first and then in order Cd(II), Pb(II) and Zn(II).

In certain bacteria, there is an actual pumping mechanism to expel the toxic metal ion. For example, the export of Cd(II), from inside the cell, is controlled by a Cd/H$^+$ antiport. A proton gradient is created across the cell membrane, due to hydrolysis of ATP, and thus, an active pump system is generated for the efflux of Cd^{2+}.

Similarly, arsenate and arsenite are ejected outside the cell by anion pump, driven by ATP hydrolysis.

Another mechanism, of detoxification of the metal ion, is to convert them to volatile forms. Many microbes detoxify the heavy metal ions by converting them to the methylated form, which has low solubility in water and is very volatile. Exact mechanism of detoxification, due to volatile compound formation, will be discussed under mercury toxicity.

Effect of Deficiency and Excess of Essential Metal Ions

i. **Zinc:** Zinc is an essential element and occurs in several enzymes. The normal requirement of zinc for an adult is 15 mg/day. Deficiency of Zn(II) can be caused by binding of Zn(II) with phytates and organic phosphates in food, so that it cannot be absorbed by the body. Zinc deficiency leads to skin disease, impaired development of gonads, dwarfism, loss of appetite, anaemia and loss of body hair. It is not known how zinc deficiency causes the above conditions. However, it can definitely be said that zinc deficiency may affect the activity of the zinc-containing enzymes.

Zinc deficiency can be reversed by administrating zinc sulphate. However, excess of zinc cannot be given, as excess zinc can have toxic effect. Zinc excess can also be caused to the workers in zinc smelters or due to inhalation of industrial fumes. The symptoms of excess Zn(II) are abdominal disorders, metallic taste, respiratory irritation and cyanosis (zinc fume fever). Excess zinc also causes deficiency of copper ion and calcium. Large excess of zinc can cause convulsion, paralysis and even death.

ii. **Copper:** Copper is a constituent of many enzymes, involved in electron transfer, oxygenation and oxidation processes. Hence, deficiency of copper causes deactivation of these processes, leading to anaemia (ceruloplasmin deficiency), and loss of hair pigment (tyrosine deficiency). Deficiency of Cu(II)-containing enzyme, cytochrome c oxidase, causes reduced arterial elasticity and stunted growth in adults and Menkes disease in children, resulting in kinky hair, retarded growth, and respiratory problems, severely limiting life span.

If synthesis of ceruloplasmin is hindered, the mechanism of the control of copper level in the biological system is damaged. This leads to accumulation of copper in liver, kidney and brain. Thus, the central nervous system is damaged, leading to tremors, rigidity and abnormality of the brain. Accumulation of copper in liver leads to cirrhosis and ultimate death. This physical abnormality is called Wilson's disease.

External intake of small excess of copper causes gastrointestinal irritation and vomiting. Serious toxic effect is observed, if more than one gram of copper is taken at one time or there is continuous intake of 250 mg/day, for a period of time. The toxic effect occurs because of strong affinity of Cu(II) for the –SH group of different enzyme proteins. The enzymes get deactivated, due to copper binding, and thus, their specific biochemical activities are inhibited, leading to physical disorders.

iii. **Manganese:** Trace amount of manganese is essential for the activity of enzymes in animals and of chlorophyll in plants. However, inhalation of excess manganese, in the industrial areas, causes damage to the brain, leading to cramps and tremors in limbs and hallucination in the brain, that is, the symptoms of Parkinson's disease. Excess manganese can also cause the degeneration of the kidneys.

iv. **Cobalt:** It is an essential metal in vitamin B$_{12}$ and coenzyme. Cobalt deficiency, causes vitamin B$_{12}$ deficiency and leads to a wasting disease in Australia and New Zealand, called bush sickness. However, ingestion of excess of cobalt in drinks or through intravenous injection, over a period of time, affects haemoglobin and causes cardiac insufficiency, leading to cardiac failure. Excess cobalt also causes goitre.

v. **Vanadium:** Though essential, in trace amount for normal growth, excess vanadium inhibits enzyme tyrosinase, nitrate reductase and also sulphhydryl activity of enzymes. It inhibits synthesis of biochemicals like amino acids, lipids and cholesterol. It precipitates serum proteins and inhibits liver acetylation process.

vi. **Nickel:** This is also essential in trace amount. Excess nickel ingestion may occur because of exposure to hazardous industrial fumes. It causes skin and respiratory disorders. It can produce bronchial cancer. It deactivates cytochrome c oxidase and also the enzymes, assisting dehydrogenation process, and thus inhibits biochemical processes.

vii. **Iron:** Though essential, excess iron intake can cause acidity, vomiting and coma conditions. Excess iron normally comes from the utensils, used for cooking, or from excessive intake of iron tablets. Excess metal gets deposited in different parts of the body, such as liver, kidney and brain, and can lead to their failure.

viii. **Molybdenum:** It is the only element of the second transition series, which is essential for biological systems, specially plants, for nitrogen fixation. But excess of molybdenum can cause gout in human beings.

ix. **Calcium:** The level of calcium in the body is usually controlled by vitamin D and parathyroid hormones. But, if there is a metabolic imbalance of calcium regulation, it gets deposited in the tissues, leading to their calcification. Formation of stone and cataract are due to calcium salt deposition.

Toxicity Due to Nonessential Metals

i. **Barium:** Soluble barium salts are toxic. One gram of barium chloride, taken at a time, can cause death. It causes vomiting, diarrhoea and even haemorrhage of stomach and intestine. It affects the central nervous system and damages the kidneys.

ii. **Beryllium:** Industrial smokes may cause beryllium poisoning, leading to acute pneumonia, damage of skin and mucous membrane, cancer formation in lungs and bones and a disease of alveolar wall, called berylliosis. Beryllium is not excreted from mammalian tissues. It exhibits toxicity, due to the inhibition of some enzymes, because of the replacement of the essential metal ions by beryllium. Manganese in the enzyme alkaline phosphate is replaced by beryllium, and thus, ATP hydrolysis is inhibited. It also deactivates the enzyme DNA polymerase.

iii. **Cadmium:** This is a very toxic metal, the lethal dose being > 350 mg at a time. Cadmium enters the body system, mainly through the food chain (40 mg/day) and industrial fumes. As tobacco contains 1 μg of cadmium per 1 g, cadmium is inhaled by the smokers.

Cadmium poisoning can also be caused from plating, pigment and alkaline battery-manufacturing factories. Itai-itai disease, which occurred along Jintu river in Japan, was due to cadmium poisoning from a mine. The disease is always accompanied by renal dysfunction.

Cadmium cannot pass through the placental membrane, and hence, new born babies are free from cadmium. Its concentration increases in the body in the later life, due to ingestion through lungs or intestine. Cadmium does not affect the central nervous system.

Excess cadmium can displace zinc from the enzymes. These enzymes get deactivated, resulting in metabolic disorder. Intake of zinc can, therefore, give protection against cadmium poisoning.

There is also a mechanism in the body for protection against excess cadmium. The absorbed cadmium is taken up by a thionein-like protein. Cadmium gets bound to its –SH groups. Thus, cadmium is sequestered, and its toxic effect is controlled. However, the protein-bound cadmium gets accumulated in the kidneys. With increasing level, of 200 μg/g of the kidney, the kidneys get damaged. The damaged kidneys cannot retain plasma protein, calcium and phosphorous. They are excreted in excess and lead to the formation of renal stones.

iv. **Lead:** This nonessential metal enters the biochemical system, through food, water and air. Lead pollution can be caused from factories, manufacturing batteries. Lead intake can be caused due to the use of lead packing for the storage and transport of food. Lead in antiknock compound, lead tetraethyl, used in fuel, ends up in the exhaust gases of the automobiles. Industrial fumes are also the sources of lead in the atmosphere. Thus, there can be accumulation of 200–300 μg of lead per day in human body.

Lead gets deposited in the softer tissues. From there, the reversibly fixed lead passes to the blood stream. Like transition metals, lead has strong affinity for the –SH group of the enzymes, and so it gets bound to the enzymes strongly and deactivates them. In the blood stream, lead is known to inhibit the activity of several enzymes, involved in the synthesis of haem. Excess lead lowers the formation of δ-aminolevulinic acid, its conversion to porphobilinogen and also the conversion of protoporphyrinogen to protoporphyrin IX. Thus, the biosynthesis of haem is inhibited, leading to anaemia.

Lead also affects the biosynthesis of bones, because divalent lead replaces calcium in bone. Deposition of lead in brain results in its reduced activity, leading to depression, nervousness and lack of concentration. Excess lead causes damage of kidney, liver and intestinal track, with consequent loss of appetite, muscle and joint pain, weakness and tremors. Excess lead also causes dental caries and abnormalities in female reproductive system.

v. **Mercury:** Mercury may enter biological systems as metallic mercury vapour, mercury salts or alkyl or aryl (organic) mercury compounds. The sources of mercury poisoning are mainly the industrial processes, involving mercury. The organomercurials, used as fungicides, are also sources of mercury in the atmosphere.

The toxicity of mercury depends on its chemical form. Liquid mercury is not absorbed by human system and passes through it without any toxic effect, though mercury vapour is absorbed into the blood. Hg(II) ion is toxic. However, the doubly charged ion cannot pass through the membrane and hence cannot enter the cell. Hg(I) salts, being less soluble than Hg(II) salts, are less toxic. Nevertheless, Hg(I) can enter the cell and is oxidised in the tissues to Hg(II) and shows high toxicity.

Hg(II), like Cd(II) and Pb(II), has affinity for –SH group of the enzyme proteins and can deactivate them. It also inhibits δ-aminolevulinic acid dehydratase, which in turn inhibits the synthesis of δ-aminolevulinic acid, with consequent decrease in synthesis of haem. It can also form bonds with haemoglobin and serum albumin with –SH groups.

Mercuric nitrate was used in the preparation of felt and many people employed in those factories suffered from mercury toxicity.

More toxic are the organomercurials, $CH_3–Hg^+$ being most toxic. Mercury salts are converted into methyl mercury in the biochemical system by the action of vitamin B_{12} (as discussed in Chapter 7). Methyl mercury is taken up by the aquatic plants and fishes, and finally, mercury enters the animal and human bodies.

The higher toxicity of the organomercurials is because they being non-polar in nature, are soluble in fats. They can pass through the biological and spread all over the body. The covalent Hg – C bond does not break easily, and so, alkyl mercury is not excreted easily and is retained in the cells of kidney, brain, heart, lung and muscle tissues. $CH_3–Hg^+$ can also penetrate through the placenta and can get accumulated in the foetal tissues, causing serious problems.

Methyl mercury gets attached to the cell membrane, thereby lowers, the transport of sugar through the membrane, whereas external passage of K^+ is increased. Thus, the brain cells do not get sufficient energy, and K^+-controlled nerve impulse transmission is also impaired, leading to walking tremors and mental instability. Such effects are observed distinctly, when the concentration of mercury in the brain exceeds 20 μg/g. The effect may be more serious in babies.

Babies born of mothers, with CH_3–Hg^+ poisoning, may have irreversible damage to brain, resulting in cerebral palsy and mental retardation.

The toxic effect of mercury was first realised, when 121 persons in Japan were affected by a disease causing tremors and mental imbalance. They were residents, near the Minamata bay, where the effluent of a factory, containing methyl mercury, was discharged. Though the concentration of CH_3–Hg^+ discharged was low, but it got accumulated in higher concentration in the fishes, whose source of food was the water of the bay. The disease called "Minamata disease" afflicted the heavy fish eaters the most. A number of babies were born, with cerebral defects, as their mothers ate fish, contaminated with methyl mercury.

vi. **Arsenic:** Arsenic poisoning is caused through pollution in air and water. It is a potent proto-plasmic poison and gets accumulated in the tissues.

Elemental arsenic is not absorbed through the intestinal wall and hence is not toxic. Arsenic is more poisonous in trivalent state than in pentavalent state. The poisoning causes gastroen-teritis. Long exposure to arsenic causes skin lesions and sometimes skin cancer.

As(III) shows toxicity due to binding with the –SH group of the enzyme protein, con-sequently deactivating them. As(III) adversely affects the generation of ATP because it binds with the two –SH groups of the enzyme pyruvate dehydrogenase. This deactivates the enzyme essential for ATP synthesis so ATP synthesis cannot take place. The poisonous effect of Lewisite gas, containing arsenic, used during World War II, showed poisonous effects, due to above reasons.

At higher concentration, As(III) causes coagulation of proteins, probably by binding with the –SH groups of the proteins. Arsenic also binds with the sulphur of the protein keratin in the hair and thus affects its texture.

Excess arsenic affects the biosynthesis of ATP in another way, also. Arsenic is chemi-cally similar to phosphorous, and hence, it interferes in the stepwise biosynthesis of ATP. In the presence of arsenic, in the step leading to the formation of 1,3-diphosphoglycerate from glyceraldehydes-3-phosphate, there is a formation of 1-arseno-3-phosphoglycerate. The latter compound undergoes immediate hydrolysis (arsenolysis), to form 3-phosphoglycerate. Thus, further oxidative phosphorylation, as in the case of 1,3-diphosphoglycerate, does not take place, and ATP is not formed.

vii. **Selenium:** Selenium, in trace amounts, is essential for plants and animals (see Chapter 13 for the role of selenium in biological systems). In the plants, selenium replaces sulphur in the bio-chemicals, and selenium analogues of cystine and cysteine, which is useful in the biochemistry of plants, are formed.

Higher level of selenium is toxic, because it binds with the plasma proteins and gets distrib-uted in the tissues. It replaces sulphur in the amino acid residues of the enzyme protein, thus altering their roles. It causes irritation of the mucous membrane and can lead to throat cancer and degeneration of liver and kidney. It causes deformation of hair and nails, nervousness and giddiness.

viii. **Tin:** Excess tin can enter the biological systems as salt, through food and as organotin, through the atmosphere.

Ionic tin cannot penetrate through the intestinal wall, but neutral organotin gets deposited in the brain, causing headache, giddiness and other disorders. It also affects the urinary system.

ix. **Thallium** Monovalent thallium compounds are highly toxic. Tl(I) though slightly bigger than K^+ still can penetrate the cell membrane. However, it binds strongly with the membrane. It accumulates in the erythrocytes, kidneys, bones and tissues. Excess Tl(I) causes fall of hair, gastroenteritis and peripheral neuropathy, and can lead to death.

x. **Aluminium:** It is a nonessential metal. It was considered to be a harmless metal and is used extensively, as a material of cooking utensils, by the poorer section of the society. Thus, large amount of aluminium enters human body, through the corrosion of the metal and through the action of acidic food on it. Another source, of aluminium intake, is in the process of dialy-sis, where the parts of the equipment are made of aluminium. Excess aluminium can also be ingested in the form of anti-acid tablets. Excess of aluminium causes degenerative neurosis in the brain, resulting in Alzheimer's disease, a kind of dementia in old age. However, the fact that aluminium is responsible for Alzheimer's disease is not fully established.

Aluminium has also been shown to bind with long chain fatty acids, and this may have toxic effect.

Reactive Oxygen Species

Reactive oxygen species (ROS) are formed as a result of various biochemical processes involving molecular oxygen, O_2. Various species formed are hydrogen peroxide (H_2O_2), hydroxyl radical ($OH^{\bullet}$), superoxide anion (O_2^-) and singlet oxygen ($^1O_2$). For example, in mitochondria, ROS are formed as products of respiratory chain. Other biochemical processes causing their formation are photochemical and enzymatic reactions, exposure to UV light, ionising radiation or toxic metal ions. Reduction of O_2 yields superoxide which is then converted to H_2O_2 by dismutase. Fenton reaction in the presence of Fe^{2+} can result into formation of hydroxyl radical from hydrogen peroxide.

$$H_2O_2 + Fe^{2+} \rightarrow OH^{\bullet} + OH^- + Fe^{3+}$$

Hydroxyl radical is the most reactive form of oxygen and can cause damage to DNA, cell membranes, etc. Apart from Fe^{2+}, other metal ions are also known to lead to formation of ROS, e.g. Cu^{2+}, Cr^{3+}, Cr^{6+}, V^{3+}, Co^{2+}, Ni^{2+}, Cd^{2+} and Zn^{2+}. In the presence of a reductant like ascorbate, Cu^{2+} is reduced to Cu^+, which can react with H_2O_2 in a Fenton-like manner to produce Cu^{2+}, OH^- and $OH^{\bullet}$:

$$H_2O_2 + Cu^+ \rightarrow OH^{\bullet} + OH^- + Cu^{2+}$$

There are several other pathways resulting into formation of ROS.

The ROS are produced at low concentration and are important in various processes such as signal transduction, host defence and gene expression. These are mainly produced in mitochondria. Under physiological conditions, there is cellular balance between production of ROS and their removal by antioxidants and enzymes so that they do not cause cellular damage. If, however, there is an excess of ROS generation compared to the capacity to clear them, *oxidative stress* is said to occur, which results in damage to biomolecules of cells and tissues. Alternately, if the antioxidant mechanism is dysfunctional, there can be a build-up of ROS in the cell, causing damage to cell organelle. Oxidative stress can also result into cell death or mutations in DNA leading to cancer or other neurodegenerative diseases.

The five major types of intracellular enzymes acting in antioxidative defence mechanisms are manganese superoxide dismutase (Mn-SOD), Cu/Zn superoxide dismutase (Cu/Zn-SOD), catalase, glutathione peroxidase (GPx) and glutathione reductase (GR). The superoxide is converted to oxygen and hydrogen peroxide by superoxide dismutases, whereas hydrogen peroxide is converted to water and oxygen by catalase and GPx.

Cell Death Due to Iron

Redox-active metal ions or metal complexes can interact with DNA, either directly or via other chemical species produced by them, such as ROS, causing damage to DNA or other cell organelle, which may ultimately result into cell death. Damage to DNA can also be caused by interaction with high valent metal-oxygen active species.

Traditionally, cell death in biological systems has been considered to be either as necrosis or as apoptosis. Necrosis is considered as premature cell death, caused by factors external to the cell, such as infection, toxins or trauma. Apoptosis, on the other hand, is programmed cell death, triggered by normal, healthy processes in body. Apart from these, newer processes of cell death have recently been identified such as autophagy – process in which the cell degrades its own internal structure via lyso-somes. Another process, proposed by Dixon in 2012, is ferroptosis – an *iron* dependent, non-apoptotic mode of cell death.

Ferroptosis is characterised by the accumulation of lipid ROS which causes cell death. Thus, ferroptosis is different from necrosis, apoptosis and autophagy. Unlike the features characterising necrosis (swelling of the cytoplasm and organelles and rupture of the cell membrane) or apoptosis (cell shrinkage, chromatin condensation, formation of apoptotic bodies and disintegration of the cytoskeleton), or autophagy (formation of classical closed bilayer membrane structures, called autophagic vacuoles), ferroptosis exhibits shrinkage of mitochondria with increased membrane density and reduction in or vanishing of mitochondrial cristae, but the cell membrane remains intact and nucleus is of normal size. Also, there is a decrease in the amount of intracellular glutathione (GSH) and decreased activity of glutathione peroxidase 4 (GPX4). Ferroptosis follows different pathways, but eventually iron-dependent ROS accumulation is involved in all pathways (mitochondrial cristae are the folds in the inner membrane of mitochondria, which give the inner membrane wrinkled shape and also provide large surface area for chemical reactions to occur on).

As mentioned in Chapter 5, iron in body has various stages of absorption, transport, storage etc., aided by different proteins. In short, intestinal absorption or erythrocyte degradation releases Fe^{2+}, which is oxidised to Fe^{3+} by ceruloplasmin. Fe^{3+} then binds to transferrin (TF) on the cell membrane to form TF-Fe^{3+}, which forms a complex through membrane protein TF receptor 1 (TFR1) to endocytose. This bound Fe^{3+} is reduced to Fe^{2+} and stored in unstable iron pool and ferritin. Excess of Fe^{2+} is oxidised to Fe^{3+} by ferroportin (FPP), a protein that transports iron out of the cell into blood stream. In this way, internal iron is recycled, strictly maintaining iron homeostasis in cell, with very little free iron. If, however, this cycle gets affected, there is an excess of Fe^{2+}, available for reactions with ROS.

Major biochemical reaction causing the ferroptosis is process of lipid peroxidation, that is, oxidative addition of O_2 to lipids, such as polyunsaturated fatty acid tails present in phospholipids making up the cell membranes, lipoproteins and other structures.

Excess of intracellular iron is necessary for ferroptosis. Iron dependence for ferroptosis is confirmed by observing the factors that can inhibit ferroptotic cell death. Thus, lipophilic antioxidants (e.g. α-tocopherol/vitamin E), iron chelators (like deferoxamine, DFOA – a drug administered for removing excess of iron and aluminium from body), inhibitors of lipid peroxidation (a selenium containing hydroperoxide, viz. phospholipid hydroperoxide glutathione peroxidase 4′ (PHGPX or GPX4)) and depletion of polyunsaturated fatty acyl phospholipids (PUFA-PLs) which are primarily the target of lipid peroxidation can all prevent cell death due to ferroptosis. Also, factors causing deficiency of iron, such as, increased expression of iron transporter TF, or its receptor also cause inhibition of ferroptosis.

Recent studies have shown that ferroptosis plays an important role in the occurrence and development of many diseases – nervous system diseases such as neurodegenerative disorders (e.g. Alzheimer's disease), cardiac diseases, pancreatic disorders (e.g. diabetes), certain types of cancers, e.g. gastrointestinal cancer, and some liver- and kidney-related ailments. A lot of current research is therefore focused on understanding the mechanism so that it can be used for treatment of such diseases, either by activating or by blocking the ferroptosis pathways.

Metal Ions and DNA Damage

The phosphate backbone of DNA, its bases and deoxyribose sugar contain donor atoms O and N enabling strong binding with several metal ions. Of many aspects of metal ion binding with DNA, the one that causes damage to DNA is an active area of research. Understanding this interaction can throw light on mechanisms of several diseases and also can potentially lead to treatment pathways. Other areas being explored are metal-dependent DNAzymes for catalytic cleavage of DNA substrates and DNA-based sensors for metal ions.

Metal ions bind to the purine bases G and A through nitrogen atoms (N_1, N_7) and oxygen atom of G (O_6), whereas binding to pyrimidine bases occurs through N_3 and O_2 (Figure 11.3).

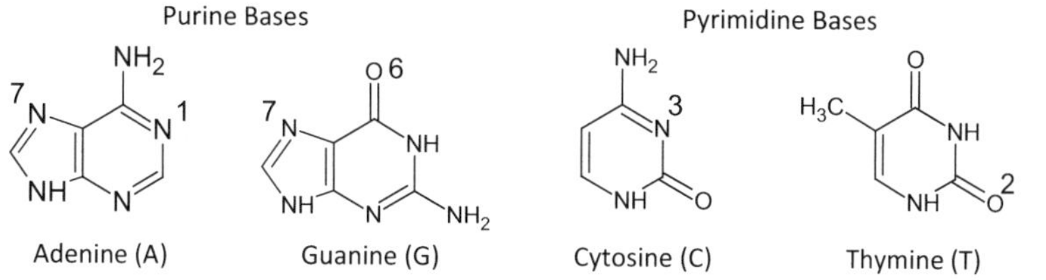

FIGURE 11.3 Sites of DNA binding with metal ions.

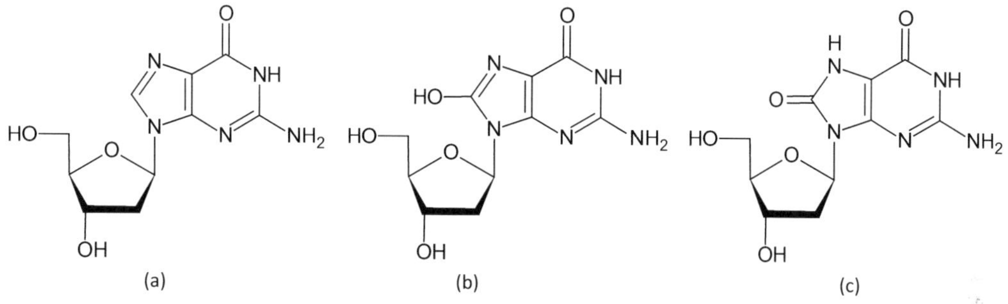

FIGURE 11.4 (a) dG. (b) 8-OH dG. (c) 8-oxo-dG.

These bases in DNA can displace water molecules from the coordination sphere of transition metal ions in aqueous systems and bind with them strongly. Bound metal ions affect the nature and stability of the secondary and tertiary structures of DNA, and DNA is subject to modifications from further reactions of these metal ions with other species in the system. The subsequent outcome of the metal ion–DNA binding depends on factors like nature of metal ion, its oxidation state, other ligands bound to it, its redox activity and actual site(s) of binding with DNA.

ROS produced in proximity of DNA can cause oxidative damage to DNA resulting into several diseases and disorders such as kidney disease and diabetes, certain types of cancer, hypertension and aging. However, such damage is desired in tumour cells to treat cancers when DNA replication is to be prevented, for example, in chemopreventive and chemotherapeutic applications.

Redox-active metal ion-mediated oxidative damage to DNA can occur in two ways: either ROS produced by the metal ions free in solution (not bound to DNA) or ROS produced by metal ions bound to DNA cause damage. Whereas former typically causes single strand breaks in DNA which are not very site specific, the latter results into more site specific modifications of bases and double-strand breaks in DNA. The base modifications can then lead to DNA strand scission. Among the four bases that make up the DNA, guanine (G) is the most easily oxidised base. Damage to DNA is substantially at base G sites which results into many derivatives of guanine like hydroxyl and keto forms. For mononucleoside 2′-deoxyguanosine (dG), Figure 11.4 shows corresponding hydroxyl and keto derivatives produced due to oxidative attack. Of these, 8-hydroxy-2′-deoxyguanosine (8-OH-dG) is considered as marker compound for such DNA oxidative damage (Figure 11.4b). Thus, Fe^{2+}, Cu^{2+}, Cr^{3+} and V^{3+} produce significant levels of 8-OH-dG in the presence of H_2O_2, and the Cu^+/H_2O_2 system also yields significant 8-OH-dG levels. Furthermore, binding of the metal ions with phosphate groups leads to other reactions as well.

Model studies have been conducted to differentiate the interactions of the metal ions Fe^{2+}, Cu^{2+} and Cr^{3+} with dG base and/or the phosphate group, leading to formation of 8-OH-dG. Mononucleoside dG and mononucleotide 2′-deoxyguanosine-5′-monophosphate (dGmp) were used as models. It has been observed that level of 8-OH-dG formed depends on the nature of metal ion and on whether the metal ion interacted with base only (the mononucleoside dG) or the phosphate group only or both (dGmp). Fe^{2+}/H_2O_2 produced much larger level of 8-OH-dG for the mononucleotide (dGmp) than for mononucleoside (dG). This shows that Fe^{2+} interacts with both N7 of the guanine (G) base and the adjacent phosphate on the mononucleotide. In the case of Cu^{2+}/H_2O_2 system, levels of 8-OH-dG were found to be same for both dG and dGmp, whereas Cr^{3+}/H_2O_2 system yielded higher levels of the marker compound from dG. This proved that Cu^{2+}– N7 and Cr^{3+}–N7 interactions with the guanine base favoured formation of the marker 8-OH-dG.

Studies involving DNA–metal interactions are being pursued for improving the efficacy of anticancer, antiviral and antibacterial drugs.

QUESTIONS

1. Under which circumstances metal ions in biological systems are considered as toxic?
2. Can all metal ions cause toxicity in biological systems?
3. Describe various ways metal ions can exhibit toxicity in biosystems.
4. Does toxic effect of a metal depend on its chemical nature (species)?
5. Do biological systems have a way to detoxify in case of metal toxicity?
6. Which factors can cause deficiency of zinc in humans? What are symptoms of zinc deficiency? How can the deficiency be reversed?
7. What are the toxic effects of cobalt, nickel, vanadium, iron, molybdenum and calcium?
8. What are the toxic effects of the following nonessential metals in humans? Ba, Be, Cd, Sn and Tl
9. How does human body protect itself against excess of cadmium?

12

Metal Complexes in Therapeutics

As seen in the previous chapters, metal ions have very important roles in biochemical systems. Metal ion deficiency or excess can result in inhibition of metabolic processes and consequent diseases. Hence, metal complexes find various applications in medicine. There may be direct administration of the metal salt, or metal complex, to compensate the deficiency of a metal ion. The externally administered metal ion gets bound to the relevant biomolecule, at the site specified for the deficient metal ion, and thus revives the normal activity of the biomolecule. Alternatively, the drug may be a chelating agent, which gets bound to the excess toxic metal ion, externally ingested, or to the excess essential metal ion, deposited at various improper parts of the body, due to the failure of some metabolic process. The chelating agent forms a metal chelate in vivo, and thus, the excess metal ion is sequestered.

Various roles of metal complexes in medicine can be classified as follows.

Metal Salts and Metal Complexes as Drugs in Treatment of Metal Deficiency

Deficiency of essential metal ions is widespread in human population. Following are the examples of essential metal ion deficiency and the suggested treatment, using metal salts and metal complexes.

a. **Iron deficiency:** Deficiency of iron can occur due to continued blood loss from ulcers, infections or cancerous growth, and this leads to anaemic condition. The patient gets exhausted easily and suffers from headache and gastrointestinal disturbances.

   The condition can be treated by intake of iron rich food, e.g. spinach, corns, egg yolk and liver. However, only a part of the iron (10%) is absorbed from the food by human system. Hence, soluble iron salts like ferrous sulphate are administered, along with ascorbic acid, to increase the absorption of iron. A preparation of iron (III) chloride, with alkali modified dextran, is also commonly used.

   In case of severe anaemia, as in leukaemia, thalassaemia or sickle-cell anaemia, blood transfusion is carried out.

b. **Zinc deficiency:** The deficiency of zinc is caused during pregnancy or due to the presence of chelating compounds in the food (high phytic acid content of cereals) or by use of sequestering agents for the removal of excess of other metal ions. Zinc also gets sequestered in the process.

   As discussed in Chapter 11, deficiency of zinc may cause tiredness, depression and formation of skin lesions. Zinc deficiency can affect carbohydrate and protein metabolism, mental alertness and reproductive maturity.

   Zinc salts are administered orally in case of zinc deficiency. Zinc sulphate is given orally for the treatment of cysts, ulcers, burns and sickle-cell anaemia. With the observation that rheumatoid arthritis patients are deficient in zinc, zinc sulphate therapy is being attempted for the rheumatoid arthritis patients, with significant benefit. This anti-inflammatory effect of zinc is due to the inhibition of the function of calcium protein, in the presence of zinc.

c. **Copper deficiency:** The deficiency of copper can be caused due to malnutrition, loss of copper due to metabolic disorder or genetic disease, like Menkes disease.

   Normally human beings get adequate supply of copper in food, but the presence of some metal ions like cadmium, mercury, silver and zinc or ascorbic acid reduces absorption of copper.

Powdered copper metal and copper salts like sulphate, acetate and carbonate have been used, since ancient times, to treat anaemia, for stimulating heart and for improving general strength.

In Menkes disease, where copper deficiency occurs, due to the failure of the process of absorption of copper, l-histidine complex of Cu(II) is administered. This complex can be transported better in blood serum, than simple copper salt, and hence is more effective in the treatment of Menkes disease.

d. **Cobalt deficiency:** Cyanocobalamin (vitamin B_{12}) is required in human body, for synthesis of haemoglobin. Vitamin B_{12} is absorbed from the food by the stomach, assisted by a glycoprotein, called the intrinsic factor. If the glycoprotein is deficient, vitamin B_{12} is not efficiently absorbed, leading to pernicious anaemia.

Vitamin B_{12} coenzymes are not stable in the presence of platinum metal. This is because vitamin B_{12} coenzyme transfers alkyl group to Pt(II), thereby vitamin B_{12} is becoming ineffective in assisting haemoglobin formation. Hence, those cancer patients, who have undergone cisplatin therapy, become vitamin B_{12}-deficient and suffer from pernicious anaemia.

In such cases of pernicious anaemia, vitamin B_{12} is administered to the patients in physiologically active form.

Metal Salts and Metal Complexes as Drugs for Other Diseases

Besides the use in treatment of metal deficiency, metal salts and metal complexes are also used as drugs for the control of diseases.

Following are the examples:

a. **Metal salts as drugs:** Aluminium, magnesium and bismuth hydroxides are used as antacids. Mercuric chloride and phenyl mercuric nitrate are used as antiseptic. The organomercurial compounds are used as diuretics:
   - Magnesium salts are used as purgatives.
   - Silver salts are applied in the treatment of burn injuries in patients.
   - Osmium carbohydrate polymers are known to exhibit anti-arthritis activity.
   - The polyoxy anions of tungsten are being tried in the treatment of AIDS.

   Zinc oxide is used externally for healing of wounds and is also used in the toothpaste. Zinc salts are also used as ointment for the treatment of herpes. The metal ion binds with DNA of the herpes and thus destroys it.

b. **Gold compounds as drugs:** The use of gold compounds, as medicine, can be traced way back to 2,500 B.C. There is a record of systematic use of gold salts for the treatment of tuberculosis, syphilis and other germ diseases, since 1890. Since 1929, Au(I) compounds, such as $Na_3[Au(S_2O_3)_2]$ (sanocrysin), sodium and calcium salts of Au(I) thiomalate, (Figure 12.1a), colloidal gold sulphide and polymeric gold thioglucose (Solganol) (Figure 12.1b), are used as intramuscular injection, for the treatment of rheumatoid arthritis. As the injection is painful, now it is administered as triethyl phosphine gold(I) tetraacetyl thioglucose (auranofin) (Figure 12.1c).

Various studies are being carried out to understand how auranofin works to stop the progress of arthritis. The mechanism is not definitely known. A probable pathway is that gold gets coordinated to the –SH groups of cysteine – 34 of the albumin, and thus, the gold drug is carried in the blood. Gold gets deposited in the lysosome. This inhibits release of the enzyme hydrolase, which destroys the tissues, around the joints.

It has also been suggested that gold thiomalate stimulates liver and kidney. This boosts the distribution of Zn(II) and Ca(II) in the body. As these metal ions are essential for some of the processes involved in the inhibition of the progress of rheumatoid arthritis, progress of the disease is checked, by administration of gold drugs.

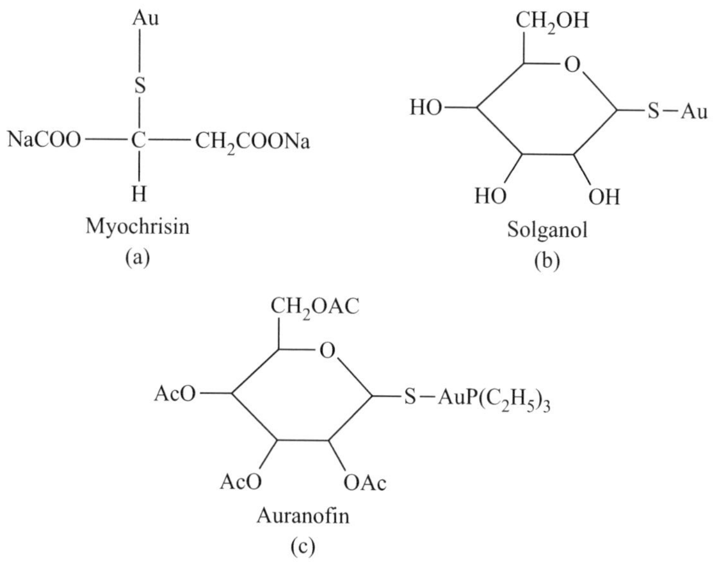

FIGURE 12.1 Gold compounds as drugs (a) myochrisin, (b) solganol, and (c) auranofin.

c. **Lithium compounds:** Lithium carbonate is known to be a psychotropic drug and is administered in doses of 250 mg to 2 g for the manic disorders, and in symptoms of periodic aggression and depression.

The exact mechanism of the therapeutic role of Li^+ ion is not known. Three probable mechanisms have been suggested:

i. Lithium (I) has small ionic radius and hence is highly polarising. It has strong affinity for O^- and N coordination sites. It may displace Na^+, K^+ or Mg^{2+} in the phosphatase enzyme and thus deactivate it. Thus, the breakdown of inositol phosphate into inositol is inhibited. Consequently, the concentration of inositol containing phosphate in the brain is reduced. This deficiency causes disruption of the neurotransmission pathway. Thus, neuron communication is reduced in the brain, and the hyper activity, leading to manic aggression, is controlled.

ii. However, the above mechanism does not explain how lithium carbonate controls depression. Hence, in an alternative mechanism, it has been proposed that Li^+ reduces the formation of cyclic adenosine monophosphate. Cyclic AMP is a signalling molecule between the cells in the brain. Its deficiency affects neurotransmission process. However, this also does not explain how depression is controlled by Li^+.

iii. Recently, it has been suggested that the protein bound to guanosine triphosphate, (G – protein) in the brain is involved in the information transduction in the brain. It has been suggested that the administered Li^+ may get bound to G protein, by displacing Mg^{2+}, bound at the guanosine triphosphate site. Alternatively, Li^+ may bind at another vacant site of G protein, required for the reactivity of G protein, and thus block it. Hence, the activity of G protein is inhibited on administration of Li^+. This affects the information transduction, and thus, manic disorders, either aggression or depression, are controlled.

d. **Arsenic and antimony compounds:** Though arsenic is highly poisonous, its compounds have been used for the treatment of parasitic infections and syphilis. The compounds used are phenyl arsonic acid (Figure 12.2a) and substituted phenyl arsonic acid (Figure 12.2b).

More effective drug for syphilis is arsphenamine (Figure 12.3). A very small dose of the drug needs to be administered.

Antimony tartrate is used for the treatment of trypanosomiasis and allied diseases.

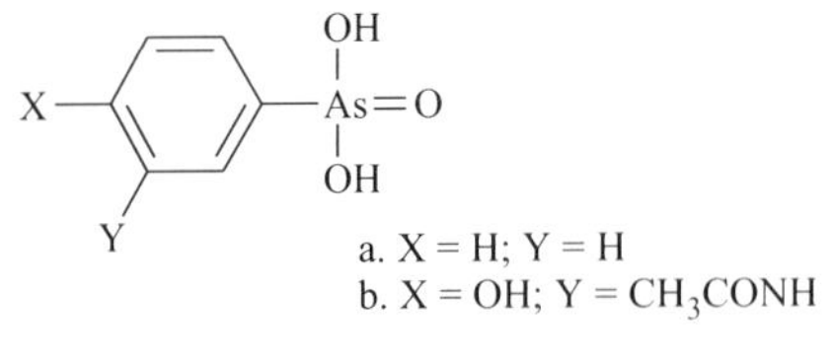

a. X = H; Y = H
b. X = OH; Y = CH₃CONH

FIGURE 12.2 (a) Phenyl arsenic acid and (b) some derivatives as drugs.

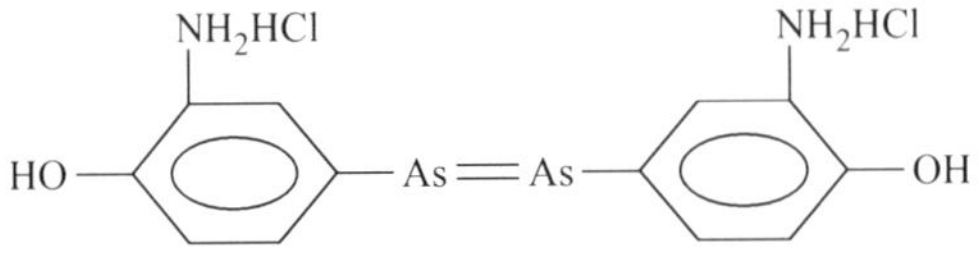

FIGURE 12.3 Arsphenamine.

Both antimony and arsenic act as drug, by binding with the thiol groups of the protein. This may be of help in arresting the disease.

e. **Anticancer drugs:** The observation of Barnett Rosenberg that cis-dichlorodiamine platinum(II), cisplatin (Figure 12.4), exhibits anticancerous activity was a land mark in the use of metals in medicine. Cisplatin was found to be very effective in the treatment of testicular cancer. It is also active in inhibition and eradication of ovary, lung, neck and head cancers.

Since the chlorides in cisplatin are labile, in an aqueous solution, the chlorides are replaced by $-H_2O$ or $-OH^-$. Hence, the active species is $[Pt(NH_3)_2(H_2O)_2]^{2+}$ or $[Pt(NH_3)_2H_2OOH]^+$. In exhibiting anticancerous activity, the platinum (II) of cisplatin gets bound to the nitrogen of the seventh carbon atom of the two guanine bases of one DNA molecule of the cancerous cell. It is interesting that cisplatin has greater affinity for the DNA of cancerous cells.

The water or OH^- of cisplatin is replaced by the N atoms of the bases of DNA molecules of cancerous cells. The distance between the two Pt(II) coordinated, base N atoms of the DNA, is comparable to the distance (3.3 Å), between the two displaced cis groups of cisplatin.

The EXAFS study of cisplatin, bound to DNA, shows the absence of Pt–Cl bonds and the presence of four Pt–N/O linkages. (EXAFS cannot distinguish between nitrogen and oxygen.) This shows that the chlorides are displaced, and the positions are occupied by the two nitrogens of the guanine base of DNA. The ^{195}Pt NMR study also shows binding of four nitrogens with Pt in $Pt(NH_3)_2$ DNA.

The binding of Pt(II) with the two bases of DNA can be of the following types:

i. Intrastrand cross-attachment of Pt(II) with two guanine of the same DNA (Figure 12.4a).

ii. Interstrand cross-attachment of Pt(II) with guanine bases of the two DNA strands of the same double helix (Figure 12.4b).

iii. DNA protein crosslink–Pt(II) may bind with one guanine base nitrogen of a DNA and another protein molecule (Figure 12.4c).

iv. Pt(II) may get bound to one bidentate guanine from N_7 and adjacent O (Figure 12.4d).

Initially, it was thought that there is interstrand binding of $Pt(NH_3)_2$ with the DNA double helix. However, now it is established that cisplatin exhibits anticancer activity, through intrastrand cross-linkage with two guanine bases in the guanine-rich area of the same DNA, that is binding of type(i). As a result of intrastrand crosslink of Pt(II), the DNA duplex undergoes bending (Figure 12.4e). Electrophoresis studies have shown that on binding with cisplatin, the DNA undergoes bending of 40° and there is uncoiling of the double helix by 13°. Thus, the number of helical turns is reduced. There is a change in the super helix density. Even after the binding of Pt(II), the H bonding between the base pairs of the double helix persists. The mode of binding of Pt(II) with DNA and its bending is shown in Figure 12.4e.

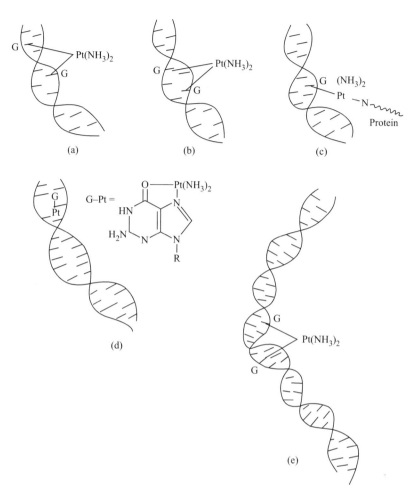

FIGURE 12.4 Anti-cancer activity of cis-platin. (a) Intrastrand cross-attachment; (b) Interstrand cross-attachment; (c) DNA protein cross link; (d) Guanine N7 and O bidentate binding; and (e) DNA duplex bending.

The exact mechanism of the anticancerous activity of cisplatin is not known. Probable reason is that on binding of cisplatin, the replication of DNA is blocked. An alternative explanation is that cisplatin-bound DNA gets bound to damage recognition protein, and thus, its interaction with other cellular components is inhibited and replication is retarded.

It is interesting to observe that trans $Pt(NH_3)_2Cl_2$ does not exhibit anticancer property. Probably, cis orientation may be specifically required for intrastrand crosslinking in the DNA of the target molecules. An alternative suggestion is that before binding with DNA, there may be some specific cellular receptor of the drug, which requires the cis orientation of the chlorides, for binding at the sites of the two groups of the receptor, to be replaced by the drug.

The necessary conditions for a metal complex to have anticancerous activity are as follows:

i. The complex should be neutral, so that it can diffuse through the hydrophobic membrane of the cell.

ii. The complex should have square planar structure. The two groups, to be substituted by DNA, that is leaving groups (chloride in case of cisplatin), should be at cis position.

iii. The groups, which are not substituted, should have low trans effect, like ammonia, heterocyclic amines or diamines.

iv. The leaving groups (chlorides) should be labile, so that they can be quickly substituted.

However, there are exceptions of complexes, with anticancerous property, which are charged and have only one labile group, e.g. $[Pt(NH_3)_2Cl\,L]^+$, where L=pyridine, 4'-methyl or 4'-bromopyridine.

Though cisplatin has been found very effective in cancer therapy, it has some toxic side effects like giddiness, vomiting and damage to kidney. Hence, it is given as an intravenous injection, nearly 5 mg/ 1 kg of the body weight, with a large amount of saline solution, to reduce the damage to the kidneys.

Alternatives to Cisplatin

Some platinum(II) complexes, with anticancerous activity, comparable to cisplatin, but with lesser toxicity have been tried.

In one of the complexes, called spiroplatin (Figure 12.5a), two ammonias of cisplatin have been substituted by a long-chain diamine, and instead of two chlorides, there is one sulphate and one water molecule.

In another complex, called carboplatin (Figure 12.5b), two chlorides in cisplatin have been substituted by dicarboxylate group.

These second-generation anticancer drugs are being investigated for activity against tumours, resistant to cisplatin.

Octahedral complexes of Pt(IV) (Figure 12.6a) and Ru(III), with two ammonias and four other anionic ligands (Figure 12.6b), have also been shown to exhibit anticancer activity.

Other Ru^{2+} (Figure 12.7a), Ru^{3+} (Figure 12.7b) and Pd^{4+} (Figure 12.7c) complexes also exhibit some anticancer activity.

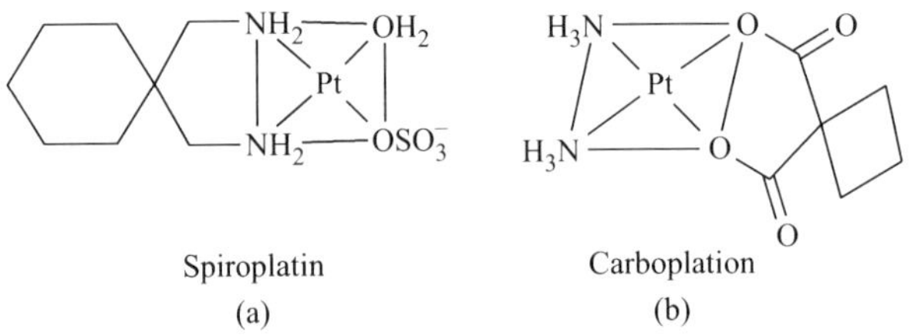

FIGURE 12.5 Anticancer drugs (a) spiroplatin and (b) carboplatin.

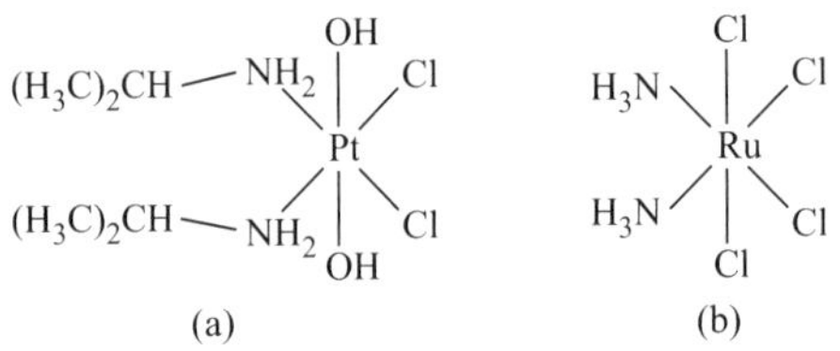

FIGURE 12.6 Octahedral complexes of (a) Pt and (b) Ru as anticancer drugs.

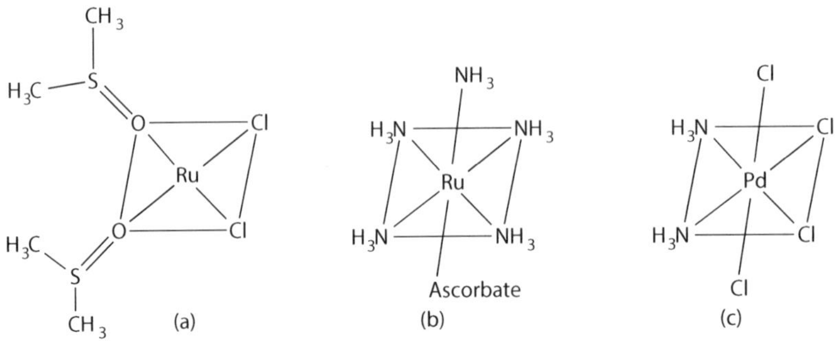

FIGURE 12.7 Complexes of (a) Ru(III) (b) Ru(IV), and (c) Pd (IV) as anticancer drugs.

In all these complexes, there are cishalide or ammonia groups, and they may be getting displaced for the binding of the DNA with the complex, as in case of cisplatin.

A binuclear rhodium (II) complex, with four bridging carboxylate groups (Figure 12.8), also exhibits anticancer property, due to the coordination of the Rh(II) to the DNA bases and the formation of H bond, between the DNA base and the carboxylate group of the complex.

The distance, between two Rh(II) centres, is suitable for the formation of interstrand crosslinking, and this arrests the tumour growth.

Some more transition metal complexes, without ammonia and amine ligands, have been shown to have anticancer activity. Organometallic compounds such as $(C_5H_5)_2M\ Cl_2$, M=Ti(IV), Nb(IV) or Mo(IV) are active against tumours in animals. It has been suggested that these metallocenes may not bind to the DNA base by chloride displacement. They may bind to DNA through metallointercalation. The Pt(II) complex [Pt (terpyridyl)SCH_2CH_2OH]$^+$ also binds to DNA by intercalation. The planar molecule, with delocalised π electron cloud over terpyridyl, inserts between the base pairs of DNA, due to hydrophobic and stacking interaction. As a result, there is unwinding of the double helix of DNA by ~26°, per molecule of the complex inserted. This leads to the arrest of the replication of DNA.

The phosphine complexes of Au(I) like aurinofin (Figure 12.1c) and also phosphine complexes of Ag(I) and Cu(I), exhibit limited anticancer activity. Activity is more in case of diphosphine bridged complex of Au(I) (Figure 12.9), and is comparable with cisplatin.

However, it is observed that phosphines, themselves, have anticancer activity, comparable to their Au(I) complexes. It is, therefore, suggested that the phosphine part of the complex acts as the anticancer agent. The binding of phosphine with the metal ion only helps in preventing its oxidation to phosphine oxide, which does not exhibit anticancer activity.

Besides transition metal salts, non-transition metal salts, such as $Ga(NO_3)_3$ and tin complexes (Figure 12.10), also show activity against leukaemia in mice.

Besides the deactivation of the cancerous cell, the metal complexes have another role in cancer prevention. Many of the organic carcinogens have strong ligating sites, with the possibility of their coordination

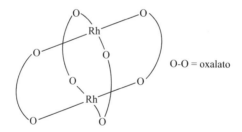

FIGURE 12.8 Binuclear Rh(II) oxalate complex.

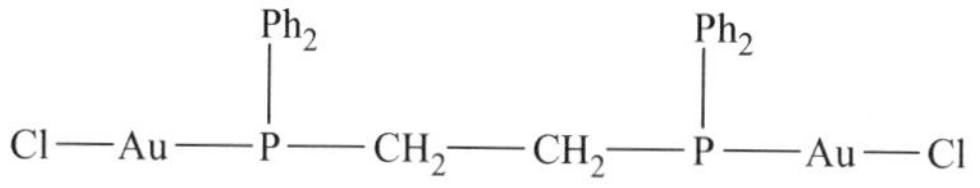

$$Cl-Au-\overset{\overset{\displaystyle Ph_2}{|}}{P}-CH_2-CH_2-\overset{\overset{\displaystyle Ph_2}{|}}{P}-Au-Cl$$

FIGURE 12.9 Diphosphine bridged complex of Au(I).

R = alkyl phenyl
L = pyridine
X = halide

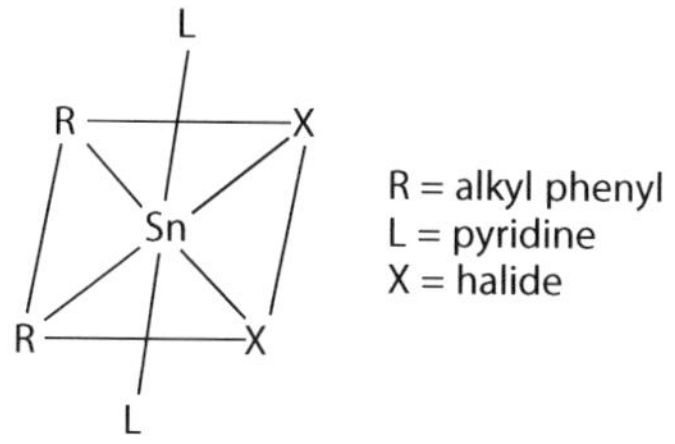

FIGURE 12.10 Tin complexes as anticancer drugs.

with the active metal bio sites leading to cancer. The metal complex drugs can deactivate the carcinogen, by binding with it. Thus, the cancerous effect of the carcinogen is inhibited.

Adverse Effect of Anticancer Drugs

Treatment of cancer, by metal containing drugs, does reduce malignant tumour growth, but may cause mutation of DNA, that is, change in the sequence of bases of DNA. It has been observed, in case of cisplatin therapy, that the cell repair enzyme protein avoids that region of DNA, which is bound to the metal drug. This results in the production of DNA, with one or more incorrect nucleotide. This mutagenesis can lead to serious complications in health.

It is necessary that after the deactivation of the cancerous cell, the platinum should be extruded from the DNA. It is done by a process, called excision repair. In this process, the phosphates on DNA, on both sides of the platinated part, get hydrolysed and the DNA gets separated into two parts. These platinum free parts develop into two new DNA. From the, $Pt(NH_3)_2$ bound, small part of the DNA, platinum is flushed out and is removed from the biological system in steps. If the excision repair is not effective on some platinum-bound cells, deposition of the platinum takes place in the body of the cancer patients, undertaking platinum complex chemotherapy. This produces adverse effects.

Chelating Ligands as Drugs

There can be an excess of toxic metal ion, created in the body, in case of poisoning, or there can be an accumulation of excess of an essential metal ion, in different parts of the body, due to impaired metabolic activity. In either case, the excess metal ion is not desirable and should be removed at the earliest. This is done by administering a strongly metal binding ligand, which sequesters the metal ion, and the resulting metal complex is excreted from the blood stream, through kidney, along with urine or through liver, along with faeces.

This method of removal of metal is termed chelation therapy. The chelating drug can be taken orally. However, if it is not permeable, through the intestinal wall, it can be given as an injection. The ligands administered and the metal complex formed, in chelation therapy, should have following characteristics:

i. The biochemical systems have multidentate coordinating molecules, like proteins and nucleic acids, which bind strongly with the metal ion. Hence, the chelating drug should form highly stable complex with the metal ion to extrude it out.

   The ligand should have strong coordinating sites (O or N coordinating hard base sites to bind with class A metal ions, and soft base sites S, P or As, to bind with class B metal ions), and it should be polydentate, so that it can bind at all coordination sites of the metal ion, without leaving any one free. Thus, no other ligand in biochemical system can remain bound with the metal ion and thus the metal ion cannot interfere with the biochemical system.

ii. The ligand should not have any toxic effect. Its lethal dose should be greater than 400 mg/kg of the body weight. Thus, the ligand should not cause damage to other biomolecules, while taking out the metal ion.

iii. The ligand should be specific for the target toxic metal ion, so that it does not get bound to the other essential metal ions. Thus, if the dose of the chelating drug is controlled, it can circulate in the blood and arrest the toxic metal ion, without causing any damage to the essential metal ion.

iv. The chelating drug should remain unchanged in the biochemical system, without undergoing any other reaction, so that the metal binding coordination sites are not lost.

v. As the chelating drug has to act on a metal ion, inside the cell, it has to pass through neutral hydrophobic lipid membrane of the cell. Hence, the ligand should have such groups in its structure, such that it is hydrophobic in character.

vi. However, the complex of the chelating drug, with the toxic metal ion, should be water-soluble, so that it can be excreted through the kidney.

Following are the more commonly used chelating drugs:

i. **Polyamino polycarboxylic acids:** They form very stable chelates with class A metal ions, coordinating through amino and carboxylate sites and forming stable five-membered rings. Hence, in the case of lanthanide or actinide metal ion poisoning, disodium calcium salt of ethylene diamine tetracarboxylic acid (Na$_2$CaEDTA) is used as sequestering agent.

Amino polycarboxylic acids, with more coordinating sites, like diethylene triamine pentaacetic acid, form more stable complex and hence act more efficiently in the removal of lanthanide and actinide metal ions, particularly industrial poisoning.

ii. **British anti-Lewisite (BAL) and analogues:** This was the first chelating agent used for controlling metal poisoning. The origin of the name of the chelating agent is due to its use against the poisonous arsenic containing gas ClHC=CHAsCl$_2$, called Lewisite.

BAL is chemically 2,3-di-mercapto propanol and gets bound with the arsenic of the Lewisite from two mercapto sites and removes it from the body in the form of the complex, as shown in Figure 12.11a. Thus, BAL protects the enzymes containing mercapto sites from attack by Arsenic.

BAL is also used as an antidote for poisoning, due to other metal ions, such as Hg, Sb, Bi, Cu, Ni, Zn and Au. These metal ions coordinate strongly to the mercapto sites of BAL, and the resulting water-soluble complexes are excreted through urine.

BAL, however, has toxic effects. Hence, alternative compounds unithiol (sodium salt of 2,3-dimercapto ptopane-1-sulphonic acid) (Figure 12.11b) and DMSA (2,3-dimercapto succinic acid) (Figure 12.11c), which are less toxic and more soluble in water, are being used for the detoxification.

iii. **Cysteine derivates:** Since cysteine is a constituent amino acid of the enzyme proteins, it is acceptable to the body. It binds strongly to the metal ions. But it undergoes biodegradation easily. However, its dimethyl derivative, D-penicillamine (Figure 12.12) has been found to be an effective chelating agent for the removal of Pb, Hg, Au, Pt and Sb. It is a better alternative of BAL, as it has very low toxicity and can be taken orally. Now it is being used as an effective drug for the removal of excess copper, deposited in the body parts of the patients of Wilson's disease. It forms a soluble complex of copper which it is excreted from the body.

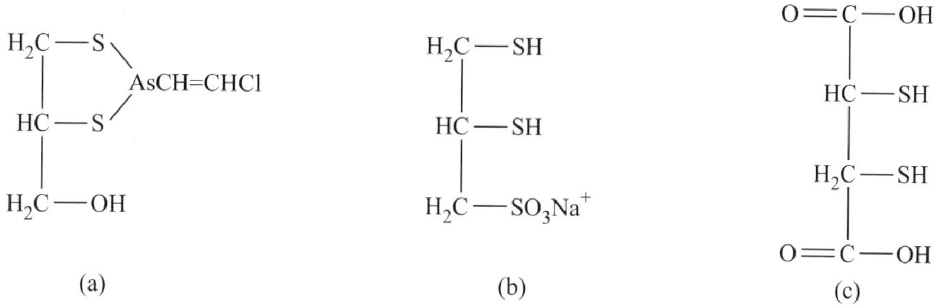

FIGURE 12.11 Chelating ligands as drugs for metal toxicity (a) BAL complex of lewisite, (b) unithiol, and (c) DMSA.

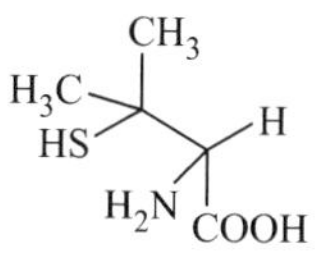

FIGURE 12.12 D-penicillamine.

iv. **Chelating ligands for removal of excess iron:** In the case of following diseases, the haemo-globin of the patient's blood is defective or deficient and hence cannot carry oxygen effectively.

   a. **Cooley's anaemia:** This is a hereditary disease, in which due to a genetic disorder, there is no generation of haemoglobin in the patient's body.

   b. **Thalassaemia:** In this genetic disease, the rate of synthesis of globin chain in haemoglobin is reduced, causing disorder in its structure. There is premature destruction of the red blood cells, and hence, oxygen transport cannot take place.

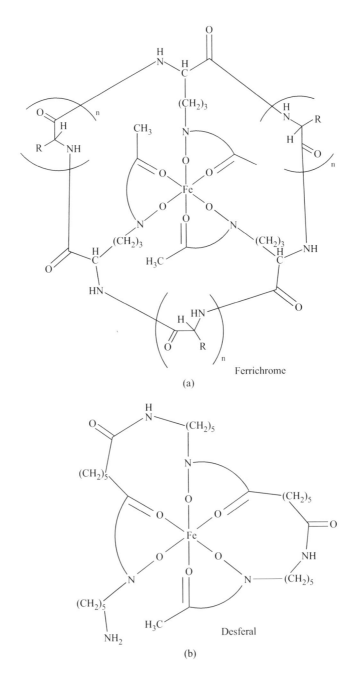

FIGURE 12.13 Complexes of chelating ligands with iron (a) ferrichrome and (b) desferal.

c. **Sickle cell anaemia:** In the patients of this disease, glutamic acid substitutes the valine at the sixth position of the globin part of the B units of haemoglobin. This mutant haemoglobin has a sickle shape. Since, in this sickle-shaped haemoglobin, the globin part is normally bound to the haem part, and hence, the capacity of the iron centre to bind with O_2 should not be affected. However, there is lowering in the capacity of sickle-shaped haemoglobin to carry O_2 to the tissues. This is because the sickle-cell haemoglobin polymerises into fibrous structure and hence its all four units cannot, effectively, carry O_2 to the tissues.

In all the above diseases, causing anaemic disorders, the patient has to be given blood transfusion. However, the body cannot excrete the large amount of iron in the transfused blood. Thus, iron loading takes place in the heart, liver, brain and endocrine glands.

This problem can be overcome by using such chelating drugs that bind selectively with iron resulting in sequestering out the excess iron.

Nature has a very effective iron chelator, called ferrichrome (Figure 12.13a). This is a cyclic protein, with three anchored, potentially bidentate, oxamine coordinating sites.

A synthetic drug has also been in vogue, for a long time, for trapping iron. This is called desferrioxamine or desferal (Figure 12.13b). It is a hexadentate ligand, with three bidentate oxamine coordinating sites, as in ferrichrome. These hexadentate ligands bind with all the six coordination sites of the excess iron in the biochemical system, and deactivate it. Finally, the iron complex is extruded from the body.

QUESTIONS

1. How are the following metal deficiencies treated? Iron, zinc, copper, cobalt.
2. Give examples of metal salts and metal complexes used as drugs for diseases.
3. What is the active species of cisplatin in aqueous solution?
4. How does cisplatin act against cancerous cells?
5. Why trans-$Pt(NH_3)_2Cl_2$ does not exhibit anticancer property?
6. If some metal complexes other than cisplatin are to be tried as anticancer agents, what should be possible conditions they must fulfil?
7. Though cisplatin is very effective as anticancer drug, what are concerns related to its use?
8. Is it necessary that the metal complex being tried as anticancer agent should be square planar, like cisplatin?
9. How is metal poisoning taken care of?
10. What does the abbreviation BAL stand for? What are its uses?
11. Describe the ligands which are used to cure excess of iron in body.

13

Role of Trace Nonmetals in Biochemical Systems

The biochemical systems have been shown to contain some nonmetal elements, other than the normal ones, C, H, N, O, Cl and S, which are the constituents of the biomolecules. They are discussed next.

Role of Silicon in Biochemistry as an Inorganic Nonmetal Element

Silicon is an inorganic nonmetal and is the eighth most abundant element in the solar system and makes up to 27.7% of the earth's crust by weight, and second only to oxygen (46.2%). However, in spite of great abundance of the element in nature, its occurrence in animals is scarce. It is found in higher amount in the plants, in which silica is used for the strengthening of the cell wall. In marine algae and diatoms silicon is used for the construction of frustules[1], which surround the organisms.

However, in the higher animals, silicon occurs in traces and forms only <0.01% of the human body weight. This is because silicon occurs in nature mainly as silicates or silica (as crystalline quartz or amorphous SiO_2). These water-insoluble forms of silicon cannot be digested by the animals, and hence, silicon cannot enter the human system easily. The water-soluble compound of silicon, orthosilicic acid (H_4SiO_4), is the only dietary form of silicon that is biologically available to the higher animals. This is formed in nature by the action of the acidic water on silica or silicate rocks. Soluble silicic acid is absorbed in plants. Silicon enters the human body through the plants eaten by human beings.

During the preparation of beer, barley husks, containing large quantities of silicon containing phyths, are macerated for a long time in acidic water, during which silicic acid passes into solution. Average content of silicic acid in beer is 19.2 mg/L, whereas in drinking water, it is <10 mg/L. Thus, beer is a rich source of silicon for human beings.

Fruits and vegetables also contain silicon, though 100% is not absorbed by human beings. Though banana contains 5 g/100 g silicon, only 2% is absorbable. However, 50% of silicon in beans (2.5 g/100 g) is in absorbable form. Availability of silicon from meat is very low, because silicon is present in tissues of animals, which are not eaten.

In lower organisms like sponges, transport of silicon as orthosilicic acid (H_4SiO_4, OSA) and its ultimate conversion to SiO_2 is enzyme mediated. In frustules, orthosilicic acid first enters in the membrane compartments (called silica vesicles), where it polymerises to form silica. In both the cases, OSA has to enter the membrane against concentration gradient, as concentration of H_4SiO_4 inside the cell is already high. This reverse osmosis is assisted by H_4SiO_4 binding to special transporter molecules. Furthermore, the glycoprotein, present in the cell, also binds with OSA and assists its transport. However, such transporter molecules are not present in the cells of higher animals. So, the mechanism of silicon transport in human beings is not definitely known, and hence, its role in biochemistry is not known with certainty.

In the early stages of investigations, the role of silicon on biological systems was mainly concentrated on the lung disease, silicosis, caused by the inhalation of crystalline silica in the mining of feldspar rocks, in agate polishing industry and stone cutting process in making of statues. Records available indicate that there is an increase in the cases of silicosis in Gujarat (India) over last five years, due to advancement of agate cutting and polishing industry, causing many deaths. The mechanism by which silica causes the lung disease (silicosis) has invited the attention of scientists. It has been suggested that the effect depends on silica particle size and shape and the duration for which the particles stay in the lung. There is, of course, some effect of the chemical reaction of H_4SiO_4 with the lung membrane surface.

---

[1] **Frustules:** Hard and porous cell wall of the diatoms.

However, the effect is mainly physical as silica inhaled in crystalline form like quartz can cause silicosis and even cancer, whereas amorphous form of silica is considered to be safe. Amorphous silica is used in several food and cosmetic products. It is also one of the constituents of antacid preparations.

Thus, initially silicon was considered to have mainly adverse effects on human health. However, now the scientists are looking at silicon in a more positive way, indicating that the element can aid biological reactions and may be helpful for human health.

About twenty years ago, a group of scientists emphasised on the significant biochemical role of silicon. It was considered essential for the synthesis of DNA, protein and chlorophyll. It was also suggested that Si controls mammalian hormone activity and protects human beings from heart disease. However, these claims were rejected as there was no evidence of biomolecules having Si–C or Si–O bonds. However, there is a growing realisation that silicon is equally important to life, as it is to geology, as a rock-forming element. Silicon can be considered as an essential element to produce strength in many life forms, such as frustules in diatoms, strengthened cell wall in certain higher plants and bone and cartilage in higher animals. Though the mechanism, by which silicon enters biochemical system, is not definitely known, there is no doubt that its presence in diet is essential.

In diatoms, the presence of silicon-containing precursors is essential. For several metabolic activities within the organisms, silicon is required. Deficiency of silicon has inhibiting effects on protein, DNA and chlorophyll synthesis.

It is interesting to note that in higher plants, the toxicity due to heavy metal ions is reduced, in the presence of excess silicon. The presence of silicon is also found to improve photosynthesis in plants, due to strengthening of leaves and improvement in their chlorophyll content.

The primary cell wall of plants contains 35% pectin (a polysaccharide rich in galacturonic acid residues). The pectin polymers are crosslinked by bridging with Ca. In this Ca-rich environment of the cell wall, silicon mineralisation occurs. This helps in strengthening of leaves.

It has been observed that the stinging nettles lose their capacity to sting, if there is deficiency of silicon in their body, showing that Si strengthens the body parts of the insects.

By feeding weaning rats with diet, containing range of Si and Ca compounds, it was observed that silicon increased calcium content of the bone and also increased the rate of bone mineralisation.

Biogeochemical Cycle of Silicon

It is observed that there is deficiency of biologically available silicon, because the biogeochemical cycle does not facilitate high abundance of Si in the biosphere. This is because a lot of silicon available in soluble form, from the earth's crust, as orthosilicic acid, is removed, by the formation of insoluble compounds. This phenomenon is called abiotic drain. However, this loss of silicon in biosphere by abiotic drain is compensated by the large abundance of silicon in the earth's crust. The silica minerals, such as quartz, alkali feldspars and plagioclase, are leached out by the weathering of the silicate minerals. Thus, soluble orthosilicic acid passes into water and to the biosphere. The rate of weathering is dependent on various factors such as the composition and the pH of the leaching solvent. The reaction, which leads to the weathering of the silica and silicate rock, can be shown as follows:

i. Quartz is mainly weathered by carbonic acid formed by the dissolution of CO_2 in water. This is a slow process.

$$2SiO_2 + 4CO_2 + 2H_2O \longrightarrow 2H_4SiO_4 + 4H_2CO_3 \tag{13.1}$$

ii. Feldspar rocks are also weathered by carbonic acid, formed as above. It is also a slow process.

$$2KAlSi_2O_8 + 2CO_2 + 11H_2O \longrightarrow Al_2Si_2O_3(OH)_4 + 2K^+ + 2HCO_3^- + 4H_4SiO_4 \tag{13.2}$$

The dissolution of feldspars is faster than that of quartz, and hence, it is a better source of H_4SiO_4, which is soluble in water.

This process of dissolution of silica and silicates absorbs a lot of the greenhouse gas CO_2 and thus has global cooling effect. Hence, it can be argued that the recent global warming caused by increase of CO_2 may be considered as the net result of the CO_2 pumped into the atmosphere by the human activity and dependent rate of use of CO_2 for leaching out OSA from the silicate mineral.

Inhibition of the Role of Aluminium by OSA

Initially, it was suggested that OSA may inhibit the activity of the biochemically essential elements by binding with them and forming insoluble silicates. However, it is now known by in vivo studies that reactivity of the σ base OSA, with the enzyme activity of class B metal ions, is less. Moreover, OSA is seen to activate the role of biochemically essential metal ions. Hence, it has been suggested that OSA has an indirect role and that it reacts with Al^{3+} present in the biological systems and eliminates it, so that it does not inhibit the activity of the biochemically important elements Fe, Mn, Cu and Zn, by binding with active sites, as discussed in the section on Al, earlier in Chapter 10. Hence, when there is an increase in the amount of OSA in the environment and in turn in the diet of animals, the biochemical roles of the essential metal ions are better expressed as Al^{3+} is eliminated and the coordination sites on the biochemicals are not blocked by aluminium and the essential metals are better taken up by the biochemicals and there is enhancement of their activity.

Therefore it will not be wrong to suggest that OSA has an indirect effect by undoing the adverse effect of Al^{3+} on the activity of the essential metal ion.

Another suggestion is that Al^{3+} may affect by inhibiting the absorption of these metal ions in the intestine by blocking the sites of absorption, that is mucin glycopeptides of the intestine wall. Al^{3+}, being a strong class A metal ion, binds strongly with the σ base sites of mucin, thus blocking the binding of the essential metal ions with mucin of the mucous membrane of intestine and thus disrupting their absorption through the intestine wall. An alternative suggestion has also been extended for Al^{3+} blocking the activity of essential metal ions. It may not be binding of Al^{3+}, with either the mucin or enzyme site preferred by the essential metal ions. Instead, Al^{3+} may bind with another σ base site of mucin or enzyme, leading to change in the conformation of the ligand at the essential metal binding site and thus reducing the uptake of the metal ion by the enzyme or mucin, leading to lowering in absorption of the metal ion through the membrane or reducing the role of the metal ion as a cofactor of the enzyme.

Furthermore, it has been observed that orthosilicic acid, by eliminating Al^{3+} as insoluble silicate, prevents it from forming insoluble aluminium phosphate. Thus, phosphate remains available in biochemical system in free form for several biochemical processes, like synthesis of nucleotides and nucleic acids.

It has also been observed that on addition of silicic acid in the soil, there is an improvement in the growth of iron-deficient plants. This is because iron may form aggregate of colloidal Fe–O-containing particles, which are insoluble and become unavailable to the plant. If silicic acid is present in the soil, aggregation of Fe is prevented, and free iron ions are available to the plant for its growth.

Thus, there is emphasis, mainly on the indirect role of silicon in the biological system, of eliminating Al^{3+} and its adverse role. No direct role of silicon was visualised because no specific site has been identified in the biochemistry for the binding of silicon. Moreover, study of the effect of silicon deficiency has been considered more in the small organisms. In the case of human beings, the effect of silicon deficiency has not been studied exhaustively. However, recent studies have shown the direct role of silicon in biological systems, which are discussed next.

Direct Role of Silicon in Biological Systems

i. **Role of silicon in bone formation:** Silicon deficiency causes changes in bone and cartilage formation. Chicks, fed on silicon supplemented diet, as metasilicate, exhibit better growth of cartilage and strong bone formation. Silicon possibly gets deposited in the growth areas. There is also an increase in bone collagen[2] with an increase in orthosilicic acid in the osteoblasts.

---

[2] **Collagen:** A protein of which 30% of body parts are made.

This explanation, of bone and collagen growth, presumes polymerisation of OSA, leading to formation of SiO_2. However, for polymerisation of OSA an optimum concentration of OSA is required, which is not obtained inside the bone, even when diet is rich in OSA.

An alternative suggestion has been extended for the role of Si in bone and collagen formation. Interaction of silicon with iron in the bone has been suggested as the mechanism, by which silicon helps in the formation of hard and soft tissues and collagen.

It is considered that Al^{3+} in the biochemical system can block the sites in the biomolecules which normally bind with iron and provide the base for the formation of Si–Fe bond, leading to formation of collagen, essential for bone growth. Silicates present in the biological system also combine with the free Al^{3+}, to remove it as insoluble aluminosilicate, thus inhibiting Al from blocking sites of collagen formation.

Another probable explanation for the concentration of silicate in the cell is by binding with ascorbic acid, through Si–O bond formation. Ascorbic acid is an antioxidant, which is essentially present in the cells. It binds with silicic acid, and thus, Si gets accumulated in the cells of collagen, assisting its growth.

Bone is mainly composed of hydroxyapatite, a mineral containing phosphate and calcium. In this, there is a substitution of silicon and some other ions such as Mg^{2+}, Na^+, CO_3^{2-}, Cl^- and F^- by ion exchange. This substitution by other ions affects the quality of the bone. Increase of silicon in the diet of human beings increases the density of hip bone, but it does not affect the growth of the bones of vertebral column. This indicates that silicon promotes the formation of bone, as required for the increase in bone density of hip bone, but does not act as inhibitor of osteoblastic[3] resorption. Osteoblastic resorption retards the strengthening of the vertebral column. As this process is not inhibited by Si, it (Si) does not help in the strengthening of the vertebral column.

ii. **Silicon in extracellular matrix:** It is now known that silicon is necessary for the normal development of glycosaminoglycan (GAG) network in the extracellular matrix. Silicon assists the stabilisation of the structure of the polysaccharide GAG by forming crosslink of R–O–Si–OR. This serves as an organic matrix, which binds with calcium and starts the process of calcification. Phosphorylation of GAG and its reaction with collagen leads to the formation of hydroxyapatite and the fully mineralised bone.

The fact that silicon has an important role in the formation of the organic matrix is supported by the observation that there is higher concentration of silicon in the body areas of mice, where there is active growth of bone. As the bone attains maturity, the level of silicon goes down in the bone and there is an increase in the calcium content.

iii. **Silicon in teeth formation:** In the initial stage, teeth are only sacs, made of organic material, with no mineral content. Small amount of silicon may be present in the organic sac, as a structural component. This takes up iron for mineralisation, just as in the case of collagen formation. There is another phenomenon of controlled mineralisation in the formation of teeth of the small animal limpet. In this case, the two minerals, goethite (α Fe–O–OH) and silica, are both formed in the teeth of the limpet in the same environment. This provides sufficient strength to the teeth which helps the animal to cut even the rocks.

In conclusion, it can be said that silicon can form bonds with other biometals and this is of biological importance. However, to establish the nature of reaction between Si and biometal, detailed study of species, containing both silicon and the biometal, has to be carried out.

Silicon in Orthopaedic Implants

Silicon is also used in the preparation of bioactive glass, which has composite and favourable structure, to form strong bonds with the bone. These are used in orthopaedics as implants. In implants, the silica

---

[3] **Osteoblast:** Cells from which bones are synthesised.

content is less than 60% and the ratio of calcium and phosphorous is same, as in hydroxyapatite in bone. When implanted on wounded bone, bioglass is incorporated in the healing callus of the bone. This process, of regeneration of the bone, assisted by bioglass, involves twelve stages of reactions. From the bioactive glass, orthosilicic acid gets collected around the surface of the implant. Calcium and phosphate ions are absorbed from the solution around the implant, and deposits of hydroxyapatite are formed covering the wounds in the bone. The bioglass strengthens the callus of the bone against internal movement and sheering force. Furthermore, the layers of silica and hydroxyapatite can react with other biomolecules like proteins, which have important healing effect at the wounded bone site.

As the bioactive glass has low concentration of silicon, it has an open network structure. This enables Si to interact with the surrounding metal ions, and thus facilitating the release of silicon after the precipitation of hydroxyapatite.

Selenium in Biochemistry

Selenium is also an inorganic nonmetal element and is present in the biochemical systems in trace amount. The requirement of selenium varies in men (70 µg/day) and women (60 µg/day). Deficiency of selenium can cause an endemic cardiomyopathy, called Keshan's disease, other heart diseases and even cancer.

Occurrence: Selenium, with outer electronic configuration $s^2 p^2$, is a member of the sixth group of the periodic table, and has properties similar to that of sulphur. It occurs in the soil as the anions selenide, Se^{2-}, and selenate, SeO_4^{2-}. It enters the vegetation growing on the selenium rich soil and further in the animal and human systems through the food chain.

In food, selenium exists in small amount, in inorganic form Se^{2-} and SeO_4^{2-}. Mainly, it exists in food in the form of organic compounds, containing selenium, which are better absorbed in the human systems. The main sources of selenium are cereals, sea fishes and animal meat. However, the content of selenium is less in vegetarian food and milk. Hence, there is a possibility of Se deficiency among vegetarians. In view of this, in some countries, selenium compounds are added to the fertilisers, through which selenium-fortified food is obtained.

Though selenium is a later member of the VI A (16) group and is more basic and with metallic properties, it exists in the biochemical system, like other nonmetals (N, P, O, S) as a constituent of the biomolecules. This is unlike the metals of biochemical importance, which are coordinated externally to the biomolecules.

Selenium is available to the animals through diet, mainly in organic form of Se, that is Se-containing amino acids. However, selenium exists in biological systems mainly as selenocysteine, which is the main component of the selenium-containing proteins and enzymes. Selenomethionine and selenocysteine have structure similar to methionine and cysteine, respectively, with only the –SH group in the amino acid being replaced by –SeH group.

In the proteins, selenium exists mainly as selenocysteine residue at the 21st position in the protein chain.

Early studies showed that selenium exists in three enzymes, glutathione peroxidase, thioredoxin reductase and iodothyronine deiodinase, which act as antioxidants. These enzymes have very important role in biochemical processes. This prompted the search for more selenium-containing enzyme proteins. Now, it is known that mammals have in their system up to thirty seleno proteins. In human systems, 25 seleno proteins have been detected.

Biosynthesis of Se Proteins and Se Enzymes

Selenium has properties similar to sulphur. But selenium, being bigger in size, is more basic in nature and hence –Se–H bond tends to dissociate more readily than –S–H bond. That is why selenocysteine is more acidic than thiocysteine and forms more stable cystein selenate ion than cystein thiolate. For the same reason, selenocysteine Se–H gets dissociated at lower pH and is in dissociated cystein selenate form at biological pH (~6).

Seleno proteins are formed in biological systems by the introduction of selenocysteine in different proteins at the 21st amino acid residue assisted by UGA codons. A codon is a sequence of three DNA or RNA nucleotides that corresponds with a specific amino acid or stop signal during protein synthesis. UGA codon is one of the three termination or stop codons that causes release of new polypeptide chain from ribosome.

As stated earlier, selenium in the diet is mainly in the form of selenomethionine with small amount of selenocysteine, and inorganic selenites and selenates. In the biochemical system, these Se compounds get decomposed, giving ultimately hydrogen selenide, H_2Se. This forms the base for reformation of selenocysteine, which is incorporated in the different proteins at specific position, guided by UGA codon. This results in the formation of different seleno proteins and seleno enzymes. H_2Se also forms methyl selenol (CH_3–SeH) and dimethyl selenide ((CH_3)$_2$Se), which are not useful and are excreted through breath, perspiration and urine.

The main selenoenzymes are six different glutathione peroxidase (GPx 1 to 6) and three thioredoxin reductase (Trx R 1 to 3). These selenoenzymes exhibit special role as antioxidants and protect the cells from damage caused by active oxygen species, such as superoxide (O_2^-), peroxide (O_2^{2-}) and hydroxy radical ($OH^\bullet$), and reactive nitrogen species such as NO and peroxynitrite.

Biochemical Roles of Seleno Enzymes

i. **Protection of heart disease:** If there is accumulation of excess active oxygen species, this can result in the oxidation of low-density lipoprotein (LDL), which may lead to atherogenesis and heart disease.

   Seleno enzymes, which are antioxidants, reduce the possibility of HDL oxidation. Seleno enzymes also assist in the formation of HDL, which is a good cholesterol, thereby improving the lipid profile. That is why, proper intake of selenium is helpful in protection from heart disease.

   Though there is no direct evidence of selenium deficiency causing heart disease, it has been observed that the organoselenium compound based drug ebselen, which shows glutathione peroxidise (GPX) activity, is cardioprotective for myocardial ischemia.

ii. **Protection from cancer:** It has been observed that in human beings, the chances of cancer occurrence due to carcinogenic chemicals and radiation are less, if there is excess of selenium in the body. Se compounds act as antioxidant chemopreventives for cancer. Intake of yeast containing selenomethionine reduces the risk of prostrate, intestinal and urinary bladder cancer. There are, however, exceptions that men and women, showing high selenium content in nails and blood plasma, are not necessarily protected from prostrate and ovarian cancer, respectively.

   The exact mechanism of the chemopreventive activity of the seleno organic compounds is not known. Following three probable mechanisms have been suggested:

   a. Selenium acts as an antioxidant and thus suppresses the activity of carcinogenic materials, which inhibits the damage of DNA. It has been observed that the anticarcinogenic activity is dependent on the chemical form of selenium. For example, selenocysteine is more effective in preventing damage of DNA, due to caner, than selenomethionine.

   b. Selenium helps in inhibiting the multiplication of cells and also assists in the repair of DNA, thus inhibiting the growth of cancer.

   c. Selenium in the human body reduces the activity of the enzyme kinase C and controls expression of hormones oestrogen, and androgen. This reduces the possibility of occurrence of breast and prostate cancer in females and males, respectively.

iii. **Improvement of immunity and control of thyroid activity:** It has been shown that deficiency of selenium, in animals, causes lowering in immunity, which causes various disorders. These can be controlled by proper intake of selenium. With the improvement of the immune system, there is a reduction in autoimmune thyroiditis, a condition caused by excess iodine intake, viral infections or immune therapeutic agents, and this condition leads to hypothyroidism.

Selenium is present in the largest amount in thyroid glands as selenium-containing enzymes and oxidants glutathione peroxidase and thioredoxin reductase. It has been suggested that the possible role of Se in arresting autoimmune thyroiditis is the antioxidant activity of the Se enzymes, GPx.

During the synthesis of the thyroid hormones by thyrocytes in the thyroid glands, there is production of the reactive oxygen species (ROS) and H_2O_2. These oxidants can cause damage to thyroid cells. However, the selenoenzymes present in the thyroid gland have antioxidant activity and can arrest the role of the oxidant and thus avoid thyroid cell damage. Furthermore, selenoenzyme (deiodinase) present in thyroid gland assists the thyroid hormone metabolism. Hence, the deficiency of selenium can cause defective thyroid hormone metabolism, leading to damage to the thyroid gland.

However, the exact role of selenium as antioxidant in preventing cell damage and controlling malignancy, heart disease and hypothyroidism is not known. Hence, at the present stage, use of selenium supplements in food and of selenium-containing drugs in order to increase the level of selenium in the body, should be done under strict expert supervision.

Role of Iodine in Biochemistry

Iodine is the last but one member of the VIIA 17th group of the periodic table and the last halogen. Though halogens are known to be highly electronegative, iodine, with bigger atomic radius, has low ionisation potential. It can lose electrons easily and exhibit basic properties. This characteristic of iodine has special role to play in the biochemistry of iodine. Like selenium, iodine is a nonmetal and exists as a constituent of the biochemicals.

Iodine is obtained in the plants grown in iodine-rich soil. Other sources of iodine are the sea weeds and sea fishes, as sea water is rich in iodide salt. The residents near sea coast get food rich in iodine. But in places distant from sea, specially on hills, soil and food are deficient in iodine.

Biochemistry of Iodine

The daily required intake of iodine by an adult human body is 500 µg, out of which only 150 µg is absorbed. This trace amount of 150–200 µg of iodine is present in human body. Through blood in salivary glands, stomach and intestine, iodide reaches the thyroid glands, which have a number of follicles on the cell membrane. The iodide enters the thyroid cells through the follicles. The transport of iodide inside the cell is against concentration and electrical gradient, and is aided by a transmembrane glycol protein called sodium iodide symporter (NIS). Energy for this process is provided by the electrochemical gradient of Na^+ across the cell membrane, maintained by Na^+/K^+ pumps (Chapter 3).

Inside the follicles of the thyroid membrane cells, there is a colloidal glycoprotein called thyroglobulin. This is formed in steps. First, a polypeptide unit with several tyrosine residues (molecular weight 330 K daltons) is formed. Two of such units undergo dimerisation and also combine with a carbohydrate, forming thyroglobulin in the small vesicle of the follicle, and are further released in the follicular lumen. Thyroglobulin is produced and consumed entirely in thyroid gland.

The iodine absorbed by the thyroid (described above) enters the follicular lumen and gets easily oxidised to molecular iodine, I_2. This happens because of larger size of iodide ion which loses electron easily and forms neutral I, two of which unite to form I_2.

The tyrosine residue of the thyroglobulin reacts with I_2 to form monoiodotyrosine (MIT) thyroglobulin by uptake of I at position 3. There is further formation of diiodotyrosine (DIT) thyroglobulin by iodination of the second tyrosine at position 5.

In the next step, the enzyme thyroid peroxidase catalyses the coupling of two diiodotyrosine thyroglobulin to form tetraiodo thyronine (T_4), also known as the hormone thyroxine. Alternatively, one molecule of MIT and one of DIT couple together to form triiodothyronine (T_3). These hormones T_3 and T_4 remain stable in the follicular cells of the thyroid for three months in the form of colloid. The colloid passes from follicular cells to the epithelial cell and stays in the cytoplasm. The enzyme protease digests the colloid and MIT, DIT, triiodothyronine and thyroxine are liberated in the cytoplasm. T_4 and T_3 gradually diffuse into the blood, whereas MIT and DIT get deiodinated and iodine is liberated for further cycles.

The hormones T_3 and T_4 are released in the blood as per requirement of the body. T_4 hormone has longer life than T_3 and exists in blood in larger amount ~ 20 times more than T_3. However, T_3 is three to four times more active than T_4. T_4 loses one iodine assisted by enzyme iodinase to form T_3. T_3 and T_4 hormones released in blood are mostly in bound form, bound to proteins globin, prealbumin and albumin.

Steps in the synthesis of T_3 and T_4 hormones are shown in Figure 13.1.

The thyroid hormones T_3 and T_4 in the blood are metabolised in the liver. First, the thyroid hormones T_3 and T_4 lose the binding proteins releasing more active free hormones. The structures of the free form of T_3 and T_4 are shown in Figure 13.2.

T_3 and T_4 release iodine in the extracellular cell region for use in the body and part of iodine goes to the bile for excretion in stool.

The normal thyroxine (T_4) level in the blood has to be maintained for proper supply of iodine to the body. If there is deficiency, thyroxine is released in blood from the thyroid gland. This release of T_4 from thyroid gland is controlled by the secretion of thyroid stimulating hormone (TSH) by the pituitary gland. If there is fall in the thyroxine level in the blood, the pituitary gland is stimulated to release TSH, which in turn stimulates the thyroid gland to release thyroxine.

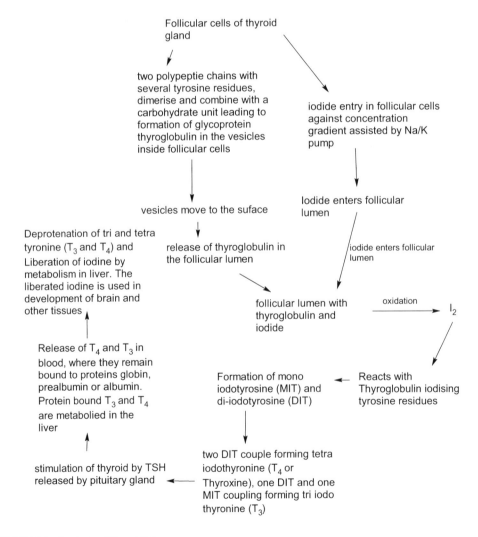

FIGURE 13.1   Synthesis of T_3 and T_4 hormones.

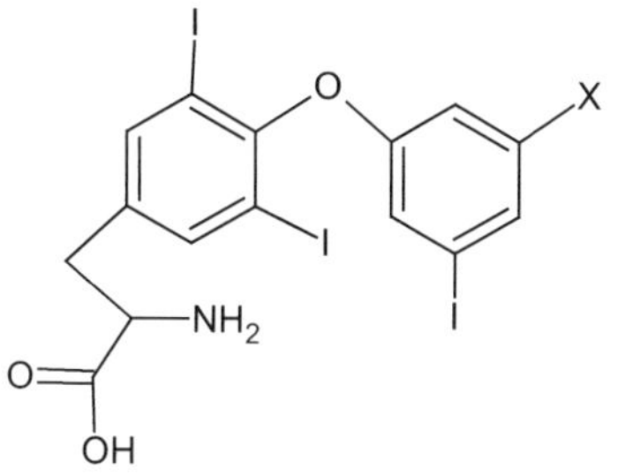

X = H triiodo thyronine (T$_3$)

X = I tetraiodo thyronine (T$_4$)

FIGURE 13.2 Free form of T$_3$ and T$_4$.

In case of inadequate supply of iodine, formation of T$_3$ and T$_4$ in the thyroid is less, and hence, the thyroxine level in the blood goes down. This is compensated by the excess release of TSH from the pituitary gland, which activates the thyroid to release the stored thyroxine. The thyroid can bear the strain for some time, but if there is continued acute iodine deficiency, the release of TSH also cannot stimulate the thyroid to release thyroxine. In case of the people residing in iodine deficient areas, the thyroid must continuously release hormones and there is increased TSH release to stimulate the thyroids. Due to this strain, the thyroid increases in size. This condition caused by low thyroxine in blood due to low activity of thyroid is called hypothyroidism. In this, there is less of thyroxine and more of TSH in blood. There is large enlargement of the thyroid gland, and it bulges as a lump outside the throat. This condition is known as goitre.

In coastal areas, people who eat lot of sea weeds and sea fishes which are rich in iodine, have greater intake of iodine. In Japan, individual intake of iodine is nearly ten times more than the prescribed upper limit (0.1 mg) of iodine. It is seen that the occurrence of breast cancer in females and prostate cancer in males is less among Japanese. Thus, excess iodine does not have any adverse effect for persons with normal thyroid activity, as thyroid stores the excess iodine and releases thyroxine as per requirement. However, in case of persons with higher thyroid activity, the thyroxine content increases in blood with excess supply of iodine. The TSH requirement is low and hence, TSH content in blood becomes low. This condition is called hyperthyroidism. For people suffering with this condition, intake of excess iodine has harmful effects.

As discussed earlier, nonhormonal iodine in the extracellular region, outside the thyroid exhibits antioxidant property, helps in development of immunity in human beings and prevents the occurrence of cancer, especially breast cancer, in females and prostate cancer in males.

With the awareness of the significant role of iodine in overall development and well-being of human population the administration in areas with iodine deficiency has become alert. To compensate for the iodine deficiency, the table salt, most used by people, is "iodised" by mixing it with calculated amounts of potassium iodate. This is called iodised salt. Iodine can also be mixed with edible oil. However, in India, fortification of edible salt with iodine is preferred and sale of uniodised salt is prohibited.

Role of Fluorine in Biological Systems

Fluorine is the first member of the halogen family (group 17th or VIIA) in the periodic table. Chemically, it is the most electronegative element among all elements. Fluorine occurs in nature in the form of minerals such as fluorspar (CaF_2), cryolite (Na_3AlF_6) and fluorapatite ($Ca_5(PO_4)_3F$). However, these minerals are insoluble in water. This was considered as a reason for fluorine not being involved in biochemical processes in nature during evolution. Also, it is very strong oxidising agent and has very high oxidation potential, and therefore, biomolecules may not be able to make it participate in biochemical reactions. Its high hydration energy prevents it from acting as good nucleophile in aqueous systems; hence, it does not participate in nucleophilic substitution reactions. Fluorine is not required for development or growth of body in humans. In fact, studies have indicated that it has the potential to accumulate in body inciting toxic effects. Fluoride overload in body can result into dental and skeletal fluorosis.

Commercially important fluorine-containing products are chlorofluorocarbons (Freons) used as refrigerants, polytetrafluorethylene (Teflon) and fluoride-containing toothpastes, among many others.

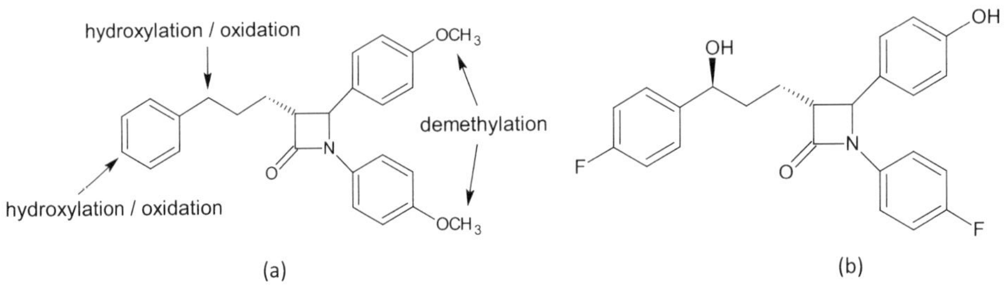

FIGURE 13.3 Drug used for treatment of hypercholesterolaemia (a) original non-fluorinated molecule and (b) fluorinated molecule developed as drug ezetimibe.

Some chlorofluorocarbons play a role in depletion of ozone, their manufacture has been restricted and many of such compounds are being replaced by hydrofluorocarbons. Chlorofluorocarbons have unique properties – chemical and thermal stability. Another important application area of fluorine-containing organic compounds is pharmaceuticals and agrochemicals.

About 30% of the pharmaceutical drugs available in market at present are organic molecules containing one or more fluorine atom. Fluorine-containing drugs are the most prescribed drugs – these include anticancer agents, antidepressants, anti-inflammatory agents, anaesthetics and central nervous system drugs. And the expected trend is that more of such drugs will soon come in the market. Fluorine-containing agricultural products are also very popular for their useful properties. The following properties of fluorine make this possible:

Fluorine substitution in organic molecules as drugs results into enhanced biological activity and increased chemical or metabolic stability. Important factors that contribute to this are: relatively small size of fluorine atom, electron withdrawing property, greater stability of C–F bond compared to C–H bond and greater lipophilicity imparted to the organic molecule, which affects the distribution of the fluorinated compounds through membranes. Smaller size of fluorine causes minimal steric hindrance effects for the molecules while bonding with target sites in enzymes or receptors. Electronegativity of fluorine changes significantly the physiological response of the drug molecule in most cases, thereby increasing efficacy of the drug. This is because introduction of fluorine changes pKa and dipole moment of the drug molecule and also stability of other functional groups present in the drug molecule.

An example of these effects is the development of drug ezetimibe, which inhibits cholesterol absorption in intestine. The original molecule (Figure13.3a) that promised to be a drug during studies was substantially affected by metabolic attack on four primary sites by demethylation, hydroxylation and/or oxidation, decreasing its effectiveness. Incorporation of fluorine atom in the molecule yielded the drug ezetimibe (Figure 13.3b) which is much more stable to such metabolic attacks. The drug was approved in 2002 by the USFDA to reduce cholesterol levels in patients with hypercholesterolaemia.

Fluorine-containing compounds are also important as diagnostic medicines in positron emission tomography (PET). PET is an imaging technique based on the use of radioactive nuclei which are incorporated in the compound (as radio tags) to be administered to patient. The radionuclides commonly used are ^{11}C, ^{13}N, ^{15}O and ^{18}F. These radionuclides emit positrons (β^+) during decay, which in turn produce γ rays which are monitored to produce a three-dimensional tomographic image. ^{18}F is the most important radionuclide used in the imaging technique. This is because of its longest half-life (110 minutes), higher percentage of β^+ emission and relatively low positron energy. Due to this, image quality is much better compared to when using other radionuclides. An important radiopharmaceutical in use is ^{18}F-fluorodeoxyglucose (FDG) which can show glucose uptake and energy consumption in various cells, and its main uses are in the fields of oncology and neuroscience. Structure of FDG is shown below.

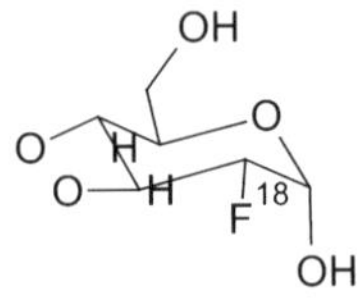

With increased use of fluorinated pharmaceuticals, there are concerns about increased fluorine load in human body. Some pharmaceuticals had to be withdrawn from market, or research work to develop them as drugs was terminated due to toxic effects of fluorine-containing metabolites produced in body. Notable examples of drugs withdrawn are heavily fluorinated anaesthetics such as methoxyflurane, isoflurane and others.

Ground water contamination by fluoride is a global challenge, especially in developing countries. Certain industries can cause contamination of ground water by fluoride, for example, brick kilns, aluminium and steel industry. Some phosphatic fertilisers also cause fluoride to leach into water from soil. Leaching of soluble fluoride from minerals in soil to ground water is possible especially in arid regions where ground water movement is less (providing more contact time), temperature is higher and water has more of sodium bicarbonate and less of calcium and magnesium ions, that is water having higher alkalinity and lower hardness. Volcanic activity also leads to increased fluoride concentrations in ground water. In such cases, fluoride concentration can go as high as 10 mg/L. WHO guidelines for drinking water limit the fluoride concentration to 1 mg/L. Above 1.5 mg/L, possibility of mottling of teeth increases. Higher concentration of 6 mg/L causes skeletal fluorosis while continued consumption of fluorine-containing water may cause very severe form of the disease. Higher intake of calcium and vitamin C is recommended in the regions having ground water contamination of fluoride. Dilution by artificial recharge techniques like flooding of groundwater with surface water is also suggested. Finally, treatment of water before consumption should be undertaken.

It is estimated that approximately half of the fluoride ingested is removed from human body, rest gets absorbed in calcified tissues like bones and teeth. Bio-accumulation of fluoride is known to occur upon continued intake. Fluoride is incorporated into teeth and bones by replacing the hydroxyl ion in hydroxyapatite to form fluorohydroxyapatite or simply fluorapatite. This binding, however, is reversible. That means, when fluoride intake is stopped, the fluoride can go back to plasma from bones during bone remodelling and can get excreted by kidneys. Protective role of fluoride at concentration less than 1 mg/L is by virtue of the same reaction. If there is a cavity in tooth enamel (which is made up of hydroxyapatite) due to caries, fluoride replaces hydroxide and forms fluorohydroxyapatite, which is denser than hydroxyapatite and more resistant to acid attack – the first step in tooth decay. Though some studies have contended that the beneficial effect of fluoride in dental decay is outweighed by its harm to bones and other negative effects, it can be said that the topical use of fluoride in the presence of tooth cavities is certainly beneficial. The bacteria like *Streptococcus mutans* (cariogenic bacteria) in the mouth convert sugars to mineral-attacking acids. Fluoride at higher concentration kills these bacteria by interfering with some of their enzymes, like inhibition of enzyme-mediated glycolysis, adversely affecting polysaccharide metabolism in bacterial cells. In some countries, therefore, community water fluoridation is considered and practised as an effective way of controlling tooth decay.

Effect of fluoride on controlling cariogenic bacteria thus proves that fluoride does participate in some biochemical processes and can interact with enzymes. Due to its small size (comparable to that of hydrogen) and very high electronegativity, fluoride can bind strongly with metal ions in enzymes and also can disrupt hydrogen bonds at active sites. Several enzymes, including P-450 oxidases, are affected by fluoride ingestion. A recent study of the effects of inorganic fluoride compounds on human cellular functions revealed that fluoride can interact with a wide range of enzyme-mediated cellular processes and genes modulated by fluoride including those related to the stress response, metabolic enzymes, the cell cycle, cell–cell communications and signal transduction. Detailed mechanistic pathways are being investigated.

There are some studies indicating that fluoride toxicity can adversely affect cognitive development. Also, some data indicates that excess fluoride can negatively affect working of thyroid gland, leading to hypothyroidism. Mechanism for fluoride-induced hypothyroidism is considered to be competitive binding with iodine and obstruction in synthesis of T^3 and T^4. This has been the basis for certain studies undertaken for a dose of 5 mg/L of fluoride to be administered as a treatment for hyperthyroidism.

Thus, fluoride chemistry in biological systems is expanding.

Role of H⁺ Ion in Biochemical Systems

Hydrogen is the first element in the periodic table with electronic configuration $1s^1$. It can lose one electron to form H^+ ion with stable $1s^0$ configuration, or gain one electron to form H^- ion with $1s^2$ configuration. However, H^+ is of significance in biochemical reactions. It is present in the gastric acid in the stomach, which is essential for the digestion of protein in food.

For the secretion of gastric juice in the stomach, there is a proton pump on the surface of stomach. It secretes the H^+ in the gastric acid.

Proton pump is a protein pump, which builds up proton gradient across a biological membrane because there is more proton outside the membrane. They can be said to catalyse the following reaction:

$$H^+ \text{ (on one side of membrane)} \xrightleftharpoons{\text{energy}} H^+ \text{ (another side of membrane)}$$

The proton pump uses energy to transport protons from the outside of the mitochondria to the space inside membrane. It works against concentration gradient and is an active pump like Na^+ or Ca^{2+} pumps.

Proton pump is H^+/K^+ ATPase, and energy is liberated by the breakdown of ATP–ADP. There are other mechanisms of proton transport.

In case of release of excess of acid in the stomach, there can be two possible remedies.

  i. Antacids – they neutralize acid in the stomach. They are bicarbonates of Al, Ca, Mg or Na, which act as soft bases, and so neutralize the stomach acid. But this effect is temporary.
  ii. Proton pump inhibitors (PPIs) – they are a class of medicines that inhibit the action of the proton pump and reduce the release of acid. The PPIs work by blocking the H^+/K^+ APA enzyme, which is mainly responsible for the acidity of proton pump. However, there are side effects of PPIs such as headache, diarrohea and vomiting. There are also serious side effects of kidney disease, infections and vitamin deficiency, though, associated with long-term use.

Therefore, in the treatment of hyperacidity, there should be a judicious choice between of antiacid or PPI.

QUESTIONS

  1. Are nonmetals important in trace amount in biochemical processes? Give some examples.
  2. How is silicon important in biological systems?
  3. How does silicon pass from geosphere to biosphere?
  4. What is the source of silicon in human body?
  5. Does silicon cause any disease in humans?
  6. What is role of orthosilicic acid (OSA) in biochemical processes in human body?
  7. How much selenium is needed for an adult human?
  8. What is Keshan's disease?
  9. In what chemical form does selenium exist in food?
  10. Why is there greater possibility of selenium deficiency in vegetarian population?
  11. How does selenium exist in animals?
  12. Name the main enzymes containing selenium and their function.
  13. Why iodine has special role in iodine biochemistry?
  14. How does iodine occur in nature?
  15. Why do the people in hilly areas suffer from iodine deficiency syndrome (IDS)?
  16. Why is beer considered a rich source of silicon for humans?

Answers

Chapter 1

1. Nucleus.
2. Lipids and proteins.
3. DNA, RNA and proteins.
4. RBC.
5. The cell membrane is made of phospholipids, coated on both the sides with proteins. The general structure of the lipid molecule is shown in Figure 1.2. The phospholipids have a hydrophobic end and a hydrophilic end, and exhibit surfactant properties. In the cell membrane, the lipids have a bilayer structure, with the polar heads of the lipids pointing outwards. Depending on the hydrophobic or hydrophilic nature of the protein, it occupies different parts of the lipids. The protein is considered to float like icebergs, in the "fluid mosaic model" of the cell membrane.
6. Oxidation of glucose, in the respiration, liberates energy and sustains life.
7. Glycosidic.
8. Ribose and deoxyribose, respectively.
9. O in hydroxyls are not strong donors, and stable complexes with hard acids like Ca^{2+} are known.
10. Ca^{2+} ion.
11. By coordination through amine N and carboxylate O, resulting into formation of chelate rings.
12. A ligand with coordinating group, that can coordinate with metal ions using different (sets) of donor atoms; simple ligands that can act as ambidentate ligands are thiocyanate (SCN^-) and nitrate (NO_3^-). Both are monodentate ligands can bind with metal ion through either S or N (SCN^-), and N or O (NO_3^-).
13. Cysteine and histidine.
14. Both C=O and C=N bonds involve sp^2 hybridisation.
15. The fact that the coordination site changes at higher pH has been confirmed by studying the electronic spectrum of Cu-glycylglycine (1:1) solution at varying pH (Figure 1.14). At low pH, the band appears at high $\lambda > 800$ nm, due to CO and NH_2 coordination. With increasing pH, there is an increasing N^- coordination, creating a stronger field, and a higher energy absorption band is observed at 625 nm.
16. If the side chains are very large, for accommodating them, each peptide chain becomes coiled in the form of helix like a spiral stair case. There is a hydrogen bonding between different coils of the same chain, and this holds the helix together (Figure 1.17a). This structure is found in the α-keratin of unstructured wool, hair, horn and nails. It was suggested by Pauling and Corey in 1951 that the five coils in the helix of the polypeptide chain of keratin are made of 18 amino acids; that is, there are 3.6 amino acid residues per turn of the helix. There is a repeat distance of 1.5 Å between the amino acid residues, and this provides room for the side groups. This is called a helix structure and is of significance in protein chemistry.
17. Phospholipids.

Chapter 2

1. It depicts stability of a metal complex MLn, formed between metal ion M and ligand L, n being number of ligands bound to metal centre. Greater the value of this constant, more stable is the complex.

2. Nature of metal ion, nature of ligand(s), thermodynamic factors.

3. Enthalpy change, which should be negative large value for a metal complex to be more stable.

4. Entropy change.

5. The ionic size of the metal ion is inversely proportional to the charge density, and hence, reciprocal of ionic radius is proportional to the electrostatic attraction of the metal ion for the ligand, that is its Lewis acidity. On this basis, the bivalent metal ions of the first transition series are arranged in the following order, as per their tendency to form the complexes with one specific ligand.

    Ti (II) < V (II) < Cr (II) > Mn (II) < Fe (II) < Co (II) Ni (II) < Cu (II) > Zn (II)

    The above order is called Irving–Williams order.

6. The anomalous positions of Cr(II) d^4 and Cu(II) d^9 in the Irving–Williams order are due to Jahn–Teller effect, which brings additional crystal field stabilisation energy. Cr(II) and Cu(II), octahedral complexes, being doubly degenerate in ground state, undergo distortion and get stabilised, and hence, Cu(II) and Cr(II) form more stable complexes than the metal ions on both sides.

7. Cavity size of macrocyclic ligands with respect to that of metal ions.

8. This term applies to greater stability of metal complexes of macrocyclic ligands. Busch and coworkers have attributed the greater stability of the macrocyclic ligand complexes to the fact that in such complexes, the ligand atoms are suitably placed around the metal ion and the metal ion is simultaneously bound to number of donor atoms. Hence, the M–L bond is less susceptible to dissociation, leading to the greater stability of the complex. This factor of the stabilisation of the macrocyclic ligand complex was termed "multiple juxtapositional fixedness" (MJF).

9. It has been observed that in ternary complexes of the type [M(dipy)L], the mixed-ligand formation constant K_{Mdipyl}^{Mdipy} is much higher than expected from statistical considerations. 2,2'-Dipyridyl (dipy) is a bidentate ligand bound to the metal ion through $N \rightarrow M$ σ bonds. Besides this, there is also $M \rightarrow N$ π bond formation, by the back donation of electrons from the metal $d\pi$ orbitals to the vacant delocalised $p\pi$ orbitals of the ligand. This, $d\pi$–$p\pi$ interaction, does not allow the concentration of electrons on the metal ion to increase significantly, on the binding of dipyridyl. In other words, electronegativity of $[M(dipy)]^{n+}$ is almost same as that of $[M(H_2O)x]^{n+}$. Hence, the tendency of L to combine with $[M(dipy)]^{n+}$ is almost same as to combine with $[M(H_2O)x]^{n+}$, and K_{Mdipyl}^{Mdipy} is nearly equal to K_{ML}^{M}.

10. In case of MAL type of complexes, where A is a π-acidic ligand, $M \rightarrow A$ π interaction results in increase in class A character of the metal ion in MA. This brings in discriminating behaviour of MA towards the secondary ligand L, coordinating through N–N, N–O⁻, O⁻–O⁻. The tendency of binding is in the order O⁻–O⁻ > N–O⁻ > N–N. The order can also be explained in terms of electron repulsion concept. In the formation of M–L complex, where L coordinates through O⁻, there is a repulsion between metal d electrons and O⁻ lone pair of electrons. In the ternary complex, MAL, $M \rightarrow A$ π bonding reduces π electron density over the metal ion, and hence, the lone pair of electrons over O⁻ has to face less repulsion from d electrons, while O⁻ coordinating L combines with MA, than when combining with M. The effect is more significant in the complexes of copper(II), so that the value of $\Delta logK$ in $[CuA(Catecholate)^{2-}]$ complex is positive. Substitution over the π-acidic ligand, which increases its π-acidic character, makes the $\Delta logK$ of MAL complexes more positive. In case of O⁻–O⁻ coordinating bidentate ligand L, substitution over L, which increases the charge density over O⁻, stabilises the ternary complex further, making $\Delta logK$ more positive.

11. The study of the effect of a π-acidic ligand, in stabilising the ternary complex, is important, as in biological systems, there are instances of formation of ternary complexes, involving imidazole type of ligands with possibility of $M \rightarrow N$ π interaction. In the metalloenzymes, the electron density around the metal ion does not increase significantly on binding with the π-acidic imidazole nitrogen of the histidine residue of the apoenzyme protein. Hence, the metal–apoenzyme can effectively bind with the substrates with $O^-–O^-$ (ATP) or O–N (peptide) coordinating sites, resulting in the formation of stable apoenzyme–M–substrate ternary complexes.

12. Charge on the ligands, steric hindrance due to bulky groups, tridentate character of one ligand, intramolecular interligand interactions.

13. The most predominant interaction which stabilises the ternary complexes in biological systems is intramolecular interligand interaction. These are of two types:

    a. **Rigid interaction:** One of the ligands is coordinated rigidly with the metal ion, and the other coordinated ligand has a flexible non-coordinating group interacting with the rigid ligand. For example, there is a rigid interaction between a coordinated aromatic diamine and the non-coordinated base of a nucleotide as in [Cu(2,2′-dipyridyl)-adenosine triphosphate] (Figure 2.11b).

    b. **Flexible interaction:** If both the ligands have free non-coordinated group, there can exist flexible interaction between them. For example, in [Cu–ATP–tryptophan] (Figure 2.11a), there is stabilisation due to interaction between the non-coordinated indole part of tryptophan and adenine base part of ATP. Similar interaction, between non-coordinated side groups of the two amino acids, stabilises the ternary complex, [Cu(phenylalanine)(tryptophan)].

14. Hydrogen bonding, ionic or electrostatic interactions, stacking interactions, hydrophobic interactions.

15. Electrostatic, stacking, hydrophobic.

16. In complex of type (Figure 2.12a), where the aliphatic or aromatic non-coordinated group of one ligand cannot sufficiently overlap with the second coordinated ligand, dipyridyl or analogue, there can be partial hydrophobic interaction between the two ligands, leading to the stabilisation of the ternary complex. Alternatively, it can be considered that the non-coordinated aliphatic or aromatic part of the ligand tends to occupy space near the metal ion, rather than out in solution (Figure 2.13), due to interaction between the axial metal d orbital electron and the delocalised π electron cloud over the ligand side group. This bonding tendency is referred to as ligand side group–metal interaction.

17. Nucleophilic substitution is the reaction of an electron pair donor (the nucleophile) with an electron pair acceptor (the electrophile). Ligands have electron pairs which they use for formation of coordinate bonds with metal ions; thus, ligands act as σ bases, they are nucleophiles, and hence, the ligand substitution reactions are nucleophilic substitution reactions.

18. a. The metal ion should be a weak Lewis acid, and the ligand should be a weak base.

    b. In the substitution of X by H_2O, in cis and trans $Co(en)_2LX$, it has been observed that when L is electron donating group (s or both s and p donor) such as OH^-, NH_3, Cl^-, Br^-, NCS^-, the reaction proceeds by SN_1 mechanism.

    c. If the ligands in ML_6 are bulky in nature, they produce steric hindrance to each other and prefer to undergo dissociation into five coordinated intermediate, for substitution to take place.

19. The trans effect is related to substitution reactions in square planar metal complexes.

20. According to Taube, if a substitution reaction over a metal complex is fast and is complete in less than one minute, for 0.1 N concentration of the reactants at room temperature, the system can be termed labile. If the substitution reaction takes more time, the complex is said to be inert. However, the boundary between these two types of reactions is not rigid.

21. No. The lability and inertness are not related to the thermodynamic stability of the system. Thermodynamically stable systems may be kinetically labile or vice versa, because the

ease with which substitution reaction takes place does not depend on the strength of M–L bond. For example, the complex $[Mn(CN)_6]^{3-}$ with lower stability constant (log K = 27) has a relatively less and measurable rate of substitution reaction than $[Ni(CN)_6]^{4-}$ with higher log K = 30. Though $[Ni(CN)_6]^{4-}$ has high stability constant, the system is fast to substitution. $[Fe(CN)_6]^{4-}$ (log K = 37) is very slow, whereas $[Hg(CN)_4]^{2-}$, with higher log K = 42, is very fast to substitution.

22. In case of d^0, d^1, d^2 metal ions, there is an octahedral complex formation, due to d^2sp^3 hybridisation. However, one or more low-lying t_2g orbitals are still vacant. Thus, one low-energy d orbital is available for d^3sp^3 hybridisation, and thus, the intermediate hepta-coordinated structure MX_6Y can be easily formed, and hence, the substitution reaction can proceed through SN_2 mechanism, and the system is labile.

23. Isomerisation reactions.

24. It was observed by Ray and Dutt that though the complex $[Co(biguanide)_3]^{3+}$ is both thermodynamically and kinetically very stable, it undergoes racemisation reaction. Hence, they argued that the racemisation must be by the distortion of the structure, rather than by the breaking of biguanide at one end. They suggested rhombic or tetragonal twist, leading to an intermediate planar hexadentate structure (Figure 2.31). This can orient back to original octahedral structure with the retention of original optical active form or change to another form, leading to racemisation. This mechanism is popularly known as rhombic or tetragonal twist.

25. An alternative mechanism was suggested by Bailar and is called triangular twist mechanism. In this process, one of the trigonal planar faces of the octahedron undergoes a twist through an angle of 60°, along the C_3 axis, resulting in an intermediate trigonal prismatic structure. When it orients back to the original octahedral form, there may be change to the opposite optical isomeric form leading to racemisation.

26. Bailar's mechanism is different from Ray's twist, as well as bond rupture process, suggested by Werner, in the fact that the latter two can give simultaneous racemisation and isomerisation reaction, whereas Bailar's twist does not permit cis–trans isomerisation.

27. The formation of the outer sphere complex involves three factors:
    i.   The thermodynamic driving force of the reaction.
    ii. The work involved in bringing together of the two reactants and separating them after electron transfer. The forces which hold the reactants in the outer sphere complex are the electrostatic interaction, van der Waals forces and H bonding between the reactants.
    iii. The reorganisation of the coordination and solvation spheres of the two reactants during the reaction.

Chapter 3

1. The four cations are widely distributed in all living organisms. Mainly, they have role as constituents of the body parts or body fluids. They also play other roles like transmission of nerve impulses and muscle contraction, through transmembrane concentration gradient. They stabilise the structures of proteins, cellular membranes and skeletal mass, and also act as enzyme activators.

2. In the modern periodic table, group 1 metals (elements excluding hydrogen) are called alkali metals, while group 2 metals are called alkaline earth metals. Alkali metals are called so because they form alkaline solution when they react with water. Alkaline earth metals, on the other hand, exist in nature as their alkaline salts in the crust of earth, hence the name.

3. Both Na^+ and K^+ act as cofactors for ATPase enzyme. However, roles of Na^+ and K^+ are, mainly, antagonistic (opposite to each other), and this is a controlling factor in metabolism. K^+ activates respiration in muscles, kidney, adipose and erythrocytes, helps protein synthesis and acetyl choline synthesis, whereas Na^+ opposes all these reactions.

4. The normal osmotic transport of metal ions, through semi-permeable membrane, is from higher concentration to lower concentration, leading to equal concentrations on both sides. This is called passive transport. This process of ion transport in the direction of concentration gradient is a passive transport, governed by the osmotic phenomenon. In living cells, a concentration gradient (difference in concentration of ions) is maintained, across the semi-permeable membrane, defying passive transport.

   The transmembrane concentration gradient of Na^+ and K^+ plays an important role in a variety of transport mechanism, used by cells to accumulate and expel nutrients and toxic ions. Further, the concentration gradient, across the membrane, causes the generation of an electrical potential of about 60 mv in the membranes of the excitable tissues. This transmembrane potential inhibits the flow of ions, from one side of the membrane to another. When there is an excitement of the tissue, the permeability of the alkali metal ion increases. There is a diffusion of Na^+ inside the cell and of K^+ outside the cell, and there is lowering in the transmembrane potential. Thus, the flow of ions in the channels, in the membrane, is voltage-gated (regulated by voltage). Electrical signals are transmitted between nerve cells by a mechanism involving ion channels.

5. The ionophores are present in cell membranes and are usually monobasic carboxylic derivatives of ethers. Their molecular weights range from 500 to 2,000 daltons. These ionophores form channels for the positively charged ion to pass through. The negatively charged ionophore forms a neutral complex with the metal ion. However, if the ionophore is a neutral species, it forms a positively charged complex with the alkali metal ion, and it needs a permeant anion, to move through the membrane. Thus, there is, carrier ionophore-assisted, passive transport of the metal ion, from the region of higher concentration to the lower concentration region, across the cell membrane, and this explains the phenomenon of passive transport.

6. There is a higher concentration of Na^+ outside the cell membrane, whereas there is a higher concentration of K^+ inside the cell; that is, there exists concentration gradient across cell membrane.

7. The energy required for the transport of metal ions against the concentration gradient is released due to, Na^+ and Mg^{2+} promoted, enzyme ATPase catalysed hydrolysis of adenosine triphophate (ATP) to adenosine diphosphate (ADP) and inorganic phosphate. Hence, this phenomenon of active transport is called "sodium pump mechanism".

8. Calcium has, mainly, the role as the constituent of extracellular structure of the body. Calcium is essential for the development of extracellular structures such as egg shells, bones and teeth. It is stored in the form of small crystals in the tissues, so that deposition and reabsorption are rapid and calcium can be mobilised and deposited, for the formation of extracellular structure. Besides these extracellular roles of calcium, it is also an important constituent of the intracellular material and has an important role in muscle contraction, hormone secretion, glycolysis (glucose breakdown), gluconogenensis (glucose synthesis) and cell growth. The changes in the concentration of intra- and extracellular calcium also govern the metabolic regulations known as the calcium messenger system.

9. See the "Other Biochemical Roles of Calcium" section.

10. The concentration of Mg(II) is high, inside the cell, whereas it is low in the plasma. As the intracellular concentration of K(I) and Mg(II) are high, ribosomal RNA and proteins in the cells are bound to these metal ions, more to Mg(II), with higher Lewis acidity. Mg(II) binds at the phosphate end of the nucleic acid and reduces the electrostatic repulsion between the phosphates and thus stabilises the interaction between the base pairs in RNA. This results in the increase in the melting point of RNA.

11. Ribozymes are the RNA enzymes which exhibit catalytic activity to promote hydrolysis of RNA phosphodiesters. They are bound to five $[Mg(II)(H_2O)_6]^{2+}$, which retain their solvent environment and are H-bonded to O and N atoms on the bases, phosphates and sugar rings of RNA. These five Mg^{2+} help to stabilise the structure of RNA. One more Mg(II) is bound to the RNA and has catalytic role in phosphodiester hydrolysis.

12. A phosphodiester bond is formed when two of the hydroxyl groups in phosphoric acid react with hydroxyl groups on two other molecules to form two ester bonds. These make up the backbone of strands of nucleic acids.

13. Mn^{2+}.

14. Kinases are Mg^{2+}- or Mn^{2+}-containing enzymes required for the hydrolysis of ATP molecules for generating energy for biological functions. Hydrolysis of an ATP produces ADP, inorganic phosphate and 7.3 Kcal of energy per mol. Some examples of kinases are creatine kinase, pyruvate kinase, hexokinase and phosphofructose kinase.

15. Calcium, in millimole concentration, is necessary for muscle contraction. Muscle contraction is governed by the release of calcium from sarcoplasmic reticulum and the Ca^{2+}-binding proteins, present on the muscle fibre. When an impulse comes in the nerve ending in the muscle fibre, there is release of Ca^{2+} from sarcoplasmic reticulum. This Ca^{2+} interacts with the regulatory proteins present in the muscle. In the higher vertebrates, the skeletal muscles contain the regulatory protein, troponin.

Chapter 4

1. Though present in biological systems as a trace metal, zinc is indispensable for the growth and development of plants and animals. Zinc is necessary for the synthesis of nucleic acids and for the maintenance of their conformation. Deficiency of zinc affects cell growth, division and differentiation and thus leads to abnormalities in the compositions and functions of cells.

2. Zn(II) is absorbed in the small intestine, the absorption being minimum in the stomach and long intestine. Zn(II) absorption is dependent on the biochemical ligands present in the diet, which form complex with Zn(II) and increase or decrease its uptake. The biochemicals, with cysteine and histidine coordinating sites, increase Zn(II) uptake. Zn(II) is more efficiently absorbed from food from animal sources than from plant sources. Phytate, present in plants, forms strong chelate with Zn(II) and makes its absorption by the membrane difficult. However, in a balanced food, where other more effective Zn(II) binding ligands are present, along with phytate, its effect may be nullified, and Zn(II) is easily absorbed by the membrane. Infants can absorb Zn(II) better from breast milk than from cow's milk. The presence of minerals, like calcium, lowers down the zinc absorption. Similarly, a large excess of Cu(II) in the diet reduces Zn(II) absorption. This antagonism is due to various factors, involved in the absorption of trace metals. Another metal ion of nutritional significance, reducing zinc absorption, is iron.

3. Phytate, also called phytic acid, is principal storage form of phosphorous in several plant tissues. Phytic acid is a sixfold dihydrogen phosphate ester of inositol, also called inositol hexakisphosphate or inositol polyphosphate. At physiological pH, the phosphates are partially ionised, resulting in phytate anion, which is a strong chelating ligand for several nutrient metal ions such as Ca, Fe, K, Mg, Mn and Zn, affecting their absorption.

4. In the enzymes, Zn (II) may have four different roles:
   i.   The first type of Zn(II) is directly involved as a Lewis acid catalyst.
   ii. In some enzymes, there may be two Zn(II) centres present. One is of the first type and has catalytic activity. The second type of Zn(II) is not itself catalytic, but regulates the enzymatic activity of the first type of Zn(II). It may act as an activator or inhibitor.
   iii. The third type of Zn(II) stabilises the overall conformation of the enzyme protein, mainly stabilising its quaternary structure, so that the protein acts as an efficient catalyst. Thus, this type of Zn(II) acts as a coenzyme.
   iv. There is a fourth type of Zn(II) in the enzymes, with none of the above three types of roles. This is called a noncatalytic type of Zn(II).

5. The mammalian zinc aminopeptidases are oligomeric (molecular weight 200,000 daltons), whereas corresponding microbial enzyme is monomeric (molecular weight 40,000 daltons). The former have two Zn(II) per mole, whereas the latter may have one or two. Zn(II) can be removed from the enzyme by 1,10-phenanthroline, without affecting the hexameric structure. Zn(II) can be replaced by Co(II), retaining the enzyme activity. In the mammalian aminopeptidase, replacement of one Zn(II) by Mg(II) or Mn(II) enhances the catalytic activity, but the replacement of both the Zn(II) ions, by these metal ions, does not retain the catalytic activity. This shows that at least one Zn(II) is essential for the catalytic activity. The other one may have regulating role, which can be maintained by the substituted Mg(II) or Mn(II). Hence, the substitution of the second Zn(II), with catalytic role, by other metal ions, makes the enzymes catalytically inactive.

6. These are Zn(II)-containing proteins which cause hydrolysis of monophosphate esters of primary alcohols, secondary alcohols, phenols and mono-substituted phenols (Eq. 4.1). They also catalyse trans-phosphorylation reaction. Besides Zn(II), the enzymes also have Mg(II) in the structure. They are called alkaline phosphatase as they are catalytically active at around pH 8.

7. These enzymes catalyse hydrolysis of monophosphate esters of primary alcohols, secondary alcohols, phenols and mono-substituted phenols. The proposed reaction mechanism involves deprotonation of H_2O, forming OH^-. The deprotonation of water bound to Zn(II) occurs only at a high pH, and hence, the formation of $-OH^-$ takes place at a high pH only. This explains why alkaline phosphatase is active at high pH only.

8. Carbonic anhydrase, a Zn(II) protein

9. The hydration reaction is important, because CO_2 is generated by the oxidation of carbohydrate and it has to be converted into carbonic acid in the blood, to be carried back by deoxyhaemoglobin to the lungs. In the lungs, the enzyme carbonic anhydrase (CA) catalyses the dehydration of HCO_3^-, so that CO_2 is exhaled. It may be noted that the nomenclature of the enzyme is based on this reverse reaction of removal of water from carbonic acid.

10. A Zn (II)-containing protein known as alcohol dehydrogenase.

11. Co(II).

12. The exchangeability of Zn(II) and Co(II) may be due to following reasons:

    i. Since there is no CFSE in d^{10} Zn(II) ion, structure of its complex is solely dictated by electrostatic forces. In case of high spin Co(II) with d^7 configuration, there is a small difference in stabilisation energies in tetrahedral (−8 Dq) and octahedral (−12Dq) forms, and hence, in Co(II) complexes, one form can change into other easily, as in Zn(II) complexes.

    ii. In high spin tetrahedral complex of Co (II), the electronic configuration is $e^4t_2^3$, and it behaves like a spherical ion, as d^{10} Zn(II).

    iii. Further, Co(II) and Zn(II) have similar ionic radius. Hence, Co(II) is the best substituent for Zn(II) in the enzymes.

13. Carbamoyl-*l*-aspartate(III), formed by the condensation of carbamoyl phosphate(I) with *l*-aspartate(II); the reaction being catalysed by a Zn(II) protein known as aspartate transcarbamoylase.

14. Zn(II) has an important role in the biochemistry of nucleic acids. It has been recognised that Zn(II) is essential for the transfer of genetic information. The enzymes DNA and RNA polymerases, which catalyse the replication and transcription of nucleic acids in prokaryote and eukaryote, are nucleotide transferases, with stoichiometric amount of intrinsic Zn(II). Catalytic role of Zn(II) is not yet confirmed.

15. It has been observed that Zn(II)-bound haemoglobin shows increasing tendency to bind with oxygen. Zn(II) has greater affinity for oxyhaemoglobin than haemoglobin, showing that oxygen-bound haemoglobin is more suitably oriented to Zn(II) binding, or Zn(II) binding orients haemoglobin to bind with oxygen more favourably.

16. It has been observed that in the formation of nucleic acid and protein assemblies, important in biological reactions, metal ions bind with the protein, to help it in getting into an active

conformation, to interact with nucleic acid, efficiently. The metal ion selected by nature is Zn(II), because of its abundance and also because there is no redox chemistry, so that the damage of the DNA, due to oxidative degradation, is avoided. A class of zinc proteins, involved in genetic factors, was discovered in 1983. It was observed by Klug and coworkers that such proteins have finger-like structural domain and hence were called zinc fingers. Unlike the other Zn proteins, which bind with DNA in a three-dimensional way, utilising the double-helix structure of the DNA molecule, zinc finger proteins recognise the nucleic acid sequence of different length by binding together in a two-dimensional way.

17. See the "Other Roles of Zinc Proteins" section.

Chapter 5

1. Haem b is the Fe(II) complex of the most common porphyrin in nature, called protoporphyrin IX with $CH=CH_2$ at X and Y positions, and $-CH_3$ at position Z as shown in Figure 5.1b. This forms the prosthetic group in a large number of iron proteins, like haemoglobin.

   Haem a has a formyl group at position Z, a long phytol chain at position X (Figure 5.1b) and $CH_2=CH$ at position Y, in the porphyrin ring.

   In haem c, the porphyrin part is covalently bound to a protein through S sites of the cysteine residues at positions X and Y (Figure 5.1b). There is a CH_3 group at position Z. Haem b and haem c prosthetic groups are present in cytochromes. The porphyrin rings in haem a, b and c consist of four pyrrole rings, linked through $-CH$ groups.

   In haem d, the double bond of the pyrrole ring I is dihydro-reduced.

2. In the formation of haem, the four nitrogens of protoporphyrin IX get coordinated to Fe(II) through σ bonds. There is deprotonation of two NH groups. Thus, the 16-membered macrocyclic ring gets two negative charges and forms a neutral haem complex with Fe(II). The resultant chelate has 4 six-membered rings with double bonds, leading to a stable structure. Furthermore, there is a macrocyclic effect (favourable enthalpy, entropy and MJF effects, see Chapter 2), resulting in further stabilisation of the macrocyclic chelate ring. Besides the formation of N → Fe(II) σ bonds, there is an Fe → N π bond formation, due to the interaction between metal dπ orbitals and the delocalised π orbitals of the ligand ring. Thus, a very stable haem complex is formed.

3. Cytochrome P450, a protein with Fe(III) protoporphyrin IX (haem b type) as the prosthetic group, is responsible for conversion of water-insoluble, aliphatic or aromatic hydrocarbons into soluble alcohols or phenols and of olefins into epoxides. It is found in the liver, intestine and lung tissues of mammals. It is bound to the membrane of the tissue and assists in transferring one oxygen atom of the molecular oxygen to a substrate, resulting in its monooxygenation.

4. The different forms of microsomal cytochrome P450, in human beings, have the same prosthetic group, but differ in their protein backbone and assist the hydroxylation of various substrates such as drugs, steroid precursors, pesticides and halogenated hydrocarbons (like DDT), making them water soluble, so that they are excreted through urine. Thus, the enzyme is a part of the body's detoxification system.

5. Cytochromes are proteins containing haem as cofactor. In the enzymatic activity of cyt P-450, the substrate (RH) gets bound to the enzyme in a protein pocket, close to the haem, and the water or any other group, present at the sixth position, is replaced. As RH binding is noncovalent and weak, effectively Fe^{3+} enzyme–RH complex is penta-coordinated, and Fe(III) is in high spin state. Due to the binding with RH, the reduction potential of Fe^{3+} enzyme–RH becomes more positive, and it is more easily reduced to Fe^{2+} enzyme–RH, by receiving electron from the biochemical reducing agent NADH or NADPH. The electron is transferred from NADH to Fe^{3+} enzyme–RH, through electron transfer iron proteins. The Fe^{2+} enzyme–RH species binds with O_2 at the sixth position. If CO is supplied, it can occupy the sixth position.

In the electronic spectrum of this carboxy form, there is a characteristic absorption band at 450 nm (called Soret peak or band) and that is the reason for the nomenclature of the enzyme.

6. It is an iron(V) oxo species, which oxygenates substrate.

7. Two, Fe(III) centres.

8. Catalase and peroxidase are related iron proteins, containing Fe(III) protoporphyrin (haem b type) prosthetic group. They catalyse the reactions of hydrogen peroxide, leading to its decomposition.

9. In the enzymes cyt P-450, catalase or peroxidase, the intermediate is the same oxo cation, but in the case of the first, oxygen is transferred to the substrate and the enzyme acts as a monooxygenase, (Eq. 5.2) whereas in latter two, two electrons are transferred from the substrate to the oxo cation, resulting in the oxidation of the substrate. Thus, catalase and peroxidase exhibit oxidase activity.

10. Cytochromes, meaning cell pigments, were discovered by Keilin. These are Fe(III) haem proteins, occurring in cells, and act as electron carriers in mitochondrial oxidation process. They are involved in the terminal stage of the respiratory chain, that is carrying electron from glucose to molecular oxygen, leading to its decomposition and also transporting electron in the photosynthetic process.

11. One-step oxidation of the substrate (glucose), by transfer of electron directly to O_2, is highly exothermic process which can cause harm to the tissues. To avoid this, the electron transfer must occur in several steps, each involving smaller amount of energy. This is possible because the cytochromes differ in their reduction potentials and are arranged in sequence in the electron transport process, in the increasing order of the reduction potentials. Thus, the cytochromes affect sequential stepwise transfer of electron from the substrate to molecular oxygen, leading to its reduction to water. The electron transfer process involves reversible change of haem iron from oxidation state III to II.

12. These are non-haem iron proteins, called iron sulphur proteins, which like cytochromes carry electron in biological systems, from a molecule with lower redox potential to one with higher redox potential. They are found in anaerobic and aerobic bacteria, algae, fungi, higher plants and animals. In plants, they act as electron carrier for reduction of N_2, in nitrogenase activity, and in photosynthesis, for electron transfer from PS I to PS II. In the animals, they assist electron transfer process in the mitochondria. These proteins are similar to cytochromes, in the fact that they also involve one electron process in the Fe(III)/Fe(II) couple.

13. Iron sulphur proteins are classified into two types: rubredoxin and ferredoxin.

14. Both rubredoxin and ferredoxin are iron sulphur proteins. In the $(Fe-S)_n$ part of ferredoxin, the iron cations are bound to the inorganic sulphide ion S^{2-}, which are acid labile and are liberated as H_2S on treatment with acid. $(Fe-S)_n$ prosthetic group is bound to the protein, through the S of the cysteinyl residue. This sulphur is called organic sulphur. Rubredoxin differs from ferredoxin in the fact that it does not have labile inorganic sulphide.

15. HiPIP is abbreviation for Hi potential iron protein, a Fe_4-S_4 protein. The Fe_4S_4 proteins occur in bacteria, plant and mammals. They are known to act as electron transport system in photosynthesis and in the activity of nitrogenase, hydrogenase and other redox enzymes. They act as one electron transfer proteins and are classified into two types. The first type are the Fe_4S_4 protein, which have redox potential ~ -600 mv, and are called normal ferredoxin. The second type of proteins has high redox potential +350 mv and is called high potential iron proteins (HiPIP). Both the types have a [4Fe–4S] prosthetic group, bound to a low-molecular-weight protein (6,000 daltons).

16. Haemoglobin and myoglobin in humans and bigger animals; hemerythrins in lower organisms.

17. Figure 5.6.

18. See the "Cytochrome c oxidase" section.

19. X-Ray, magnetic moment, Figure 5.10.

20. Both are oxygen transport proteins containing iron porphyrin units. Haemoglobin has four sub-units of iron porphyrin (tetramer, molecular weight 64,450 daltons), while myoglobin (molec-ular weight 17,500 daltons) has single iron porphyrin subunit. In human beings and bigger animals, O_2 is transported from the lungs to the different parts of the body through the arteries. The transport process is carried out by haemoglobin, present in the RBC. In the muscles, the same function is carried out by myoglobin. The overall rate of binding of O_2 with haemoglobin is less than with myoglobin. Consequently, haemoglobin can take up O_2 at higher pressure, a condition available in the lungs (higher animals), gill (fishes) or skin (smaller organisms). Myoglobin can take up O_2 at lower pressure in the muscles.

21. In both haemoglobin and myoglobin, Fe(II) is in high spin state, though bound to strongly coordinating ligands. Perutz and Kendrew carried the X-ray study of deoxyhaemoglobin and showed that Fe(II) is 0.8 Å above the plane of porphyrin ring nitrogens. Thus, the distance of ligand nitrogens from Fe(II) is 2.9 Å, and hence, the four N atoms create a weak field and Fe(II) is in high spin state (Figure 5.17). When O_2 gets bound to the Fe(II) at the sixth position, the metal ion is pulled down to the centre of the plane of the macrocyclic ring. As a result Fe(II)–N distance is reduced to 2 Å, and hence, a strong field is created and Fe(II) goes to a low spin state. In high spin state, Fe(II) is bigger in size and cannot be accommodated in the porphyrin ring. In low spin state, Fe(II) is smaller in size and can be accommodated in the ring. Thus, the 16-membered porphyrin ring has very specific discrimination of high and low spin Fe(II) and is, therefore, preferred in haemoglobin, for its efficient functioning.

22. Ferritin is an iron storage protein in mammals, plants, fungi and bacteria. In mammals, iron is stored in bone marrow, liver or spleen in the form of ferritin. It is a precursor of other forms of iron storage in living systems, such as hemosiderin, an iron protein in animals, magnetite (Fe_3O_4), an inorganic oxide form of iron storage, found in magnetic bacteria, bees and hum-ming pigeons. Hemosiderin is another form of iron protein, which stores iron in human beings and animals.

23. Transferrin is an iron (III) protein in the blood serum of animals of phylum chordata. It trans-ports iron for haemoglobin synthesis. Only a small fraction of iron in the body is in transit at a time. Iron transfer protein is also present in the egg, called ovatransferrin, and in milk, called lactotransferrin. Human transferrin has molecular weight 80 K daltons, ovatransferrin 77 K and lactotransferrin 80 K daltons. Though not definitely known, this protein may have a role in sequestering the excess iron in the body. Transferrins are iron glycoproteins. In bacteria and fungi, that cannot synthesize transferrin, an alternate mechanism of iron uptake has been provided by nature. When the microorganism faces iron deficiency, an organic compound is secreted, which binds strongly with iron. The general name for such compounds is siderophore, which is a Greek word, meaning iron carrier. More than two hundred types of siderophores have been isolated from the microorganisms, and their structures have been determined.

Chapter 6

1. Copper occurs in the form of copper proteins, which exhibit stereochemistry and properties, distinct from simpler copper complexes. These special features of copper proteins are due to the protein structure, which creates a specific stereochemistry and local dielectric constant, at the site of the coordination of the metal ion. The protein coordinating site attributes specific characteristics to the copper, bound at that site. On that basis, the copper, in the biological pro-teins can be classified into the following types:

    i.   Blue copper(II) type I.
    ii. Normal copper(II) type II.
    iii. Coupled Cu(II) type III.
    iv. Cu(I) type IV.

2. The copper proteins may have one or more types of copper centres and thus exhibit different kinds of roles. The main functions of copper proteins are

    a. Catalytic-electron transfer, oxidation and oxygenation of substrates and superoxide dismutation.

    b. Dioxygen transport.

    c. Copper transport and storage.

3. The copper proteins involved in electron transfer processes in biochemical reactions are smaller proteins, with low molecular weight, and occur widely in organisms, from bacteria to human beings. They contain a single type I copper centre, which is characterised by an extremely intense (ε = 4,000–5,000 $M^{-1}cm^{-1}$) absorption band at 600 nm. Thus, these proteins exhibit intense blue colour {400 times that of $[Cu(H_2O)_6]^{2+}$} and hence are called blue copper proteins. Some examples are plastocyanin, azurin and stellacyanin.

4. The apoprotein parts in these three blue proteins show some similarity in structure, supporting the fact that they belong to the same class. For example, following amino acid residues are present in all the three blue proteins: histidine, asparagine, valine, glycine, tyrosine, cysteine, proline and methionine. It is observed that azurins have more similarity in amino acid sequence with plastocyanin, indicating that these two types of proteins have same origin. Stellacyanin does not have much similarity in amino acid sequence with plastocyanin and azurin. Furthermore, forty per cent of this protein is carbohydrate, with carbohydrate groups attached on the asparagines, at 28, 60 and 102 positions of the protein. It is observed that any of the above proteins, obtained from different sources, retains some of the amino acid residues. For example, plastocyanin from different sources has conserved amino acid residues at positions 31–44 and 84–93. Azurins from different sources have three cysteine residues. The protein, in all the blue copper proteins, is coiled, and copper(II) is located inside the cavity. In plastocyanin and azurin, the Cu(II) ion is bound to the nitrogen atoms of imidazoles of two histidine residues and to two sulphur atoms, of one cysteine and one methionine residue of the protein. Thus, in all the three proteins, Cu(II) is linked to two S ligands and two N ligands, and the four ligand atoms are oriented around the Cu(II) ion in a distorted tetrahedral geometry. The skeletal structure of plastocyanin is shown in Figure 6.1.

5. The blue copper proteins exhibit high reduction potential of the Cu(II) centres, ~0.38 V in plastocyanin, 0.33 V in azurin and 0.30 V for stellacyanin. This has been attributed to two factors: (i) the coordination of Cu(II) to softer S ligand, resulting in lowering of electron density on Cu(II). (ii) Cu(II) is predisposed in a tetrahedral geometry in the protein, and hence, its reduction to Cu(I) is easy, because Cu(I) prefers tetrahedral geometry. Thus, it is seen that in the blue proteins, the protein offers the metal ions a tetrahedral geometry and softer ligand atoms, so that the conversion of Cu(II) to Cu(I), and vice versa is facilitated.

6. These are multicopper proteins which catalyse the oxidation of the substrates by transfer of four electrons to dioxygen molecule, leading to its reduction to two O^{2-}, which form two water molecules. These proteins contain type I, II and III Cu(II) centres in different proportions in different oxidase enzymes. In these oxidases, the type I Cu(II) has electronic spectral, electrochemical and ESR spectral characteristics, similar to that of electron transfer blue proteins. Hence, these have intense blue colour and are called blue oxidases. The oxidase, which does not have type I Cu(II), is not intense blue and is termed non blue oxidase.

7. On reduction of Laccase by three equivalents of ascorbate, it is observed that type I and type III Cu(II) centres are reduced, simultaneously, as revealed by the disappearance of the electronic spectral bands at 660 and 330 nm, characteristic of type I and type III Cu(II), respectively. On addition of further equivalent of ascorbate, type II Cu(II) is reduced. According to Malmstrom, there is transfer of electron from the substrate in steps. First, the electron is transferred to the type I Cu(II). This, in turn, is transferred to the type III Cu(II) through intramolecular process, with type II Cu(II) playing some role. The type III site is thus reduced to Cu(I)–Cu(I). The oxygen molecule is bound at the binuclear type III site. Two electrons are transferred from the

Cu(I)–Cu(I) pair to the O_2 molecule, reducing it to peroxide. Simultaneous transfer of two electrons does not allow the formation of the thermodynamically unstable superoxide O_2^-. The type III site is reoxidised to binuclear Cu(II), after transferring two electrons. Binuclear Cu(II) site further accepts two electron from type I site. These two electrons are transferred to the bridging peroxide, reducing it to two oxide (O_2^{2-}) ions. These combine with $4H^+$, liberated by the substrate, and two H_2O molecules are formed. Thus, at the type III Cu(II) site, the molecular oxygen is reduced to two water molecules, and the substrate parahydroxy phenol (in Laccase activity) or ascorbic acid (in ascorbic acid activity) is oxidised by the loss of four electrons.

8. Tyrosinase, a Cu(II) protein, with monooxygenase activity, catalyses the conversion of monophenols to diphenols and thus assists in the synthesis of the pigment melanin and other polyphenolic compounds.

9. This is a binuclear Cu(II) protein, with molecular weight 40–50 K daltons. The two Cu(II) centres are strongly anti-ferromagnetically coupled through the bridges; hence, the enzyme has very low magnetic moment and is EPR silent.

10. The term "non-blue" implies the absence of type I Cu(II) site in the protein. Dopamine-β-hydroxylase occurs in the medulla of animals and causes hydroxylation of dopamine, using molecular oxygen, as oxidant. This is a monooxygenase. The enzyme obtained from bovine medulla is a glycoprotein with molecular weight 290 K daltons. The enzyme consists of two types of protein, differing only by a NH_2 terminal tripeptide. There are two atoms of Cu(II), per molecule of protein. In the activity of dopamine-β-hydroxylase, the enzyme acts as a mono-oxygenase, and ascorbic acid acts as a reducing agent and hydrogen donor. The reaction can is shown in equation 6.5.

11. These are type II Cu(II)-containing proteins, which appear to act as oxidases. However, ESR spectral study shows that Cu(II) does not undergo change in oxidation state, during the oxidation of the substrate by molecular oxygen. Nevertheless, Cu(II) is essential for the oxidation process, catalysed by this enzyme. If a chelating agent is added, which sequesters Cu(II), the catalytic reaction is deactivated. This indicates that though Cu(II) does not participate in the redox process, it gets chelated with the substrate and acts as a Lewis acid catalytic centre, and assists the oxidation process. Examples of this type of copper proteins are amine oxidase, galactose oxidase and urate oxidase.

12. Breakdown of superoxide (O_2^-), generated in the biological system, as a by-product of metabolic processes, involving molecular oxygen, is called superoxide dismutase activity. The task is performed by enzymes, called superoxide dismutases (SODs). Three types of super oxide dismutase (SOD) are known. They are characterised by having copper, manganese or iron as cofactors. The copper protein has Zn(II) as an essential component. Thus, the three types have been named, CuZn SOD, Fe SOD and Mn SOD.

13. Hemocyanin, a copper protein, is responsible for oxygen transport in invertebrate lower organisms, of the class molluscs (clams, octopus, squid), arthropods (spiders, crabs, lobsters) and annelids. In cephalopods (octopus), hemocyanin gets oxygenated at the gills and most of the oxygen is taken away, when the oxyhemocyanin passes through the tissues. Its oxy- and deoxy-forms are shown in Figure 6.12.

14. The intense band at 345 nm in the electronic spectrum has been assigned to $O_2^- \rightarrow$ Cu(II) charge transfer. As this band is absent in met hemocyanin, it cannot be due to protein $\rightarrow$ Cu(II) charge transfer. Resonance Raman spectrum of oxyhemocyanin shows $v_{o^- - o^-}$ band at 744 cm and supports the presence of μ peroxo-bridge. The peroxide is symmetrically bridged to the two Cu(II), in one of the three forms shown in Figure 6.13.

15. Yes. Ceruloplasmin is an intensely blue-coloured copper protein, present in the blood plasma of the vertebrates; 100 mL of human plasma contains 20–40 mg of the protein. Besides blood, it is also present in spinal and joint fluids and the secretions of eyes, ear and digestive system. It is a glycoprotein and has nine to ten saccharide chains, each containing different kinds of carbohydrates. It is synthesised in the liver. The molecular weight of the protein varies in different classes of animals, human beings 140 K daltons, rabbit 142 K daltons and horse 120 K daltons. The protein has tetrameric structure, and each subunit contains six to seven copper atoms.

Chemical analysis, and electrochemical and magnetic studies have shown that half of the copper is present as Cu(II), and the other half is Cu(I). Cu(II) are of type I, type II and type III. Ceruloplasmin acts as a copper transport protein. It provides copper to cytochrome C oxidase and other copper-containing oxidases. Albumin, which transports copper between intestine and liver in the portal vein, also takes copper from ceruloplasmin. Ceruloplasmin also acts as a catalyst in the oxidation of polyphenols, polyamines, adrenalin, serotonin, etc. It also has the role of sequestering excess Cu(II) and store it in the blood plasma, so that the delirious toxic effect of catalytic oxidative reaction by free Cu(II) is prevented, and ceruloplasmin bound Cu(II) is made available to the Cu(II)-containing enzymes, as per their requirement. Ceruloplasmin has also been recognised to have ferroxidase activity.

Chapter 7

1. Whereas haemoglobin contains a 16-membered tetraaza macrocyclic pyrrole ring, vitamin B_{12} has a tetraaza macrocyclic ligand complex of Co(III). The structure is similar to myoglobin, with the difference that the ligand is corrin, containing fifteen atoms in the ring, one less than in porphyrin. One methine (=C–H) bridge between one pair of pyrrole rings is absent (Figure 7.1). The peripheral carbon atoms, of the corrin ring system, are substituted by seven amide moieties. C_2, C_7 and C_{18} are substituted by acetamide, and C_3, C_8 and C_{13} bear propionamide. C_{17} has N-substituted propionamide.

2. In the cobalamin, vitamin B_{12}, the sixth coordination position of Co (III) is occupied by cyanide ion (Figure 7.1), and hence, it is also called cyanocobalamin. It is water soluble and is red in colour.

3. Though vitamin B_{12} is essential for higher animals, it is not required by plants. It cannot be synthesised by animals or plants, but is biosynthesised by certain bacteria.

4. Pernicious anaemia is caused in case of deficiency of vitamin B_{12} in the diet, or failure to absorb it from the food.

5. In the cobalamin, vitamin B_{12}, the sixth coordination position of Co(III) is occupied by cyanide ion (Figure 7.1), and hence, it is also called cyanocobalamin. The other derivatives of cobalamin are vitamin B_{12}(a), aquacobalamin, with water molecule in the sixth position and vitamin B_{12}(c), nitrocobalamin, with nitro group at the sixth position. Vitamin B_{12} can be reduced to vitamin B_{12}(r) with Co(III) converted to Co(II) and then to vitamin B_{12}(s) with cobalt in monovalent state (Figure 7.1).

6. Following are the roles of Vitamin B_{12}:
   i. Alkylation of organic compounds and metals.
   ii. Generation of methane.
   iii. Isomerase activity.
   iv. Deamination reactions.
   v. Reduction reactions.

7. Methylated cobalamin, CH_3–Co(III) B_{12}, can transfer methyl group to Hg(II) forming methyl mercury, CH_3–Hg^+ – an extremely toxic form of organic mercury. This was the key species responsible for mercury toxicity in Minamata disease.

Chapter 8

1. Iron and molybdenum.

2. The cycle of fixation of atmospheric N_2 into ammonia, its incorporation into biological system due to the formation of nitrogenous biomolecules, their decay into nitrate and the return of N_2 to the atmosphere, due to the denitrification of nitrate, is called the nitrogen cycle (Figure 8.1).

3. See the "Mechanism of Nitrogen Fixation" section.

4. The iron protein in nitrogenase is made up of two identical subunits of molecular weight 30,000 daltons. The protein contains four iron and four sulphide ions, which can be extruded as one $[Fe_4S_4]^{2+}$ cluster. This shows that there is one $[Fe_4S_4]^{2+}$ centre, bound, between the two subunits of the protein. This has been confirmed by X-ray studies also (Figure 8.3).

   Fe_4S_4 is diamagnetic, due to antiferromagnetic interaction between the pairs of Fe (II) and Fe(III) centres through sulphide bridges. It is thus similar to $[Fe_4S_4]^{2+}$ in ferredoxin with two Fe(III) and two Fe(II). The Fe_4S_4 centres of iron protein receive electron through ferredoxin chain and undergo one electron reduction. The reduced form shows EPR activity. However, the EPR spectrum of the reduced iron protein shows two spin states S = 1/2 and S = 3/2, raising the controversy earlier that there are more than one type of Fe_4S_4 sites. However, it has now been established that the reduced state of same types of Fe_4S_4 site may exist in different spin states. It has been suggested that one of the coupled Fe_2S_2 site in the reduced Fe_4S_4 [1Fe(III), 3Fe(II)] is paramagnetic. The spin coupling due to anti ferromagnetic interaction may lead to S = 1/2 or S = 3/2. However, there is no difference in the redox behaviours of the Fe_4S_4 with different spin states.

5. The enzyme nitrogenase contains Fe–Mo protein. This protein receives the electron from the iron protein, and its reduced form acts as catalyst for reduction of nitrogen. Fe–Mo–protein contains two types of subunits and has a molecular weight in the range 220,000–245,000 daltons. The two types of subunits are called P cluster and M cluster. The P cluster consists of four Fe_8S_8 units as pairs of Fe_4S_4. These Fe_4S_4 differ from Fe_4S_4 found in ferredoxin, as revealed by their electronic spectra and Mössbauer (MB) spectra. EPR shows that the P clusters are paramagnetic with S = 7/2. This indicates that there is an incomplete antiferromagnetic coupling between pairs of Fe(II). MB spectra reveal more than one isomer shift, showing that all four irons in Fe_4S_4 are not equivalent. This indicates that the Fe_4S_4 units are highly distorted, unlike the cubane structure in ferredoxin.

6. Nitrogenase also reduces other substrates with triple bond, such as acetylene, cyanides and isocyanides (Eqs. 8.11–8.15). It has been shown recently that the Fe–Mo–protein part of the nitrogenase can also cause hydrogenation of the dyes, in the presence of molecular hydrogen, and thus can act as a hydrogenase. This is the only example, where only one component of nitrogenase shows catalytic activity, independently.

7. In 1971, nitrogenase, extracted from *Azotobacter vinelandii*, was found to contain vanadium. This has small amount of molybdenum also and hence has lower activity than normal nitrogenase. However, because of the presence of vanadium, the enzyme shows selective activity for specific substrates. This shows that it may not have structure similar to normal nitrogenase with Fe–protein and Fe–Mo protein units. Because of being associated with some vanadium, it may have different structure. A nitrogenase was extracted from a mutant of A. vinelandii, which contained only iron protein and was able to fix nitrogen. This should be called Fe–Fe protein and shows the lowest activity in nitrogen fixation, the order being Fe–Mo > Fe–V > Fe–Fe. Fe–Fe protein also reduces ethylene to ethane.

8. Besides nitrogen fixation process, ammonia can also be produced by reduction of nitrate and nitrite. Nitrate is reduced to nitrite by two electron reduction. This is catalysed by the enzyme nitrate reductase. This protein has FAD (flavin adenine dinucleotide), haem b557 and Mo pterin complex, as prosthetic groups. Nitrate is bound at the molybdenum(VI) centre, replacing a ligand, possibly H_2O. The electron received from NADPH, through FAD and b557, reduces Mo(VI) to Mo(IV). Two electrons are transferred to the bound NO_3^-, reducing it to NO_2^-, and Mo(VI) is regenerated for repeating the catalytic cycle (Figure 8.7).

   Note: *Molybdopterins* are a class of cofactors found in most molybdenum-containing and all tungsten-containing enzymes. Also called MPT and pyranopterin dithiolate, the naming is misnomer. Molybdopterin *per se* contains no molybdenum; rather, this is the name of the ligand (a pterin) that will bind the active metal. After molybdopterin is complexed with molybdenum, the complete ligand is usually called molybdenum cofactor.

Pterin is a heterocyclic compound composed of a pteridine ring system, with a "keto group" (a lactam) and an amino group on positions 4 and 2, respectively. It is structurally related to the parent bicyclic heterocycle called pteridine. Pterins, as a group, are polycyclic aromatic compounds related to pterin with additional substituents.

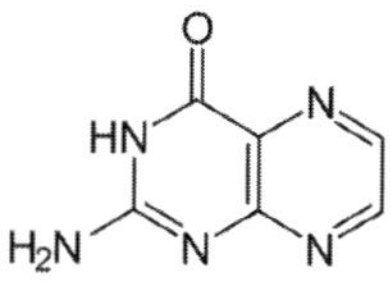

Chapter 9

1. Photosynthesis is the process of conversion of light energy into chemical energy, carried out by plants and other organisms. Photosynthesis is a bioenergetic process. It proceeds in two stages. In the first stage, occurring in the sun light, called light reaction, there is photo-induced splitting of water molecules. O^{2-} loses electron and is converted to molecular oxygen, which is liberated in the air. The light energy is converted into chemical energy and is stored in the form of ATP. The electrons liberated in the formation of O_2, reduce $NADP^+$ to NADPH. In the second stage, a series of enzymatic reactions lead to the reduction of CO_2 to glucose, by NADPH, which gets converted back to $NADP^+$. The energy for this reaction is provided by the breakdown of ATP into ADP. This reaction is called Dark reaction; that is, the reaction which does not need light and can occur in the dark. The name should not wrongly imply that the reaction occurs in dark conditions only. The net reaction can be written as

$$6CO_2 + 6H_2O \longrightarrow C_6H_{12}O_6 + 6O_2$$

Thus, CO_2 and H_2O are converted into glucose and oxygen in the presence of sun light.

2. The two steps of the photosynthetic reactions are shown in Eqs 9.1–9.3. The gross reaction is shown in Eq. (9.4).

3. The fact that O_2, liberated in photosynthesis, is from H_2O has been confirmed by using ^{18}O-labelled water. It is observed that only $^{18}O_2$ with m.w 36 is obtained. As shown in Eq. 9.2, in the normal photosynthetic process, the H used for the reduction of $NADP^+$ to NADPH is also provided by water.

4. The oxidation potential of O^{2-} is more positive; hence, liberation of electrons from O^{2-} needs solar energy, and hence, special photo system is required for "light harvesting", that is, in other words, for the capture of solar energy. The photo systems in green plants and algae have organic molecules, like chlorophyll and carotenoids, in large number of clusters, which absorb light, and thus function as molecular antenna, for capture of solar energy.

5. Chlorophylls are Mg(II) complexes of tetrapyrrole ligands of porphyrin family, with different substitutions at positions 2, 3 and 10 (Figure 9.1). There is a cyclopentanone ring E, between pyrrole ring C and D of the macrocycle, and it can exist in keto or enol form. The pyrrole ring D is in the reduced form. There is a long phytol chain attached at the carbon atom at seventh position of the D ring. This helps chlorophyll to bind with cell membrane. If X at position 3 in the pyrrole ring B (Figure 9.1) is CH_3, it is called chlorophyll a, and if X = CHO, it is called chlorophyll b. Mg^{2+} is bound to two N and two N^- of the porphyrin ring and is 0.3–0.5 Å above the plane of the ring (Figure 9.1).

6. Chlorophylls absorb at ~700 nm, due to the $\pi_b \rightarrow \pi^*$ transition, in the porphyrin ring. The photosynthetic systems also contain additional pigments, like carotenoids (β-carotene) or blue

or green phycoerythrobilins. These pigments have a broad range of absorption wavelength and thus absorb the entire range of visible spectrum from the sunlight. The absorbed light is not only useful for the photo activation of the chlorophyll, but this process, of absorption of the sunlight by chlorophyll, also protects the biochemical system of the plant from photochemical damage.

7. In photosynthesis process, there are two different sites, where light reaction occurs, and light energy is converted to chemical energy. These reaction sites are called photo system I (PSI) and photo system II (PSII). They contain the green pigments chlorophyll a and chlorophyll b in different proportions. They also differ in accessory chemicals for processing the tapped energy of the photon.

8. The photo system II has a characteristic absorption at wavelength 680 nm and hence is termed P680 (chl). It absorbs a quantum of light, followed by the transfer of one electron from the π bonding (πb) to the π anti-bonding (π^*) molecular orbital (m.o.) of the porphyrin ring. The molecule in the excited state P680 chl* is a strong reducing agent. It transfers the electron to an acceptor molecule C550, of which the composition is not definitely known. PS II is the site where O_2 is liberated; hence, photosynthetic bacteria which do not evolve O_2 do not have PS II, but have only PS I.

   The photosystem I has a characteristic absorption at 700 nm, that is at a lower energy region than PS II, and is called P700 chl. It has an oxidation potential 0.4–0.5 V and hence does not have tendency to lose electrons as such. On absorption of energy at 700 nm, one electron from π_b m.o. of the porphyrin ring is excited to the π^* orbital. In the excited state, P700 chl* transfers an electron to the electron acceptor molecule P430, and results in the formation of P700 Chl$^+$ or PSI$^+$. P430 is an Fe–S protein, different from ferredoxin. It has an oxidation potential −4V. In the reduced form, it absorbs at 430 nm, hence the name P430. The electron from the reduced form of P430 is transferred to NADP$^+$.

9. In the dark reaction, the electrons are transferred from NADPH to carbon dioxide, and it gets reduced to form glucose, as shown in Eq. (9.3). The required energy, in this dark phase, is provided by the breaking down of ATP to ADP. The dark phase of the reaction is not known to involve any metalloenzyme.

10. Model mono- and binuclear Mn(III) and Mn(IV) Schiff base complexes and mono- and binuclear Fe(III) and Ru(III) bipyridyl complexes have played an important role, in our present understanding of the role of the water-oxidising complex (WOC) and the photosystems. However, well-characterised functional models are rare. Such model complexes have direct application as catalysts in artificial photosynthesis, for splitting of water into O_2 and H_2. Hydrogen can be used in future fuel cells, for the generation of electricity. Such systems are commercially appealing, in view of the depleting energy sources and the environmental pollution due to the combustion of coal, oil or fuel gases.

11. See the "Mechanism of Photo Decomposition of Water" section.

Chapter 10

1. Following are the nickel-containing enzymes which participate in various biochemical processes:
   i. **Hydrogenase:** The enzymes, which catalyse the evolution of hydrogen from a substrate or the uptake of hydrogen by the substrate, are called hydrogenases. These enzymes may contain only iron or iron along with nickel. The latter enzymes, catalysing hydrogen uptake or liberation of hydrogen, are called nickel iron hydrogenase.
   ii. **Methanogenic bacteria:** Nickel is present in factor F430, which is a cocatalyst of coenzyme M, which catalyses the reduction of methyl into methane.

2. Vanadium has been detected in very small amounts in biological systems. Though its role is not definitely known, small amount of vanadium is required by living systems for normal health.

Based on its chemistry, it can be said that in oxidation states V and IV, it may act in biochemical reactions by deactivating certain enzymes, like Na–K–ATPase. Also, some studies suggest that activation of some enzymes by vanadium is also possible, for example, bromoperoxidase. It has now been recognised that vanadium has some role in sugar metabolism and this is an active field of research in diabetes therapy.

3. Though the toxicity of chromium(VI) is well documented as pollutant, role of chromium as micronutrient in biochemical processes is not yet well established. It has been suggested that chromium is involved in the action of the zinc protein, insulin. As the deficiency of chromium affects sugar metabolism, it is evident that it is associated with the glucose tolerance factor in the body. However, the exact role of chromium and its mode of binding with the glucose tolerance factor are not known. Chromium also acts as a cofactor, in the activity of other enzymes, and may be involved in the stabilisation of nucleic acid.

4. Aluminium occurs in the form of insoluble salts such as hydroxides, phosphates and silicate, which remain bound to the soil. Thus, in nature, free aluminium is only in traces. Hence, in the initial stages of evolution of the biological species, there was no excess of soluble aluminium salts to enter the biological system.

5. In spite of abundance of Al in the environment, it is present in vegetable and animal systems only in traces. Any excess of Al has toxic effects, leading to stunted growth of the roots of the plants and their destruction. In animals, disorders, like Alzheimer's and Parkinson's diseases, are attributed to the toxic effects of excess aluminium. Fishes are more sensitive to excess Al, and it can have fatal effects on fishes, even in very small excess. The reason for high toxicity of aluminium for animals is that the Al^{3+} ion, with three charges, is highly acidic in nature and has strong tendency to bind with hard base ligands such as amino acids, polypeptides, proteins, DNA and ATP, present in the biological systems. In the gills of the fishes, there are several anionic sites. Al^{3+} binds with these sites, stimulating the production of large excess of mucus in the gills. This causes blocking of the respiration and leads to the death of the fish.

6. With the increase in the chemical activity in the nature, there was release of acids lowering the pH of water and soil. This lead to the dissolution of the insoluble Al salts into soluble forms like chloride and nitrate, leading to abundance of free Al^{3+} ion in food and water, which in turn entered the biological system through the food chain.

7. Though there is a large amount of Al all around, its concentration in biological systems is less and thus the poisoning effect is less. The toxicity of Al^{3+} has been restricted in the evolutionary process by the presence of silicon. It is known that the oxides of Al and Si have a strong affinity, because of the similarity of the structure of silicate $(SiO_4)^{4-}$ and aluminate $(AlO_4)^{5-}$ ions. When Al and Si are together in water at pH higher than 4.5, they combine to form highly insoluble hydroxy aluminosilicate (HAS) and thus the availability of free Al^{3+} is inhibited. It has been shown that the fishes can survive, even at higher concentration of Al at pH 5, if there is excess of silicon available. This is because HAS is formed which does not affect the biological processes. Similarly, in the plants, it has been shown that the presence of excess of silicon reduces the root damage due to excess aluminium. The requirement of silicon for the inhibition of Al poisoning is more at lower pH because HAS is more soluble at lower pH and can liberate free Al^{3+} to cause poisoning. In the presence of excess silicon, the dissolution process of HAS is reversed.

8. Vanadium occurs in nature in various oxidation states. In food, it is found as the $H_2VO_4^-$ (V being in oxidation state V^{5+}). This is absorbed in the duodenum. $H_2VO_4^-$ gets reduced to V^{4+} and exists as oxycation VO^{2+} in the stomach. This has strong affinity for fibre in food. Hence, most of VO^{2+} gets bound to the fibre and is excreted out.

9. It was observed that administration of V salts to rats reduced the sugar level of rats. Since 1995, clinical trials have been carried out to see the efficacy of V as a drug for diabetes. Detailed study of the mechanism of how V compounds lower the level of sugar in the blood is being investigated. A probable mechanism is the interference of V with the phosphate signal cascade. For details, please see the section "Role of V in Sugar Metabolism" in Chapter 10.

10. Cr(III) interacts with insulin and also the insulin receptor kinase (IRK). The studies indicate that Cr(III) first interacts with insulin and boosts the mechanism of sugar entering the cell. In the second step, it activates IRK and assists in the binding of the insulin on the cell surfaces with IRK.

Chapter 11

1. Metal ions, in limited quantity, are essential for life processes. The concentration of the essential metal ions, inside the cell, is controlled by an enzyme-regulated transport mechanism, by which excess, unused, metal ion is ejected, outside the cell. However, normally the biological systems do not have transport mechanism for non-essential (impurity) metal ions, which find access inside the biological system, though they have no biological role and do not normally occur in biological systems. The excess of the impurity metal ion can, thus, be created, and they can cause disorder or poisoning effect. Such metal ions are considered to be toxic.

2. Yes. The toxic metal ions can be classified into following two types:

   i. **Essential metal ions in excess:** A metal ion, which is essential in trace amount, for the normal activity of the enzymes, may have toxic effect, when present in excess amount. For example, magnesium and manganese, essential for the activity of various enzymes and chlorophyll, are moderately toxic in large excess. Chromium, which is essential as a sugar tolerance factor, is highly toxic as Cr(VI) and moderately toxic as Cr(III). Zn(II), which is an essential component of various enzymes, is moderately toxic in excess. Iron, which is essential for all organisms, is slightly toxic in excess. Copper, cobalt and nickel, which have a significant role, as cofactors in enzymes, are highly toxic, in excess, to plants and invertebrates and moderately toxic to mammals.

   ii. **Non-essential metal ions:** These metal ions are not essential for body system and have toxic effect, if ingested. The examples are the heavy metal ions of the third transition series and f block elements. These are not essential for biochemical systems, but have strong tendency to bind with biochemical molecules and can thus have toxic effect, when introduced in the body.

3. The metal ions can cause toxic effects in following ways.

   i. The impurity metal ion can get bound to the coordinating sites of the enzyme proteins, which are the binding sites for the essential metal ions, as shown in Figure 11.1. Thus, the essential biological functional group of the enzyme is blocked, and it gets deactivated and it cannot perform its normal biochemical functions. The toxic metal ion can also bind with the ion channels, membranes and polysaccharides, which have coordinating sites. This also interferes with the normal biochemical processes.

   ii. The deactivation of the enzyme can also be due to the displacement of the metal ion, present in the enzyme and essential for its catalytic activity, by the external toxic metal ion. Heavy metal ions, such as Pb^{2+}, Cd^{2+} and Hg^{2+}, can replace the essential metal ion Zn^{2+}, from the enzyme. Normally, the essential metal ion is replaced by an external metal ion, having the same chemistry. But, since the external metal ion is biochemically inactive, it cannot catalyse the activity of the enzyme and thus exhibits toxic effect. For example, cadmium and zinc belong to the same periodic group and have similar chemical properties, and hence, Cd(II) replaces Zn(II) in the enzymes. However, Cd(II) cannot act as a cofactor for the Zn(II)-specific enzyme, and hence, its introduction deactivates these enzymes and brings metabolic disorder. Thus, Cd(II) shows toxic effect.

   iii. The entry of the new metal ion in the biomolecule can change its conformation and hence the biological activity.

   iv. The toxic metal ion can get bound to the DNA and can stimulate its replication, thus resulting in uncontrolled cell growth. Thus, the toxic metal ion can be carcinogenic, leading to the formation of malignant tumours.

v. The binding of the metal ion, with the DNA, can cause change in the base sequence. This mutagenic effect results in the transmission of wrong genetic information from DNA, leading to the production of faulty proteins and enzymes. Toxic metal, binding with DNA, can cause birth defects also.

4. The toxic effect of a metal ion depends on its chemical form. For example, Sn^{2+} is not very toxic, but the methylated form $[Sn(CH_3)_3]^{2+}$ is more toxic. This indicates that the toxicity is dependent on the whole molecule and not only on the metal ion. Similarly, Cr(VI) is more toxic than Cr(III), probably, because the former is a stronger oxidising ion and may cause oxidative degradation of the biomolecule. Methyl mercury $[CH_3.Hg^+]$ is more toxic than aqueous Hg^{2+}.

5. Though the biological systems do not have mechanism for the expulsion of the toxic metal ions, they have mechanism to inhibit the toxic effect. For example, in certain bacteria there are chemical oxidation or reduction processes, which change the toxic metal ion to less toxic form, for example, Cr(VI) to Cr(III) form. There may also be sequestering of the metal ion by binding with strongly coordinating biomolecules, like proteins or even cell membrane. The organisms contain a strongly coordinating biomolecule metallothionein, which has sulphur coordinating cysteine sites. It has special role in copper metabolism. But for other toxic metal ions, it can act as a sequestering agent, masking the metal ion and rendering it non-toxic.

6. Zinc is an essential element and occurs in several enzymes. The normal requirement of zinc for an adult is 15 mg/day. Deficiency of Zn(II) can be caused by binding of Zn(II) with phytates and organic phosphates in food, so that it cannot be absorbed by the body. Zinc deficiency leads to skin disease, impaired development of gonads, dwarfism, loss of appetite, anaemia and loss of body hair. It is not known, how zinc deficiency causes the above conditions. However, it can definitely be said that zinc deficiency may affect the activity of the zinc-containing enzymes. Zinc deficiency can be reversed by administrating zinc sulphate. However, excess of zinc cannot be given, as excess zinc can have toxic effect. Zinc excess can also be caused to the workers in zinc smelters or due to inhalation of industrial fumes. The symptoms of excess Zn(II) are abdominal disorders, metallic taste, respiratory irritation and cyanosis (zinc fume fever). Excess zinc also causes deficiency of copper ion and calcium. Large excess of zinc can cause convulsion, paralysis and even death.

7. **Cobalt:** Affects haemoglobin and causes cardiac insufficiency, leading to cardiac failure, excess cobalt also causes goitre.

   Nickel: Causes skin and respiratory disorders, and can produce bronchial cancer. It deactivates cytochrome C oxidase and also the enzymes, assisting dehydrogenation process, and thus inhibits biochemical processes.

   Vanadium: Excess vanadium inhibits enzyme tyrosinase, nitrate reductase and also sulphydryl activity of enzymes. It inhibits synthesis of biochemicals such as amino acids, lipids and cholesterol. It precipitates serum proteins and inhibits liver acetylation process.

   Iron: Excess iron intake can cause acidity, vomiting and coma conditions, and excess metal gets deposited in different parts of the body, such as liver, kidney and brain and can lead to their failure.

   Molybdenum: Can cause gout in human beings.

   Calcium: The level of calcium in the body is usually controlled by vitamin D and parathyroid hormones. But, if there is a metabolic imbalance of calcium regulation, it gets deposited in the tissues, leading to their calcification. Formation of stone and cataract are due to calcium salt deposition.

8. **Ba:** Soluble barium salts are toxic. One gram of barium chloride, taken at a time, can cause death. It causes vomiting, diarrhoea and even haemorrhage of stomach and intestine. It affects central nervous system and damages kidneys. Therefore, insoluble barium sulphate, used as contrast agent for medical diagnostic purposes, must be free from soluble salts of barium.

   Be: Industrial smokes may cause beryllium poisoning, leading to acute pneumonia, damage of skin and mucous membrane, cancer formation in lungs and bones and a disease of alveolar wall, called berylliosis. Beryllium is not excreted from mammalian tissues. It exhibits toxicity, due to the inhibition of some enzymes, because of the replacement of the essential metal ions by beryllium. Manganese in the enzyme alkaline phosphate is replaced by beryllium, and thus, ATP hydrolysis is inhibited. It also deactivates the enzyme DNA polymerase.

Cd: This is a very toxic metal, the lethal dose being > 350 mg at a time. Excess cadmium can displace zinc from the enzymes. These enzymes get deactivated, resulting in metabolic disorder. Intake of zinc can therefore give protection against cadmium poisoning.

Sn: Excess tin can enter the biological systems as salt, through food and as organotin, through the atmosphere. Ionic tin cannot penetrate through the intestinal wall, but neutral organotin gets deposited in the brain, causing headache, giddiness and other disorders. It also affects the urinary system.

Tl: Monovalent thallium compounds are highly toxic. Tl(I) is slightly bigger than K^+ and hence can penetrate the cell membrane. However, it binds strongly with the membrane. It accumulates in the erythrocytes, kidneys, bones and tissues. Excess Tl(I) causes fall of hair, gastroenteritis and peripheral neuropathy, and can lead to death.

9. There is a mechanism in the body for protection against excess cadmium. The absorbed cadmium is taken up by a thionein-like protein. Cadmium gets bound to its –SH groups. Thus, cadmium is sequestered, and its toxic effect is controlled. However, the protein bound cadmium gets accumulated in the kidneys. With increasing level, of 200 mg/g of the kidney, the kidneys get damaged. The damaged kidneys cannot retain plasma protein, calcium and phosphorous. They are excreted in excess and lead to the formation of renal stones.

Chapter 12

1. Metal ions have very important roles in biochemical systems. Metal ion deficiency can result in inhibition of metabolic processes and consequent diseases. Hence, metal complexes find various applications in medicine. There may be direct administration of the metal ion, or metal complex, to compensate the deficiency of a metal ion. The externally administered metal ion gets bound to the relevant biomolecule, at the site specified for the deficient metal ion, and thus revives the normal activity of the biomolecule. Some of the metal deficiencies are treated as follows:

   a. **Iron deficiency:** It can be treated by intake of iron rich food, e.g., spinach, corns, egg yolk and liver. However, only a part of the iron (10%) is absorbed from the food by human system. Hence, soluble iron salts like ferrous sulphate are administered, along with ascorbic acid, to increase the absorption of iron. A preparation of iron (III) chloride, with alkali modified dextran, is also commonly used. In case of severe anaemia, as in leukaemia, thalassemia or sickle cell anaemia, blood transfusion is carried out.

   b. **Zinc deficiency:** Zinc salts are administered orally in case of zinc deficiency. Zinc sulphate is given orally for the treatment of cysts, ulcers, burns and sickle-cell anaemia. With the observation that rheumatoid arthritis patients are deficient in zinc, zinc sulphate therapy is being attempted for the rheumatoid arthritis patients, with significant benefit. This anti-inflammatory effect of zinc is due to the inhibition of the function of calcium protein, in the presence of zinc.

   c. **Copper deficiency:** Powdered copper metal and copper salts like sulphate and acetate carbonate have been used, since ancient times, to treat anaemia, for stimulating heart and improving general strength. In Menkes disease, where copper deficiency occurs, due to the failure of the process of absorption of copper, *l*-histidine complex of Cu(II) is administered. This complex can be transported better in blood serum, than simple copper salt, and hence is more effective in the treatment of Menkes disease.

   d. **Cobalt deficiency:** In cases of pernicious anaemia, vitamin B_{12}, in physiologically active form, is administered.

2. Besides the use in treatment of metal deficiency, metal salts and metal complexes are also used as drugs for the control of diseases.

Following are the examples:

a. **Metal salts as drugs:** Aluminium, magnesium and bismuth hydroxides are used as antacids (antacid – a drug that prevents or corrects acidity in stomach).

Mercuric chloride and phenyl mercuric nitrate are used as antiseptic. The organomercurial compounds are used as diuretics (diuretics are medications that increase amount of water and salt expelled from body as urine).

Magnesium salts are used as purgatives (purgatives are compounds with strong laxative effects).

Silver salts are applied in the treatment of burn patients.

Osmium carbohydrate polymers are known to exhibit anti-arthritis activity.

The polyoxy anions of tungsten are being tried in the treatment of AIDS.

Zinc oxide is used externally for healing of wounds and is used in the tooth paste. Zinc salts are also used as ointment for the treatment of herpes. The metal ion binds with DNA of the herpes and thus destroys it.

Gold, Au(I) compounds, such as $Na_3[Au(S_2O_3)_2]$ (sanocrysin), sodium and calcium salts of Au(I) thiomalate (Myochrisin), (Figure 12.1a), colloidal gold sulphide, polymeric gold thio-glucose (Solganol) (Figure 12.1b), were used as intramuscular injection, for the treatment of rheumatoid arthritis. As the injection is painful, now it is administered as triethylphosphine gold (I) tetraacetyl thioglucose (auranofin) (Figure 12.1c).

Lithium carbonate is known to be a psychotropic drug and is administered in doses of 250 mg to 2 g for the manic disorders, and in symptoms of periodic aggression and depression.

Though Arsenic is highly poisonous, its compounds have been used for the treatment of parasitic infections and syphilis. The compounds used are phenyl arsonic acid (Figure 12.2a) and substituted phenyl arsonic acid (Figure 12.2b).

More effective drug for syphilis is arsphenamine (Figure 12.3). A much smaller dose of the drug needs to be administered.

Antimony tartrate as potassium salt is used for the treatment of trypanosomiasis and allied diseases.

Both antimony and arsenic act as drug, by binding with the thiol groups of the protein. This may be of help in arresting the disease.

3. *cis*-dichlorodiamine platinum(II), cisplatin (Figure 12.4), exhibits anti-cancerous activity. Since the chlorides in cisplatin are labile, in an aqueous solution, the chlorides are replaced by $-H_2O$ or $-OH^-$. Hence, the active species is $[Pt(NH_3)_2(H_2O)_2]^{2+}$ or $[Pt(NH_3)_2(H_2O)(OH)]^+$.

4. Since the chlorides in cisplatin are labile, in an aqueous solution, the chlorides are replaced by $-H_2O$ or $-OH^-$. Hence, the active species is $[Pt(NH_3)_2(H_2O)_2]^{2+}$ or $[Pt(NH_3)_2(H_2O)(OH)]^+$.

5. One possible reason is cis orientation may be specifically required for intrastrand cross linking in the target molecule's DNA. An alternative suggestion is that before binding with DNA, there may be some specific cellular receptor of the drug, and this may require the cis orientation, of the chlorides, for binding at the sites of the two groups of the receptor, to be replaced by the drug.

6. The necessary conditions for a complex to have anticancerous activity are as follows:

   i. The complex should be neutral, so that it can diffuse through the hydrophobic membrane of the cell.

   ii. The complex should have square planar structure. The two groups, to be substituted by DNA; that is, leaving groups (chloride in case of cisplatin) should be at cis position.

   iii. The groups, which are not substituted, should have low trans effect, like ammonia, hetero-cyclic amines or diamines.

   iv. The leaving groups (chlorides) should be labile, so that they can be quickly substituted.

7. Though cisplatin has been found very effective in cancer therapy, it has some toxic side effects like giddiness, vomiting and damage to kidney. Hence, it is given as an intravenous injection,

nearly 5 mg/ 1 kg of the body weight, with large amount of saline solution, to reduce the harm to the kidneys.

It has been observed that the cell repair enzyme protein avoids that region of DNA, which is bound to the metal drug. This results in the production of DNA, with one or more incorrect nucleotide. This mutagenesis (mutation of DNA) can lead to serious complications in health. It is necessary that after the deactivation of the cancerous cell, the platinum should be extruded from the DNA. It is done by a process, called excision repair. In this process, the phosphates on DNA, on both sides of the platinated part, get hydrolysed and the DNA gets separated into two parts. These platinum-free parts develop into two new DNA. From the, Pt(NH$_3$)$_2$ bound, small part of the DNA, platinum is flushed out and is removed from the biological system in steps. If the excision repair is not effective on some platinum bound cells, deposition of the platinum takes place in the body of the cancer patients, undertaking platinum complex chemotherapy. This produces adverse effects.

8. No. Several alternate metal complexes, which are octahedral, have been found to have anti-cancer activity. Octahedral complexes of Pt(IV) (Figure 12.6a) and Ru(III) complex, with two ammonias and four other anionic ligands (Figure 12.6b), have also been shown to exhibit anti-cancer activity. It is worth noting, however, that in all these complexes, there is cis halide or ammonia groups, and they may be getting displaced for the binding of the DNA with the complex, as in case of cisplatin.

9. There can be an excess of toxic metal ion, created in the body, in case of poisoning, or there can be an accumulation of excess of an essential metal ion, in different parts of the body, due to impaired metabolic activity. In either case, the excess metal ion is not desirable and should be removed at the earliest. This is done by administering a strongly metal binding ligand, which sequesters the metal ion and the resulting metal complex is excreted from the blood stream, through kidney, along with urine or through liver, along with faeces. This method of removal of metal is termed chelation therapy. The chelating drug can be taken orally. However, if it is not permeable, through the intestinal wall, it can be given as an injection.

10. BAL stands for British anti-Lewisite (BAL). This was the first chelating agent used, for controlling metal poisoning. The origin of the name of the chelating agent is, due to its use against the poisonous arsenic-containing gas ClHC=CHAsCl$_2$, called Lewisite. BAL is chemically 2,3-di-mercaptopropanol and gets bound with the arsenic of the Lewisite from two mercapto sites and removes it from the body in the form of the complex, as shown in Figure 12.11a. Thus, BAL protects the enzymes, containing mercapto sites, from attack by arsenic.

BAL is also used as an antidote for poisoning, due to other metal ions, such as Hg, Sb, Bi, Cu, Ni, Zn and Au. These metal ions also coordinate strongly to the mercapto sites of BAL, and the resulting water-soluble complexes are excreted through urine.

11. In several cases, such as Cooley's anaemia, thalassaemia, or sickle-cell anaemia, iron loading takes place in the heart, liver, brain and endocrine glands. Some chelating drugs can bind selectively with iron and are being used for sequestering out the excess iron. Nature has a very effective iron chelator, called ferrichrome (Figure 12.13a). This is a cyclic protein, with three anchored, potentially bidentate, oxamine, coordinating sites. A synthetic drug has also been in vogue, for a long time, for trapping iron. This is called desferrioxime or desferal (Figure 12.13b). It is a hexadentate ligand, with three bidentate oxamine coordinating sites, as in ferrichrome. These hexadentate ligands bind with all the six coordination sites of the excess iron, in the biochemical system, and deactivate it. Finally, the iron complex is extruded from the body.

Chapter 13

1. The biochemical systems have been shown to contain some non-metal elements, other than the normal ones, C, H, N, O, Cl, S, which are the constituents of the biomolecule. Some of the non-metal that play roles in biochemical processes are silicon, selenium and iodine.

2. In diatoms, the presence of silicon-containing precursors is essential. For several metabolic activities, within the organisms, silicon is required. Deficiency of silicon has inhibiting effects on protein, DNA and chlorophyll synthesis.

   In higher plants, the toxicity due to heavy metal ions is reduced, in the presence of excess silicon. The presence of silicon also improves photosynthesis in plants, due to strengthening of leaves and improvement in their chlorophyll content.

   The primary cell wall of plants contains 35% pectin (a polysaccharide rich in galacturonic acid residues). The pectin polymers are crosslinked by bridging with Ca. In this Ca-rich environment of the cell wall, silicon mineralisation occurs. This helps in the strengthening of leaves.

3. The biogeochemical cycle does not facilitate high abundance of Si in the biosphere. This is because a lot of silicon, available in soluble form, from the earth's crust, as orthosilicic acid, is removed, by the formation of insoluble compounds. This phenomenon is called abiotic drain. However, this loss of silicon in biosphere by abiotic drain is compensated by the large abundance of silicon in the earth's crust. The silica minerals, such as quartz, alkali feldspars and plagioclase, are leached out by the weathering of the silicate minerals. Thus, soluble orthosilicic acid passes to water and to the biosphere. The rate of weathering is dependent on various factors such as the composition and the pH of the leaching solvent.

4. The water-soluble compound of silicon, orthosilicic acid (H_4SiO_4), is the only dietary form of silicon that is biologically available to the higher animals. This is formed by the nature by the action of the acidic water on silica or silicate rocks. Soluble silicic acid is absorbed by the plants. Silicon enters the human body through the plants eaten by human beings. Availability of silicon from meat is very low, because silicon is present in tissues of animals, which are not eaten.

5. Silicon, in the form of silica (SiO_2), causes lung disease, silicosis, caused by inhalation of crystalline silica in the mining of feldspar rocks, in agate polishing industry and stone cutting process in making of statues. The harmful effect depends on silica particle size and shape and the duration for which the particles stay in the lung. There is, of course, some effect of the chemical reaction of H_4SiO_4 with the lung membrane surface. However, the effect is mainly physical as silica inhaled in crystalline form like quartz can cause silicosis and even cancer, whereas amorphous form of silica is considered to be safe. Amorphous silica is used in several food and cosmetic products. It is also one of the constituents of antacid preparations.

6. It has been suggested that ortho silicic acid (OSA) has an indirect role. It has been suggested that OSA reacts with Al^{3+} present in the biological systems and eliminates it, so that it does not inhibit the activity of the biochemically important elements Fe, Mn, Cu and Zn, by binding with active sites, as discussed earlier in Chapter 10, in the "Role of Aluminum in Biochemistry" section. Hence, when there is an increase in the amount of OSA in the environment and in turn in the diet of animals, the biochemical roles of the essential metal ions are better expressed as Al^{3+} is eliminated and the coordination sites on the biochemicals are not blocked by aluminium and the essential metals are better taken up by the biochemicals and there is an enhancement of their activity. Thus, it has been suggested that OSA has the indirect effect by undoing the adverse effect of Al^{3+} on the activity of the essential metal ion.

7. The requirement of selenium varies in men (70 μg/day) and women (60 μg/day).

8. Keshan's disease is an endemic cardiomyopathy (a fatal disease of heart muscle), caused due to deficiency of selenium.

9. In food, selenium exists in small amount, in inorganic form Se^{2-} and SeO_4^{2-}. Mainly, it exists in food in the form of organic compounds, containing selenium, which are better absorbed in the human systems.

10. The main sources of selenium are cereals, sea fishes and animal meat. However, the content of selenium is less in vegetarian food and milk. Hence, there is a possibility of occurrence of Se deficiency among vegetarians. In view of this, in some countries, selenium compounds are added to the fertilisers and thus selenium-fortified food is obtained.

11. Selenium is available to the animals through diet, mainly in organic form of Se, that is, Se-containing amino acids. However, selenium exists in biological systems mainly as selenocysteine, which is the main component of the selenium-containing proteins and enzyme. Selenomethionine and selenocysteine have structure similar to methionine and cysteine, respectively, with only the –SH group in the amino acid being replaced by –SeH group.

    In the proteins, selenium exists mainly as selenocysteine residue at the 21st position in the protein chain.

12. The main selenoenzymes are six different glutathione peroxidase (GPX 1 to 6) and three thioredoxin reductase (Trx R 1 to 3). These selenoenzymes exhibit special role as antioxidants and protect the cells from the damage done by the active oxygen species, such as superoxide (O_2^-), peroxide (O_2^{2-}) and hydroxy radical (OH$^\bullet$), and reactive nitrogen species such as NO and peroxynitrite.

13. In nature, iodine occurs mainly in the soil and in the plants grown in iodine rich soil. It exists mainly in the form of iodine salts, and partly as free iodine or as constituent of organic compounds. Human beings get iodine from vegetables, and hence, the supply of iodine to human system depends on the iodine content of the vegetables.

14. Another source of iodine is the sea weeds and sea fishes. This is because the iodine salts in the soils in high altitudes are washed down by the water, due to the erosion of the soil, and are carried to the lower altitude regions and ultimately they gets deposited in the sea and ocean. Hence, they are the biggest store houses of iodine. Sea weeds absorb the iodine, and the sea fishes feed on these plants and iodide-rich water. Hence, the sea food is the richest source of iodine.

15. The iodide in the sea water is oxidised due to oxygen and UV light and liberates molecular iodine. This being volatile is liberated in the atmosphere and is carried by air to the coastal areas. This is absorbed by the soil and gets converted to iodide form. Plants grown over the soil absorb iodide from the soil. Thus, the coastal areas are rich in iodine. As one moves up from the sea coast, there is depletion of iodine, due to loss by erosion and lesser availability of iodine back from ocean, due to greater distance and height from coast. Hence, the soil and food in hilly areas are iodine-deficient. Furthermore, there is inadequacy of iodine due to lack of sea food in the diet of people in hill areas, due to high cost of sea food and vegetarian food habits. Moreover, in general, the whole of iodine present in the food may not be available, due to the presence in the food of compounds, called goitrogens, which inhibit the utilisation of iodine in the food. Hence, the people in hilly areas suffer from iodine deficiency syndrome (IDS).

Suggested Reading

Advanced inorganic chemistry, F. A. Cotton and G. Wilkinson, 5th Edition, John Wiley and Sons, New York, 1988.

Aluminium in biological environment – a computational approach, J. I. Mujika, E. Rezabal, J. M. Mercero, F. Ruipérez, D. Costa, J. M. Ugaldea and X. Lopez, *Comput. Struct. Biotechnol. J.*, 9, No. 15, 2014, e201403002.

Biochemistry of zinc, K. Posterek, A. Horeena and J. Kogot, Annals Universitatis Marie Curie, Sklodiuska, Lablin, Polmia, Vol. XXIII, N1, 2 Section DDI, 2010.

Biochemistry, A. L. Lehninger, 2nd Edition, Worth Publishers, New York, 1974.

Bioinorganic chemistry, an introduction, E. I. Ochiai, Allyan and Becon, Boston, MA, 1977.

Bioinorganic chemistry, I. Bertini, H. B. Gray, S. J. Lippord and J. S. Valentine (Eds.), Viva Books, New Delhi, 1998.

Bioinorganic chemistry, R. W. Hay, Ellis Harwood, Chichester, New York, 1984.

Biological aspects of inorganic chemistry, A. W. Addison, W. R. Cullen, D. Dolphin and B. R. James (Ed.), Wiley, New York, 1977.

Biological cycles for elements in the environment, J. M. Wood, *Naturwiss.*, 62, 1975, 357.

The chemistry of lanthanides in biology: recent discoveries, emerging principles and technological applications, J. A. Cotruvo Jr., *ACS Cent. Sci.*, 5, 2019, 1496–1506. doi: 10.1021/acscentsci.9b00642.

Chemistry of metal chelate compounds, A. E. Martell and M. Calvin, Prentice Hall Inc., Englewood Cliffs, NJ, 1976.

The chemistry of the co-ordination compounds, J. C. Bailar, Reinhold Publishing Corporation, New York, 1956.

Coordination chemistry of macrocyclic compounds, G. A. Melson (Ed.), Plenum, New York, 1979.

Co-ordination chemistry, D. Banerjea, Tata McGraw Publishing Company, New Delhi, 1993.

Copper and its complexes – a biochemical approach, I. Iakovidis, I. Delimaris and S. M. Piperakis, *Mol. Biol. Int.*, 2011, 2011, 13.

Copper proteins, T. G. Spiro (Ed.), Wiley, New York, 1981.

Crown compounds, their characteristics and applications, M. Hiraoka, Elsevier, Amsterdam, 1982.

DNA-bound metal ions: recent developments, D. R. Morris Jr., *Biomol. Concepts*, 5, No. 5, 2014, 397.

Elements of bioinorganic chemistry, G. N. Mukherjee and U. N. Arabinda Das, Dhur and Sons Pvt Ltd., Calcutta, 1993.

Essentiality of Cr for human nutrition and health, Z. Krejpeio, *Pol. J. Environ. Stud.*, 10, No. 6, 2001, 399–404.

The fundamentals of nitrogen fixation, J. R. Postgate, Cambridge University Press, Cambridge, 1982.

Haemoglobin – an inspiration for research in coordination chemistry, J. W. Buchler, *Angew. Chem.* 17, 1978, 407–423.

Inorganic biochemistry, an introduction, J. A. Cowan, Wiley, VCH, New York, 1996.

Inorganic biochemistry, G. L. Eichorn (Ed.), Vol. I & II, Elsevier, New York, 1973.

Inorganic chemistry in biology and medicine, A. E. Martell (Ed.), A. C. S. Symposium Series No. 140, American Chemical Society, Washington, DC, 1980.

The inorganic chemistry of biological processes, M. N. Hughes, Wiley, New York, 1973.

Inorganic chemistry, principles of structure and reactivity, J. E. Huheey, E. A. Keiter and R. L. Keiter, Addison Wesley Publishing Company, New York, 2006.

Inorganic reaction mechanism, J. O. Edwards, W. A. Benjamin Inc., New York, 1964.

Inorganic reaction mechanism, M. L. Tobe, Nelson, London, 1972.

Intracellular calcium regulation, F. Bonner (Ed.), Wiley Liss, New York, 1990.

Introduction to biochemistry, W. R. Fearon, Grunne and Stratton, New York, 1947.

Invertebrate oxygen binding protein, J. Lamy and J. Lamy (Eds.), Dekko, New York, 1981.

Iodine, iodine metabolism and iodine deficiency disorders revisited, F. Ahad and S. Ganie, *Indian J. Endocrinol. Metab.*, 14, No. 1, 2010, 13–17.

Iron homeostatis, iron and heme metabolism, heme and porphyrin synthesis and metabolism, M. W. King. www. themedicalbiochemistrypage.org.

Iron sulphur proteins, T. G. Spiro (Ed.), Wiley, New York, 1982.

Iron sulphur proteins, W. Levenberg (Ed.), Academic Press, New York, 1973.

Iron transport and storage, P. Ponka, H. M. Schulman and R. D. Woodworth (Eds.), CRC Press, Boca Raton, FL, 1990.

Lanthanides: applications in cancer diagnosis and therapy, R. D. Teo, J. Termini and H. B. Gray, *Med. Chem.*, 59, No. 13, 2016, 6012–6024. doi: 10.1021/acs.jmedchem.5b01975.

Ligand substitution processes, C. H. Langford and H. B. Gray, W. A. Benjamin Inc, New York, 1965.

Long range electron transfer in biology, P. Bertrand, *Struct. Bond.*, 75, 1991, 1–47.

Long range electron transfer in multisite proteins, H. B. Gray and B. G. Malmstrom, *Biochemistry*, 28, 1989, 7479–7505.

Low-temperature magnetic circular dichroism studies of native laccase: confirmation of a trinuclear copper active site, E. I. Solomon, D. J. S. Solomon and M. D. Allenorf, *J. Am. Chem. Soc.*, 108, 1986, 5318.

Mechanism of inorganic reactions, a study of metal complexes in solution, F. Basolo and R. G. Pearson, Wiley Eastern Ltd., New Delhi, 1977.

Metal ions in biological systems, H. Sigel (Ed.), Vol. I–XXVII, Marcel Dekker, New York.

Metal-based imaging agents: progress towards interrogating neurodegenerative disease, A. C. Sedgwick, J. T. Brewster, P. Harvey, D. A. Iovan, G. Smith, X. -P. He, H. Tian, J. L. Sessler and T. D. James, *Chem. Soc. Rev.*, 49, 2020, 2886–2915, Advance Article. doi: 10.1039/C8CS00986D.

New insights into erythropoiesis: the roles of folate, vitamin B_{12} and iron, M. J. Koury and P. Ponka, *Annu. Rev. Nutr.*, 24, 2004, 105–131. doi: 10.1146/annurev.nutr.24.012003.132306.

New insights on the role of sodium in the physiological regulation of blood pressure and development of hypertension, E. Polychronopoulou, P. Braconnier and M. Burnier, *Front. Cardiovasc. Med.*, 6, 2019, 136. doi: 10.3389/fcvm.2019.00136.

New trends in bioinorganic chemistry, R. J. P. Williams and J. R. F. Desilva, Academic, New York, 1978.

The occurrence and metabolism of lanthanides. In: *Biochemistry of the lanthanides. Biochemistry of the elements*, C. H. Evans, 8, 1990. Springer. doi: 10.1007/978-1-4684-8748-0_7.

The oxygen transport and storage proteins of invertebrates, E. Words, *Essay Biochem.*, 16, 1980, 1–47.

Phtotosynthesis, J. Amsez (Ed.), Elsevier, Amsterdam, 1987.

Plastocyanin and the blue proteins, A. G. Sykes, *Struct. Bond.*, 75, 1991, 175–224.

Platinum and other metal coordination compounds in cancer chemotheraphy, M. Nicolini (Ed.), Martinus Nijhoff, Boston, MA, 1988.

Porphyrin and metal prophyrins, K. M. Smith, Elsevier, New York, 1976.

Principles of bioinorganic chemistry, A. M. Fiabane and D. R. Williams, The Chemical Society, London, 1977.

Reactivity of the laccase trinuclear copper active site with dioxygen: an X-ray absorption edge study, E. I. Solomon, J. L. Cole, G. O. Tau, E. K. Yang, K. D. Hodgson, *J. Am. Chem. Soc.*, 112, 1990, 2243.

Regulation of cellular iron metabolism, J. Wang and K. Pantoponlos, *Biochem. J.*, 434, 2011, 365–381.

Role of Cr in human health and in diabetes, W. T. Cefalu and F. B. Hu, *Diabetes Care*, 27, No. 11, 2004, 2741–2751.

The role of fluorine in medicinal chemistry, P. Shah and A. D. Westwell, *J. Enzym. Inhib. Med. Chem.*, 22, No. 5, 2007, 527–540. doi: 10.1080/14756360701425014.

Selenium, its role in human health, U. Tinggi, *Environ. Health Prev. Med.*, 13, 2008, 102–108.

Silicon in life. A bioinorganic solution to bioorganic essentiality, C. Exley, *J. Inorg. Biochem.*, 469, 1998, 139–144.

Silicon, the evolution of its use in biomaterials, J. Henstock, L. Canhan and S. Anderson, *Acta Biomater.*, 2015. doi: 10.1016/j.actbio.2014.09.025.

The sodium-potassium pump is an information processing element in brain computation M. D. Forrest, *Front. Phys.*, 2014. doi: 10.3389/fphys.2014.00472.

Spectroscopic and chemical studies of the laccase trinuclear copper active site: geometric and electronic structure, E. I. Solomon, J. L. Cole and P. A. Clark, *J. Am. Chem. Soc.*, 112, 1990, 9535.

The study of kinetic and mechanism of reactions of transition metal complexes, R. G. Wilkins, Allyn and Bacon, Boston, MA, 1974.

Synthetic analogues of O_2 binding heme proteins. K. Suslick and T. J. Reinert, *J. Chem. Educ.*, 62, 1985, 974–983.

Theoretical inorganic chemistry, M. C. Day and J. Selbin, Affiliated East West Press Pvt Ltd., New Delhi, 1977.

Threshold of inorganic chemistry, P. K. Bhattacharya, P. Samnani, Himalaya Publishing House, Mumbai, 2016.

Transition metal complexes as drugs and chemotherapeutic agents, N. Farewell, Kluwer Academic Publishers, Dordrecht, 1989.

Vanadium – an element of typical biological significance, B. Muhhrjee, B. Patra, S. Mohapatra, P. Banerjee, A. Tiwari and M. Chatterjee, *Toxicol. Lett.*, 150, 2004, 135–143.

Vanadium, a biologically relevant element, R. Wever and K. Kustin, *Adv. Inorg. Chem.*, 35, 1990, 81–115.

Vitamin B_{12} – chemistry and biochemistry. B. Kräutler, *Biochem. Soc. Trans.*, 13, Part IV, 2005, 806–810.

Water fluoridation: a critical review of the physiological effects of ingested fluoride as a public health intervention, S. Peckham and N. Awofeso, *Sci. World J.*, 2014, 2014, Article ID 293019, 10. doi: 10.1155/2014/293019.

Zinc biochemistry, from a single zinc enzyme to a key element of life, W. Maret, *Adv. Nutr.*, 4, 2013, 82–91.

Zinc coordination, function and structure of zinc enzyme and other proteins, B. L. Vallee, *Biochemistry*, 29, 1990, 5649–5659.

Zinc enzymes, T. G. Spiro (Ed.), Wiley, New York, 1983.

Index

Printed in the United States
By Bookmasters